The Role of Calcium in Biological Systems

Volume III

Editor

Leopold J. Anghileri, Dr. Chem.
Senior Scientist
Laboratory of Biophysics
University of Nancy
Nancy, France

Co-Editor

Anne Marie Tuffet-Anghileri, Dr. Sci.
Investigator
Centre National
de la
Recherche Scientifique
Paris, France

CRC Press, Inc.
Boca Raton, Florida

Library of Congress Cataloging in Publication Data
Main entry under title:

The role of calcium in biological systems.

Bibliography: v. 3, p.
Includes index.
1. Calcium in the body. 2. Calcium — Physiological effect. I. Anghileri, Leopold J. II. Tuffet-Anghileri, Anne Marie. [DNLM: 1. Calcium—Metabolism. 2. Cells—Physiology. QV 276 C143]
QP535.C2C263   574.19'214   81-15538
ISBN 0-8493-6282-2 (v. 3)   AACR2

This book represents information obtained from authentic and highly regarded sources. Reprinted material is quoted with permission, and sources are indicated. A wide variety of references are listed. Every reasonable effort has been made to give reliable data and information, but the author and the publisher cannot assume responsibility for the validity of all materials or for the consequences of their use.

All rights reserved. This book, or any parts thereof, may not be reproduced in any form without written consent from the publisher.

Direct all inquiries to CRC Press, Inc., 2000 Corporate Blvd., N.W., Boca Raton, Florida, 33431.

© 1982 by CRC Press, Inc.

International Standard Book Number 0-8493-6280-6 (Volume I)
International Standard Book Number 0-8493-6281-4 (Volume II)
International Standard Book Number 0-8493-6282-2 (Volume III)

Library of Congress Card Number 81-15538
Printed in the United States

FOREWORD

Calcium must certainly be the major bioelement of the times. Only a generation ago Ca^{2+} was known to physiologists and biochemists as a component of bone mineral and as a blood plasma constituent required in heart function and blood coagulation, but little more. Only a few, such as Baird Hastings and Walter Heilbrunn, saw more clearly into the future of Ca^{2+}, a future that was a long time coming. Then came the discovery of the role of Ca^{2+} in the contraction-relaxation cycle of skeletal muscle and the recognition that the free Ca^{2+} concentration of the resting sarcoplasm must normally be orders of magnitude lower than that in the blood plasma. Thus it was found that skeletal muscle must possess extremely efficient energy-dependent Ca^{2+} pumps. The discovery that mitochondria can accumulate Ca^{2+}, by my colleagues Vasington and Murphy, was at first regarded by many as an anomaly of in vitro conditions, since Ca^{2+} had earlier been found to uncouple oxidative phosphorylation. How could oxidative phosphorylation and Ca^{2+} transport be compatible? What possible role can mitochondria play in cellular Ca^{2+} distribution? And why does calcium phosphate form insoluble but noncrystalline granules in the mitochondrial matrix?

Answers to these and other questions came slowly at first, but in the 1970s a crescendo of Ca^{2+} research developed. Today we know dozens if not hundreds of different cellular and extracellular processes that are regulated by changes in the level of cytosolic or extracellular Ca^{2+}, in which at least three different membrane systems of the cell take part. Indeed, Ca^{2+} is now emerging as a most important and ubiquitous intracellular messenger, perhaps even broader in function than cyclic AMP, the original second messenger. What is even more remarkable is that cytosolic Ca^{2+} levels can regulate several different activities simultaneously in a single cell, raising fundamental questions regarding spatial and temporal regulatory fluctuations in cytosolic Ca^{2+} concentrations. Also remarkable are the biochemical mechanisms that keep calcium and phosphate, which occur in extracellular fluids and urine in supersaturating concentrations, from precipitating and turning us into stone. Central to all these questions is the chemistry of Ca^{2+}, its special features that endow it alone, of all the common cations, to participate in such a panoply of biological activities.

The papers in this volume address many aspects of these problems in the biochemistry and physiology of calcium and provide an important guide to recent progress.

Albert L. Lehninger
Professor of Medical Science
Department of Physiological
 Chemistry
The Johns Hopkins University
Baltimore, Maryland

PREFACE

The purpose of this review is to summarize and correlate the recent advances in several fields of scientific research related to the involvement of calcium in the structure development and function of biological systems.

Considering the general interest in calcium, this publication which is a comprehensive collection of contributions on the biochemical properties of the ion, is aimed to be of interest to workers in many fields of biology and medicine whose investigations might be related, directly or indirectly to the role of this ion in biological systems. In addition to the benefit of presenting a concise review of the state-of-the-art on each subject, it will provide a useful reference source of the work done in a wide range of scientific disciplines such as biochemistry, analytical chemistry, cell biology, physiology, nutrition, pathology, pharmacology, toxicology, etc.

The text consists of six major divisions. The first deals with the chemistry of calcium and gives both the theoretical and practical basis to interpret the role of this element in the function of normal and pathological biological systems, as described by the other subsequent divisions.

It is not the aim of this publication to provide an exhaustive compilation of all the subjects concerning the biochemistry of calcium, but to give within the limits of the present work the most important and actual highlights related to this bioelement. In most instances the given information has been made as concise as possible to make feasible the coverage of all the different subjects, but without sacrificing the updated bibliographic references which constitute a quick access to the ultimate source of knowledge. To the contributors and publisher who have made possible this publication we are very much indebted.

Leopold J. Anghileri
Anne Marie Tuffet-Anghileri

THE EDITOR

Leopold J. Anghileri, Dr. Chem. is Senior Scientist and Research Group Leader at the Laboratory of Biophysics, Service of Nuclear Medicine and Diagnostic by Ultrasons and Thermography, Medicine Faculty B, University of Nancy, France. Dr. Anghileri graduated in 1951 from Buenos Aires University where he obtained his Dr. Chem. degree in 1957. Dr. Anghileri became a member of the American Chemical Society in 1961. Until 1964 he was Investigator at the Argentine Atomic Energy Commission. From 1964 to 1968 he was a fellow at the Johns Hopkins Medical Institutions, Division of Nuclear Medicine and Radiation Health, Institut de Radium, Laboratoire Curie, Paris, and Institut für Nuklearmedizin, Deutsches Krebsforschungzentrum, Heidelberg, West Germany. During the period 1970 to 1975 he was Investigator at the Tumorforschung, Ruhr Universität, Essen, West Germany.

Dr. Anghileri has presented invited papers at international meetings. He has published more than 210 research papers. His current major research interest includes hyperthermic treatment of cancer and the calcium-magnesium membrane relationship in tumors.

THE CO-EDITOR

The late **Anne Marie Tuffet-Anghileri, Dr. Sci.** was Investigator at the Centre National de la Recherche Scientifique, Paris. Dr. Tuffet-Anghileri graduated in 1969 from Paris University. She worked in carcinogenesis at the Laboratoire Curie, Paris and in Molecular Biology (DNA-metal ions interactions) at the Paris University.

To my wife and dedicated co-worker Anne Tuffet-Anghileri (1937-1981) whose life was sacrificed for and by the Science.

L. J. Anghileri

CONTRIBUTORS

Syed N. Asad, M.D.
Attending Nephrologist
Renal Division
Department of Medicine
Nassau County Medical Center
East Meadow
Assistant Professor of Medicine
State University of New York
Stony Brook

L. Brandes, M.D.
Toronto Western Hospital
Toronto, Ontario
Canada

John Scott Carman, M.D.
Director
Adult Treatment Service
Brawner Psychiatric Institute
Smyrna, Georgia

P. Crassweller, M.D.
Associate Professor of Surgery
University of Toronto
Toronto, Ontario
Canada

A. C. H. Durham, Ph.D.
Department of Surgery
St. George's Hospital Medical School
London, England

William C. Eastin, Jr., Ph.D.
Department of Health and Human
  Services
National Toxicology Program
ResearchTriangle Park
North Carolina

W. Engelhardt, Ph.D.
University Hospital
Erlangen
Federal Republic of Germany

Andre Herchuelz, M.D.
Senior Assistant of Pharmacology
Laboratory of Pharmacology
Brussels University School of Medicine
Brussels, Belgium

Margaretha Jägerstad, Ph.D.
Assistant Professor of Nutrition
Department of Nutrition
Chemical Center
University in Lund
Lund, Sweden

Kenzo Kurihara, Ph.D.
Professor of Pharmacy
Faculty of Pharmaceutical Sciences
Hokkaido University
Hokkaido, Japan

Joseph M. Letteri, M.D.
Chief
Division of Renal Diseases
Nassau County Medical Center
East Meadow
Professor of Medicine
State University of New York
Stony Brook

Anders Linde, Odont. Dr.
Associate Professor
Laboratory of Oral Biology
Department of Histology
University of Göteborg
Göteborg, Sweden

Willy J. Malaisse, M.D.
Director
Laboratory of Experimental Medicine
Brussels University School of Medicine
Brussels, Belgium

Sachiko Moriuchi, Ph.D.
Associate Professor
Department of Food and Nutrition
School of Home Economics
Japan Women's University
Bunkyo-ku Tokyo
Japan

Robert F. C. Neher, Ph.D.
Friedrich Miescher Institut
PB 273
CH-4002, Basel, Switzerland

D. G. Oreopoulos, M.D.
Professor of Medicine
University of Toronto
Toronto, Ontario
Canada

Sushma Palmer, D.Sc.
Project Director
National Academy of Sciences
Adjunct Assistant Professor
Georgetown University School of
  Medicine
Washington, D.C.

I. Pimenta de Morais, M.D.
Professor
Faculty of Medicine of Catanduva
Department of Pharmacology
Catanduva, S. P.
Brazil

Alexandre Pinto Corrado, M.D.
Vice-Chairman
Department of Pharmacology
Faculty of Medicine of Ribeirao Preto
University of Sao Paulo
Brazil

W. A. Prado, M.D.
Assistant Professor
Faculty of Medicine of Ribeirao Preto
University of Sao Paulo
Brazil

T. J. C. Ruigrok, Ph.D.
Department of Cardiology
University Hospital
Utrecht
The Netherlands

Dieter Scholtz, M.D.
Mineral Metabolism and Endocrine
  Research Laboratory
University Hospital
Erlangen
Federal Republic of Germany

Paul Otto Schwille, M.D., V.M.D.
Professor
University Hospital
Erlangen
Federal Republic of Germany

E. W. Simon, D.Sc.
Professor
Department of Botany
Queen's University
Belfast, Northern Ireland

Allan Toguri, M.D.
Assistant Professor
Queen's University
Kingston, Ontario
Canada

Kiyonori Yoshii, Ph.D.
Research Associate
Faculty of Pharmaceutical Sciences
Hokkaido University
Hokkaido, Japan

TABLE OF CONTENTS

PHYSIOLOGICAL ROLE OF CALCIUM, PART II

Chapter 1
Role of Calcium in Steroidogenesis 3
Robert Neher

Chapter 2
Calcium and Insulin Release 17
A. Herchuelz and W. J. Malaisse

Chapter 3
Calcium in Taste and Olfaction Mechanisms 33
Kenzo Kurihara and Kiyonori Yoshii

Chapter 4
Calcium in Nutrition 45
Margaretha Jägerstad

Chapter 5
Calcium Metabolism in Dentinogenesis 55
Anders Linde

Chapter 6
Mechanism of Calcium Secretion in the Avian Shell Gland 73
William C. Eastin, Jr.

CALCIUM IN PATHOLOGY

Chapter 7
Disorders of Calcium Metabolism in Children 87
Sushma Palmer

Chapter 8
Role of Calcium Metabolism in Renal Stone Formation 103
Paul Otto Schwille and Dieter Scholtz

Chapter 9
Hemodialysis and Calcium 123
Syed N. Asad and Joseph M. Letteri

Chapter 10
The Calcium Paradox: Mechanisms and Clinical Relevance 133
T. J. C. Ruigrok

Chapter 11
Calcium and Cancer 143
Leopold J. Anghileri

Chapter 12
Calcium in Psychiatric Illness...157
John Scott Carman

Chapter 13
Calcium Deficiency in Plants 175
E. W. Simon

Chapter 14
Interactions Between Viruses and Calcium 193
Anthony C. H. Durham

CALCIUM IN PHARMACOLOGY

Chapter 15
Competitive Antagonism Between Calcium and Antibiotics 209
A. P. Corrado, W. A. Prado, and I. Pimenta de Morais

Chapter 16
The Effect of Calcium on the Toxicity of Heavy Metals 223
Sachiko Moriuchi

Chapter 17
Urinary Inhibitors of Crystallization..233
L. Brandes, D. G. Oreopoulos, P. Crassweller, and A. G. Toguri

Index 249

Physiological Role of Calcium, Part II

Chapter 1

ROLE OF CALCIUM IN STEROIDOGENESIS

Robert Neher

TABLE OF CONTENTS

I.	Introduction	4
II.	The General Role of Extracellular Ca^{2+}	4
III.	Cellular Ca^{2+} Uptake or Exchange	6
IV.	Extracellular Ca^{2+} and Hormone Binding	7
V.	Ca^{2+} and Enzyme Regulation	7
	A. Adenylate Cyclase	7
	B. Cyclic Nucleotide Phosphodiesterase	9
	C. Protein Synthesis	9
	D. Other Enzymes of Various Intracellular Compartments	9
VI.	Effects of Ca^{2+} on Translocation of Products	11
VII.	Conclusion	12
References		12

I. INTRODUCTION

When adrenocortical or gonadal tissue is stimulated by physiological concentrations of the trophic hormones corticotropin (ACTH) and lutropin (LH), respectively, a rapid neosynthesis of steroid hormones occurs. In this process, as in many similar processes of cell communication,[1] extracellular Ca^{2+} appears to be as important for the stimulatory response as the hormone itself. In discussing this particular role of Ca^{2+}, we shall refer mainly to adrenocortical tissue where most of the experimental data were obtained. These will be complemented by data obtained in gonadal tissue, considering that effects of Ca^{2+} may differ due to the use of different species or tissue preparations, such as perfused glands, tissue slices, isolated or cultured cells, or cell-free homogenates. The most reliable and abundant data are derived from work with isolated adrenal cells of the zona fasciculata-reticularis. This review will not deal with the effects of Ca^{2+} on the release of the respective trophic hormones.

II. THE GENERAL ROLE OF EXTRACELLULAR Ca^{2+}

In 1953, Birmingham et al.[2] were the first to report a requirement for Ca^{2+} in the action of ACTH on steroidogenesis in rat adrenal tissue in vitro. The Ca^{2+}-dependence of steroidogenesis in rat Leydig cells stimulated by LH was reported by Janszen et al.[3] in 1976.

Before discussing in detail the role of extracellular Ca^{2+} in the acute action of a trophic steroidogenic hormone or an equivalent agonist, it seems appropriate to review briefly the present knowledge about the mechanism of action as exemplified by the acute action of ACTH in isolated adrenocortical cells of the zona fasciculata-reticularis (Figure 1).[4-13]

According to this simplified scheme, ACTH binds to a plasma membrane receptor to be coupled to the membrane bound adenylate cyclase complex (AC) which, in the presence of its GTP-binding subunit and Mg^{2+}, converts ATP to cyclic AMP (pool 1). The intracellular level of cyclic AMP is determined by its rate of formation, its rate of degradation by phosphodiesterase (PDE) to 5'-AMP, and by its distribution into several intracellular pools and the extracellular space. The degree of steroidogenesis appears to be determined by the level of cyclic AMP in pool 3 where it is bound to the regulatory subunit (R_2) of a cyclic AMP-dependent protein kinase (R_2C_2), thus, activating the enzyme by dissociation of the free catalytic subunits C_2. The activated protein kinase phosphorylates specifically one or more preformed and relatively labile protein substrates. These P-proteins, by an unknown mechanism involving other labile proteins and possibly microfilaments and microtubuli, might promote the translocation of precursor cholesterol across the inner mitochondrial membrane to its matrix site. This process is considered to be the rate-limiting step in the stimulation of steroidogenesis. Once cholesterol has reached the cytochrome P_{450} subunit of the mixed function oxidase in the same compartment it binds immediately to it and, in the presence of NADPH and molecular oxygen, is split into pregnenolone and a C_6-fragment. Pregnenolone is subsequently converted in several steps in various compartments to corticosteroids which leave the cell. When adrenocortical cells are stimulated by cholera toxin, by extracellular cyclic AMP, or a metastable Ca^{2+} complex, the same mechanism of action exists involving the activation of a protein kinase, although the extent to which Ca^{2+} is required varies greatly as discussed below. Adrenocortical cells of the zona glomerulosa which are specialized to produce aldosterone can be stimulated, in addition to the previously mentioned agonists, by angiotensin and high K^+. When stimulated by these latter two agonists, the left part of the mechanism as shown in Figure 1 has to be replaced at least in part by another, yet unknown, mechanism since cyclic

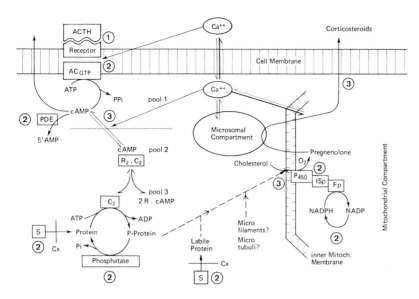

FIGURE 1. Mechanism of acute stimulation by ACTH of adrenocortical steroidogenesis. Effects of Ca^{2+} on ① hormone binding; ② enzyme regulation; ③ translocation of products. ACTH = adrenocorticotropin; cAMP = cyclic 3'5'-adenosinemonophosphate, cyclic AMP; AC = adenylate cyclase; PDE = phosphodiesterase; R_2C_2 = protein kinase holoenzyme; S = protein synthesis; Cx = cycloheximide; P_{450} = cytochrome P_{450}; ISp = iron-sulfur protein (adrenodoxin); Fp = flavoprotein (adrenodoxin reductase).

AMP does not seem to be involved.[14,15] Nevertheless, the action of these two regulators of aldosterone production is known[14-16] to be as Ca^{2+}-dependent as the steroidogenic response to ACTH.

Since 1972, various groups studying ACTH-induced steroidogenesis[14,17-22] reported that at low or physiological levels of ACTH (10^{-13} to 10^{-10} M), the Ca^{2+} requirement is absolute, but at higher levels to ACTH the Ca^{2+} requirement diminishes. The ED_{50} of ACTH for steroid formation increases for several orders of magnitude with decreasing extracellular Ca^{2+}, whereas the intrinsic activity or maximal capacity of steroid production decreases only about half, unless an excess of the Ca^{2+} chelator EGTA eliminates virtually all Ca^{2+} and abolishes any steroid formation. Interesting enough, Ways et al.[23] found that the ACTH analog $ACTH_{6-39}$ acted as a weak agonist at high Ca^{2+} concentration and as an ACTH antagonist at low Ca^{2+} concentration.

In contrast to the ACTH-induced steroidogenesis, cyclic AMP or cyclic GMP-induced steroidogenesis proved to be much less Ca^{2+} dependent.[3,17-21,24,25] This is also apparent from the fact that cyclic AMP-induced steroidogenesis is less sensitive to Ca^{2+}-antagonists, such as verapamil, ruthenium red, or La^{3+} than ACTH-induced steroidogenesis.[26-28] The stimulatory effect of cyclic nucleotides can be observed even in the presence of excessive EGTA. These findings indicate that the requirement for Ca^{2+} in ACTH action, while involved in more than one step, is greater for events preceding the formation of cyclic AMP than for those that follow. It can be concluded that Ca^{2+} has an important activating effect on the ACTH-receptor-adenylate cyclase complex (cf. Section V).

Choleratoxin is a potent pharmacological stimulator of cyclic AMP and steroid synthesis in steroidogenic tissues.[14,20,29] This stimulus appears to be less dependent on extracellular Ca^{2+} than ACTH. It is assumed that the activation of adenylate cyclase by the toxin does not involve the coupling of the hormone receptor to the catalytic subunit, but is caused by the inhibition of GTP hydrolysis.[30-32]

A few years ago, Neher and Milani[26,33-35] found that acute steroidogenesis in isolated adrenocortical cells can be triggered, in the absence of ACTH, choleratoxin, or extracellular cyclic nucleotides, by extracellular Ca^{2+} alone when this cation is presented as a metastable complex under specific conditions (cells primed for Ca^{2+}). Recently, Shima[15] and Yanagibashi[27] also observed a similar stimulatory effect of Ca^{2+} in the absence of ACTH in rat glomerulosa cells and bovine fasciculata cells, respectively. This Ca^{2+} trigger proved to be a valuable tool for the elucidation of the role of Ca^{2+} in the mechanism of action of ACTH-induced steroidogenesis.

In the following sections, the various possible sites of action of Ca^{2+} will be discussed in detail (cf. sites ①, ②, and ③ in Figure 1). It may be pointed out that most of these effects are highly specific for Ca^{2+} with the exception of Sr^{2+} which is able to substitute for Ca^{2+} at equal or slightly higher concentrations.[34,36-39]

III. CELLULAR Ca^{2+} UPTAKE OR EXCHANGE

The absolute requirement for extracellular Ca^{2+} in steroidogenesis induced by low ACTH concentrations or by the metastable Ca^{2+} trigger suggests that some Ca^{2+} has to be bound or taken up by the cell. In fact, Leier and Jungmann[40] described a stimulus-dependent net accumulation of $^{45}Ca^{2+}$ in whole adrenal glands in vivo and in vitro which was not only an increased rate of Ca^{2+} exchange. Unfortunately, it is not clear how these changes are related to the actual steroidogenetic process of the zona fasciculata cells, since Leslie and Borowitz[41] reported some evidence for a Ca^{2+} pump in plasma membranes of adrenal medulla but not adrenal cortex. Some $^{45}Ca^{2+}$ accumulation was found only in adrenocortical Golgi apparatus fraction. In contrast, Laychock et al.[42] reported some slowly increased Ca^{2+} uptake by a bovine adrenocortical 27,000 × g particulate fraction after stimulation by high ACTH concentrations in vitro or by several nucleotides of which some were nonsteroidogenic. The physiological significance of this effect remains to be elucidated.

Experiments by Jaanus and Rubin[43] with perfused cat adrenal glands produced no evidence for a net Ca^{2+} uptake by ACTH, but the rate of Ca^{2+} efflux was found to be reduced. ACTH seemed to cause an intracellular translocation of Ca^{2+} rather than an uptake of extracellular Ca^{+2}. Similarly, in isolated rat adrenocortical cells, no ACTH-induced Ca^{2+} uptake or exchange could be detected,[35] which is in agreement with the finding of Matthews and Saffran[37,44] that ACTH-induced steroidogenesis in normal Ca^{2+}-containing Krebs Ringer solution was not accompanied by a change in transmembrane potential. Nevertheless, Yanagibashi et al.[27,45] reported a dose-dependent increase in Ca^{2+} uptake in rat adrenocortical cells stimulated by 10^{-10} M ACTH which could be inhibited by 10^{-5} M verapamil. Although Ca^{2+} ionophores such as A23187 promoted a marked $^{45}Ca^{2+}$-uptake, they were found to inhibit steroidogenesis in intact cells[28,34,46,47] in the presence or absence of Ca^{2+}. It is likely that Ca^{2+} ionophores are unspecifically disturbing the accurately balanced Ca^{2+} distribution in various intracellular compartments,[48,49] thus, interfering with many enzymatic processes and inhibiting, e.g., protein synthesis.[46] However, under different conditions such as in adrenocortical slices, basal, and ACTH-induced steroidogenesis, but not cyclic AMP-induced steroidogenesis, was reported[50] to be increased by A23187 in the presence of Ca^{2+}. According to tissue preparation or to conditions, cyclic AMP formation is inhibited by A23187[35,46] or not.[47]

So far, the question of increased Ca^{2+} uptake or exchange in stimulated adrenal cells of zona fasciculata-reticularis remains controversial. This seems to be due to differing experimental procedures of cell preparation and particularly to analytical limitations in the study of Ca^{2+} metabolism in isolated cells. These cells are maintaining a Ca^{2+}-gradient between 10^{-7} to 10^{-2} M within the various intracellular compartments.[48,49]

Nevertheless, there is evidence that cell stimulation causes extracellular Ca^{2+} to be taken up but to a very limited extent and at a very specific site (cf. also Section V).

When cells of the zona glomerulosa are stimulated by high K^+, they produce aldosterone without the intervention of cyclic AMP as mentioned earlier. Although Mackie et al.[57] could not detect an increased Ca^{2+} uptake, the Ca^{2+} efflux was found to be slightly reduced. This was not seen in cells of the zona fasciculate where high K^+ has no steroidogenic effect.

IV. EXTRACELLULAR Ca^{2+} AND HORMONE BINDING

It was reported by various groups[4] that the acute stimulation of adrenocortical cells is initiated by binding ACTH to receptors on the cell surface (step ① in Figure 1). Further studies by Lefkowitz et al.[52] suggested that there may be two distinct orders of ACTH-binding sites. The ACTH-receptor interaction was also shown to stimulate adenylate cyclase activity in either intact cells or membrane fractions. In view of the dependence of adrenocortical cell stimulation on extracellular Ca^{2+}, it was of interest to know whether this very first step of binding of ACTH was Ca^{2+}-dependent. Lefkowitz et al.,[52,53] Haksar and Peron,[54] and Kowal et al.[20] found that Ca^{2+} was not necessary for binding. Similarly, the binding of angiotensin to glomerulosa cell receptors was not altered in the absence of Ca^{2+} as shown by Fakundig et al.[14] These results indicate a post-receptor location of one or more Ca^{2+}-dependent steps in the steroidogenic pathway.

V. Ca^{2+} AND ENZYME REGULATION

In view of the findings discussed in Sections II to IV it seems obvious that the main effects of Ca^{2+} in the ACTH-induced steroidogenesis have to be looked for within the cell membrane or within various intracellular compartments (cf. site ② in Figure 1). The high degree of Ca^{2+}-dependence before cyclic AMP formation and the much lower dependence after cyclic AMP formation focuses the attention to the enzymes known to regulate the intracellular cyclic AMP level, the membrane-bound adenylate cyclase (AC),[31] and the cyclic nucleotide phosphodiesterase (PDE).

A. Adenylate Cyclase

Bär and Hechter,[55] Lefkowitz et al.,[52] Kelly and Koritz,[56] Dazord et al.,[57] as well as Farese[58] and Podesta et al.,[35] found Ca^{2+} to stimulate cell-free adenylate cyclase in submillimolar concentration and to inhibit the enzyme in millimolar concentration. By contrast, in isolated adrenal cells, a millimolar concentration of Ca^{2+} is optimal. Sayers et al.[17,59] explained this discrepancy by proposing that, in the intact cell, the Ca^{2+}-sensitive site of adenylate cyclase is located in a compartment of the cell membrane wherein the Ca^{2+} concentration is low and remains little affected by high extracellular Ca^{2+}. The stimulatory effect of the latter was suggested to reflect a role for the cation in the coupling of the ACTH receptor to the catalytic unit of adenylate cyclase. Similar conclusions were drawn by Haksar and Peron,[54] Kowal et al.,[20] Ways et al.,[23] and Podesta et al.[35] The inhibition by excessive EGTA of the ACTH-stimulated adenylate cyclase complex can be reversed by adding not only Ca^{2+}, but also other cations, such as Sr^{2+}, Mn^{2+}, Co^{2+}.[60] From experiments in an even more complex system, i.e., perfused cat adrenal glands, Rubin et al.[61] draw the opposite conclusion that ACTH activates adenylate cyclase by dissociation of Ca^{2+} from its binding site at this enzyme. Similarly, Glossmann and Gips[60] concluded from the subsequent interaction of EGTA and Ca^{2+} that the hormonal activation of adenylate cyclase may be unrelated to complexing free Ca^{2+}.

Some authors believe that the primary regulation of ACTH-induced steroidogenesis may be mediated by cyclic GMP,[62,63] although this view is controversial.[64] Cyclic GMP appears to stimulate steroidogenesis less than cyclic AMP when compared by extracellular application.[4] Irrespective of this unsolved question, Perchellet and Sharma[21] found a Ca^{2+}-dependence of the hormone-activated guanylate cyclase in isolated adrenal cells which was similar to the adenylate cyclase discussed above.

In view of the evidence that pool 3 of cyclic AMP (Figure 1) which is bound to the regulatory subunit of cyclic AMP-dependent protein kinase mediates the ACTH-induced steroidogenesis,[9,10] Podesta et al.[35] studied the dependence of the level of this pool on the extracellular Ca^{2+} concentration. They found that the production of both total intracellular (pool 1 to 2) and bound cyclic AMP (pool 3) in ACTH-stimulated cells is highly dependent on the extracellular Ca^{2+} concentration and that all pools increase sharply from 0.1 to 2.5 mM Ca^{2+} with decreasing amounts of free receptor site (R_2). No change in these parameters was seen over the whole range of Ca^{2+} concentrations in the absence of ACTH. Ca^{2+} antagonists decreased the production of total intracellular cyclic AMP, bound cyclic AMP and steroid formation in ACTH-stimulated cells. This decrease in the response was reduced, but still evident in cells stimulated by cyclic AMP. This may be explained by a side effect of the Ca^{2+} antagonist verapamil or by the possibility that extracellular Ca^{2+} may be of some importance in maintaining full steroidogenesis in steps beyond cyclic AMP formation. Assuming that some Ca^{2+} is bound to the plasma membrane to a very limited extent, the Ca^{2+} dependence of adenylate cyclase activity was measured in membrane fractions from isolated cells. In the absence of ACTH, the basal activity of adenylate cyclase in the presence of GTP or the hydrolysis-resistant Gpp[NH]p was highest without Ca^{2+} (10^{-9} M or less), whereas increasing amounts of Ca^{2+} decreased the basal activity. However, in the presence of ACTH, a marked activation was seen only in the presence of about 10^{-6} M Ca^{2+} depending on the cofactor added. Ca^{2+} concentrations of less than 10^{-7} M or higher than 10^{-4} M lead to a complete loss of enzyme activation by ACTH. These results make it unlikely that Ca^{2+} supports the binding of GTP or Gpp[NH]p to the cyclase complex or affects the hydrolysis of GTP. They suggest that in the presence of ACTH and cofactors, Ca^{2+} may be involved in the coupling of the hormone-receptor complex to the adenylate cyclase complex. Mahaffee and Ontjes[65] working with solubilized adenylate cyclase preparations of whole rat adrenals came to the modified conclusion that the stimulatory effect of Ca^{2+} on ACTH action in this preparation may be exerted at a step involving guanine nucleotide interaction with the enzyme complex. Whatever the exact mode of stimulation by Ca^{2+} of the adenylate cyclase may be in the intact cell, these results suggest that one of the most important actions of extracellular Ca^{2+} in ACTH-induced steroidogenesis is to be located at the cell membrane and before cyclic AMP formation.

Complementary to this argument are the findings of Podesta et al.[35] with adrenocortical cells made responsive to a metastable Ca^{2+} complex in the absence of ACTH.[34] It was investigated whether the acute steroidogenesis in these cells could also be correlated with the total intracellular and bound cyclic AMP pool or whether a different mechanism of action is involved. The results clearly showed that the intracellular cyclic AMP pools and, in particular, the bound cyclic AMP pool increased with increasing extracellular Ca^{2+} concentration and time, in parallel to steroid production. This suggests again that adenylate cyclase is an important target enzyme for modulation by Ca^{2+}.

So far, little is known about the possible role of calmodulin in the regulation of adrenocortical adenylate cyclase. Calmodulin is a protein which interacts reversibly with Ca^{2+}, thus, regulating many of the Ca^{2+}-dependent enzymes.[66]

In line with the view that extracellular Ca^{2+} is of equal significance as ACTH for the activation of adenylate cyclase is the finding of Podesta et al.[12] that the ACTH-dependent phosphorylation of proteins is also dependent on extracellular Ca^{2+}, whereas the cyclic AMP-dependent formation of the same products is not.

B. Cyclic Nucleotide Phosphodiesterase

In isolated adrenal cells, a high phosphodiesterase activity is present which can be blocked by phosphodiesterase inhibitors.[4,9] Although cyclic nucleotide phosphodiesterase activity of different sources is known to be stimulated by Ca^{2+} by mediation of calmodulin,[66,67] there is virtually no information on the role of Ca^{2+} in the regulation of adrenal cell phosphodiesterase activity. The transient increase of intracellular cyclic AMP in ACTH-stimulated adrenocortical cells in the presence of extracellular Ca^{2+} and in the absence of a phosphodiesterase inhibitor[9] may well be explained in part by a regulation of both adenylate cyclase and phosphodiesterase through Ca^{2+} allowing a sequential stimulation of synthesis and a subsequent degradation of free intracellular cyclic AMP (pool 1 of Figure 1). The small amount of cyclic AMP bound to the regulatory subunit of the protein kinase (pool 3) is much less susceptible to degradation.[9]

C. Protein Synthesis

Protein synthesis appears to be required for the steroidogenic effect of both trophic hormones and cyclic nucleotides[13] supplying labile proteins as substrates for the protein kinase and/or further intermediates towards the rate-limiting step of steroidogenesis (cf. Figure 1, S). In contrast to Peron and McCarthy,[38] Farese[58] found that deletion of Ca^{2+} from the incubation of adrenal slices at high ACTH concentration decreased not only the formation of cyclic AMP and of steroids to a similar extent, but also the incorporation of leucine, and proposed that protein synthesis required during steroidogenesis is also Ca^{2+}-dependent. From more detailed studies,[68,69] it was suggested that the mechanism of the Ca^{2+} effect might be to speed up the transfer of the amino acids from aminoacyl tRNA to the growing peptide chain. Nakamura and Hall[46] reported A23187 to inhibit the incorporation of lysine into protein by Y_1 adrenal cells whether ACTH is present or not, presumably by inhibiting chain elongation. However, this effect could not be modified by the presence of Ca^{2+}.

Since the relatively small effects of Ca^{2+} on protein synthesis are best seen in cell-free systems at 10^{-5} to 10^{-4} M Ca^{2+}, it remains unclear whether in the intact cell Ca^{2+} by any redistribution process may be involved at the site of protein synthesis. So far, there is no evidence to assume that the role of Ca^{2+} in adrenal cell protein synthesis is more than permissive.

D. Other Enzymes of Various Intracellular Compartments

There are many Ca^{2+}- and calmodulin-dependent enzymes of intermediary metabolism whose activity is regulated by either phosphorylation or dephosphorylation,[66,70-72] a reversible process which by itself is often modulated by Ca^{2+}: Ca^{2+}-dependent protein kinases or phosphatases. In the case of angiotensin-stimulated and Ca^{2+}-dependent aldosterone production by cells of the zona glomerulosa,[14,73] one can only speculate for the present time as to the involvement of a Ca^{2+}-dependent protein kinase instead of a cyclic AMP-dependent protein kinase.

The involvement of a Ca^{2+}-dependent phosphoprotein phosphatase[70] which could counterregulate the formation of phosphoproteins was evidenced by Beckett and Boyd.[74] They showed that the cholesterolester hydrolase producing free cholesterol as potential steroid precursor (Figure 1) is activated by a cyclic AMP-dependent phosphorylation and deactivated by a Ca^{2+}- or Mg^{2+}-dependent phosphatase. In terms of more acute regulation of steroidogenesis, those phosphatases should be more interesting which dephosphorylate the phosphoproteins thought to be involved as mediators of steroidogenesis[12] (Figure 1). These phosphatases are not yet characterized. Other phosphoprotein phosphatase activities were found in various fractions of bovine adrenal cortex, the soluble portion of which appeared not to be Ca^{2+}-dependent.[75] However, it cannot be excluded that the activation of the phosphatase itself might be under the control of Ca^{2+}-dependent enzymes as pointed out earlier.

Phospholipase A_2-activity is known to be regulated by Ca^{2+}. Enzymes of this type hydrolize fatty acid ester bonds at position two of 3-sn-phospholipids and were detected in particulate fractions of cat adrenocortical cells.[76] In the same cells, Schrey and Rubin[77] found that ACTH in low steroidogenic concentration initiates specifically a rapid hydrolysis of arachidonyl phosphatidyl inositol followed by a rapid and selective reacylation of lysophosphatidyl inositol by arachidonic acid depending on the presence of extracellular Ca^{2+}. A23187 could mimic this effect and the significance of this for steroidogenesis is open to speculation. Arachidonic acid is a precursor of the Ca^{2+}-dependent formation of prostaglandins. However, prostaglandins do not appear to be obligatory intermediates in steroidogenesis,[78] although they seem to modulate steroidogenesis in some systems,[99-101] possibly also by affecting adrenal blood flow.

There are a number of mitochondrial enzymes which are Ca^{2+}-dependent and which are involved either in the intermediary metabolism leading to an appropriate supply of reducing equivalents (NADPH) or directly in the steroidogenic pathway. Thus, e.g., the pyruvate dehydrogenase complex is activated by Ca^{2+} or Mg^{2+} in various tissues,[79] although it is unknown whether the activity of these enzymes changes at all in hormone-stimulated steroidogenic cells. It seems as difficult to assess any decisive role of Ca^{2+} in the generation of NADPH through malic enzyme or transhydrogenase in intact cells as it is from experiments with isolated mitochondria.[38,80,81] It is now assumed that the intra- and extramitochondrial supply of NADPH is not a limiting factor in acute steroidogenesis.

The adrenal steroidogenic electron transport from NADPH to O_2 is thought to involve an adrenodoxin electron "shuttle", rather than a complex of reductase (flavoprotein Fp), adrenodoxin (iron-sulfur protein ISp), and cytochrome P_{450} (Figure 1). The respective dissociation equilibria seem to depend mainly on the concentration of Ca^{2+} and Mg^{2+} favoring the cholesterol sidechain split by high ion concentrations.[82] On the other hand, the studies of Kimura et al.[83] on membrane association of steroid hydroxylases in adrenal cortex mitochondria imply that a Ca^{2+}-phospholipid-sterol complex serves as a preferential substrate for the hydroxylation reaction. This is complemented by the findings of Simpson et al.[84,85] that Ca^{2+} in vitro induces changes in the difference spectra of sidechain splitting cytochrome P_{450} similar to those taking place after ACTH stimulation in vivo. In addition, data of Moustafa and Koritz[39] indicate that Ca^{2+}, apart from its swelling effect on mitochondria (cf. Section VI), stimulates mitochondrial 11β-hydroxylation by stimulating the reduction of the cytochrome P_{450} component, but not that of adrenodoxin (ISp).

Without any doubt, Ca^{2+} has many effects on enzymatic processes in the mitochondrial compartment. However, there is little evidence so far that stimulation of intact cells by ACTH involves either a shift of the Ca^{2+}-distribution between cytosol and mitochondria[86] or that Ca^{2+} is a limiting factor in mitochondria. The experiments of Farese and Prudente[87] on the mimicking role of Ca^{2+} in ACTH-induced increases in mitochondrial pregnenolone synthesis suggest that Ca^{2+} serves to amplify the steroidogenic effectiveness of other ACTH-induced factors (labile phosphoproteins?), rather than to act as the primary steroidogenic factor responsible for the ACTH-induced effect at the mitochondrial level.

VI. EFFECTS OF Ca^{2+} ON THE TRANSLOCATION OF PRODUCTS

Besides the reported actions of Ca^{2+} on various enzymatic processes in ACTH-induced steroidogenesis (cf. preceding sections), Ca^{2+} can affect the structure and physical properties of membranes, thus, changing the rate of transmembranal processes. There is little information about any direct transport effect of regulatory significance by binding extracellular Ca^{2+} on the plasma membrane of isolated steroidogenic cells. Under physiological conditions, they are permanently exposed to a fairly constant millimolar Ca^{2+} concentration. Some regulatory effects of Ca^{2+} may be expected within intracellular compartments (cf. site ③ in Figure 1). As their Ca^{2+} content differs by up to five orders of magnitude, any Ca^{2+} redistribution between them or alterations of Ca^{2+} exchange rates may well contribute to a change in enzyme activities when the cells are stimulated as discussed before.

A potential effect of Ca^{2+} on the translocation of cyclic AMP between the proposed pools 1 and 2 to 3 (Figure 1) was suggested by Podesta et al.[35] The comparison of binding kinetics of cyclic AMP in broken and intact adrenocortical cells implied the existence of an intracellular barrier between pool 1 and 2 to 3. By comparing the production of total intracellular (pool 1 to 3) and bound cyclic AMP (pool 3) after short stimulation of isolated cells by ACTH and of "primed cells" by the Ca^{2+}-trigger to a comparable degree of steroidogenesis, and at a maximal total intracellular cyclic AMP level, it was found that the ratio of these pools differed in the two sets of stimulated cells. Since in vitro Ca^{2+} had no effect on the binding of cyclic AMP to its receptor proteins, the results suggested a Ca^{2+}-dependent shift favoring pool 2.

For many years, Ca^{2+} was known to have a marked swelling effect on isolated mitochondria rendering them much more permeable to many substrates such as nucleotides or cholesterol, thus, relieving the restraints of mitochondria membranes.[38,88-92]

The exposure of mitochondria to 10^{-4} to 10^{-2} M Ca^{2+} concentrations, binding of Ca^{2+} preferentially to glycoproteins,[88,93] possible translocation of Ca^{2+} across the mitochondrial membranes, and concomitant swelling leads to the following general events of steroidogenesis as shown by many groups.[6,39,84,85,87-96]

1. Increased translocation of cholesterol from the extramitochondrial space and/or the outer mitochondrial membrane across the inner mitochondrial membrane to the matrix site
2. Increased binding of cholesterol to the cytochrome P_{450} subunit located at the matrix compartment and responsible for the subsequent sidechain oxidation
3. Increased synthesis of pregnenolone serving as a precursor for progesterone, corticosteroids, androgens, and estrogens

These effects of Ca^{2+} on isolated mitochondria, i.e., the translocation of the precursor from a metabolically inert pool to the target enzyme, are essentially those which have been pinpointed as the rate-limiting step of steroidogenesis in intact cells when stimulated by the trophic hormones. Therefore, it has been suspected that Ca^{2+} indeed could mediate the effect of ACTH in intact cells provided the hormone causes a corresponding intracellular Ca^{2+}-relocation possibly by the help of intermediary proteins. However, there is no straightforward information available whether any significant Ca^{2+} translocation takes place in intact cells stimulated by ACTH, LH, or cyclic AMP. Furthermore, since a marked effect of Ca^{2+} on mitochondria in vitro is still apparent after intact cells have been stimulated by ACTH[87,95] or LH,[92] and relatively high Ca^{2+} concentrations have to be used, it is likely that these actions of Ca^{2+} are not closely related to physiological events and that this role of Ca^{2+} is rather permissive than mediatory.

From perfusion experiments in whole cat adrenal glands, Rubin et al.[36,61,97,98] concluded that ACTH produces, in addition to cyclic AMP, an intracellular redistribution of Ca^{2+} necessary for the release of steroids from this tissue (cf. Figure 1). In other systems, such as isolated adrenocortical cells, there is no evidence for the storage of steroids in granules or an active release process by exocytoxis and a participation of Ca^{2+}. It is also unknown whether Ca^{2+} has any effect on the dissociation of steroids from their terminal enzyme complexes. The steroids appear to leave isolated cells by diffusion immediately after synthesis, whereas under stress conditions in vivo, the involvement of vesicles and microtubuli in the secretory mechanism was proposed by Bassett and Pollard.[102] Even if extracellular Ca^{2+} has a positive effect on steroid release from intact glands into the peripheral circulation, the constant extracellular Ca^{2+} environment does not seem to favor a direct regulatory role for this process.

VII. CONCLUSION

There is an obligatory interdependence of steroidogenic hormones and extracellular Ca^{2+} in order to stimulate intact cells to produce steroids under physiological conditions. Whereas the ups and downs of hormone levels regulate the degree of steroidogenesis, the constantly maintained extracellular Ca^{2+} concentration is the prerequisite of a sensitive and optimal activation of steroidogenesis within the limits of the cellular capacity. The main common target of these two diverse signals appears to reside in the membrane-bound adenylate cyclase complex responsible for the production of the intracellular messenger cyclic AMP.

It is clear that Ca^{2+} is also necessary for many later steps in various intracellular compartments, and it seems that in these processes, Ca^{2+} has primarily a permissive function and comes partly from intracellular sources. If there are other intracellular functions where Ca^{2+} exerts a mediatory or regulatory role, it is less clear whether the necessary changes in Ca^{2+} concentrations are provided for by exchange with the extracellular Ca^{2+} pool or by intracellular redistribution processes. This might be a cyclic AMP-independent hormonal action because so far there is no direct evidence that cyclic AMP in one or another of its pools causes a Ca^{2+} relocation.

Further possible regulatory functions of Ca^{2+} in intact cells may be evaluated when new techniques become available for the analysis of intracellular Ca^{2+} pools in order to follow in detail the Ca^{2+} distribution in dependence of hormone stimulation.

REFERENCES

1. Rasmussen, H., Cell communication, calcium ion, and cyclic adenosine monophosphate, *Science*, 170, 404, 1970.
2. Birmingham, M. K., Elliot, F. H., and Valere, B. H. L., The need for the presence of calcium for the stimulation in vitro of rat adrenal glands by adrenocorticotrophic hormone, *Endocrinology*, 53, 687, 1953.
3. Janszen, F. H. A., Cooke, B. A., Van Driel, M. J. A., and Van Der Molen, H. J., The effect of calcium ions on testosterone production in Leydig cells from rat testis, *Biochem. J.*, 160, 433, 1976.
4. Halkerston, D. K., Cyclic AMP and adrenocortical function, *Adv. Cycl. Nucleotide Res.*, 6, 99, 1975.
5. Berridge, M. J., The interaction of cyclic nucleotides and calcium in the control of cellular activity, *Adv. Cyclic Nucleotide Res.*, 6, 1, 1975.
6. Simpson, E. R., Cholesterol side chain cleavage, cytochrome P_{450}, and the control of steroidogenesis, *Mol. Cell. Endocrinol.*, 13, 213, 1979.

7. Mitani, F., Cytochrome P_{450} in adrenocortical mitochondria, *Mol. Cell. Endocrinol.*, 13, 21, 1979.
8. Hall, P. F., Charponier, C., Nakamura, M., and Gabbiani, G., The role of microfilaments in the response of adrenal tumor cells to ACTH, *J. Biol. Chem.*, 254, 9080, 1979.
9. Podesta, E. J., Milani, A., Steffen, H., and Neher, R., Steroidogenesis in isolated adrenocortical cells, *Biochem. J.*, 180, 355, 1979.
10. Sala, G. B., Hayashi, K., Catt, K. J., and Dufau, M. L., ACTH action in isolated adrenal cells. The intermediate role of cyclic AMP in stimulation of corticosterone synthesis, *J. Biol. Chem.*, 254, 3861, 1979.
11. Crivello, J. F. and Jefcoate, C. R., The effects of cytochalasin B and vinblastine on movement of cholesterol in rat adrenal glands, *Biochem. Biophys. Res. Commun.*, 89, 1127, 1979.
12. Podesta, E. J., Milani, A., Steffen, H., and Neher, R., ACTH induces phosphorylation of a cytoplasmic protein in intact isolated adrenocortical cells, *Proc. Natl. Acad. Sci. USA*, 76, 5187, 1979.
13. Garren, L. D., Gill, G. N., Masui, H., and Walton, G. M., On the mechanism of action of ACTH, *Recent Prog. Horm. Res.*, 27, 433, 1971.
14. Fakundig, J. L., Chow, R., and Catt, K. J., The role of Ca^{++} in the stimulation of aldosterone production by ACTH, angiotensin II, and K^+ in isolated glomerulosa cells, *Endocrinology*, 105, 327, 1979.
15. Shima, S., Kawashima, Y., and Hirai, M., Studies on cyclic nucleotides in the adrenal gland. VIII. Effects of angiotensin in cyclic AMP and steroidogenesis in the adrenal cortex, *Endocrinology*, 103, 1361, 1978.
16. Chiu, A. T. and Freer, R. J., Angiotensin-induced steroidogenesis in rabbit adrenal: effects of pH and calcium, *Mol. Cell. Endocrinol.*, 13, 159, 1979.
17. Sayers, G., Beall, R. J., and Seelig, S., Isolated adrenal cells: ACTH, Ca, steroidogenesis and cycl. AMP, *Science*, 175, 1131, 1972.
18. Haksar, A. and Peron, F. G., Comparison of the Ca^{++} requirement for the steroidogenic effect of ACTH and dibutyryl-cycl. AMP in rat adrenal cell suspensions, *Biochem. Biophys. Res. Commun.*, 47, 445, 1972.
19. Bowyer, F. and Kitabchi, A. E., Dual role of Ca^{++} in steroidogenesis in the isolated adrenal cell of rat, *Biochem. Biophys. Res. Commun.*, 57, 100, 1974.
20. Kowal, J., Srinivasan, S., and Saito, T., Calcium modulation of ACTH and choleratoxin stimulated adrenal steroid and cyclic AMP biosynthesis, *Endocr. Res. Commun.*, 1, 305, 1974.
21. Perchellet, J. P. and Sharma, R. K., Mediatory role of Ca and cyclic GMP in ACTH induced steroidogenesis by adrenal cells, *Science*, 203, 1259, 1979.
22. Shima, S., Kawashima, Y., and Hirai, M., Effects of ACTH and calcium on cyclic AMP production and steroid output by the Zona glomerulosa of the adrenal cortex, *Endocrinol. Jpn.*, 26, 219, 1979.
23. Ways, D. K., Mahaffee, D. D., and Ontjes, D. A., An ACTH analog [ACTH6-39] which acts as a potent in vitro ACTH antagonist at low Ca concentration and as a weak agonist at high Ca concentration, *Endocrinology*, 104, 1028, 1979.
24. Birmingham, M. K., Kurlents, E., Lane, R., Muhlstock, B., and Traikov, H., Effects of calcium on the potassium and sodium content of rat adrenal glands, on the stimulation of steroid production by cyclic AMP, and on the response of the adrenal to short contact with ACTH, *Can. J. Biochem.*, 38, 1077, 1960.
25. Birmingham, M. K. and Bartova, A., Effects of calcium and theophylline on ACTH- and dibutyryl cyclic AMP-stimulated steroidogenesis and glycolysis by intact mouse adrenal glands in vitro, *Endocrinology*, 92, 743, 1973.
26. Neher, R. and Milani, A., Calcium-induced steroidogenesis in isolated rat adrenal cells, in *Hormones and Cell Regulation*, Vol. 2, Dumont, J. and Nunez, J., Eds., North-Holland, Amsterdam, 1978, 71.
27. Yanagibashi, K., Calcium ion as "second messenger" in corticoidogenic action of ACTH, *Endocrinol. Jpn.*, 26, 227, 1979.
28. Warner, W. and Carchman, R. A., Effects of ruthenium red, A23187 and D-600 on steroidogenesis in Y-1 cells, *Biochim. Biophys. Acta*, 528, 409, 1978.
29. Haksar, A., Maudsley, D. V., and Peron, F. G., Stimulation of cyclic AMP and corticosterone formation in isolated rat adrenal cells by cholera enterotoxin. Comparison with the effects of ACTH, *Biochim. Biophys. Acta*, 381, 308, 1975.
30. Levitzki, A. and Helmreich, E. J. M., Hormone-receptor adenylate cyclase interactions, *FEBS Lett.*, 101, 213, 1979.
31. Glynn, P., Cooper, D. M. F., and Schulster, D., The regulation of adenylate cyclase of the adrenal cortex, *Mol. Cell. Endocrinol.*, 13, 99, 1979.
32. Abramowitz, J., Iyengar, R., and Birnbaumer, L., Guanyl nucleotide regulation of hormonally-responsive adenylyl cyclases, *Mol. Cell. Endocrinol.*, 16, 129, 1979.
33. Neher, R. and Milani, A., Mode of action of peptide hormones, *Clin. Endocrinol.*, 5, (Suppl., 29S), 1976.

34. Neher, R. and Milani, A., Steroidogenesis in isolated adrenal cells: excitation by calcium, *Mol. Cell. Endocrinol.*, 9, 243, 1978.
35. Podesta, E. J., Milani, A., Steffen, H., and Neher, R., Steroidogenic action of calcium ions in isolated adrenocortical cells, *Biochem. J.*, 186, 391, 1980.
36. Jaanus, S. D., Rosenstein, M. J., and Rubin, R. P., On the mode of action of ACTH on the isolated perfused adrenal gland, *J. Physiol.*, 209, 539, 1970.
37. Matthews, E. K. and Saffran, M., Ionic dependence of adrenal steroidogenesis and ACTH-induced changes in the membrane potential of adrenocortical cells, *J. Physiol.*, 234, 43, 1973.
38. Peron, F. G. and McCarthy, J. L., Corticosteroidogenesis in the rat adrenal gland, in *Functions of the Adrenal Cortex*, Vol. 1, McKerns, K. W., Ed., North-Holland, Amsterdam, 1968, 261.
39. Moustafa, A. M. and Koritz, S. B., Studies on the stimulation by Ca^{2+} and the inhibition by ADP of steroid 11β-hydroxylation in adrenal mitochondria, *Eur. J. Biochem.*, 78, 231, 1977.
40. Leier, D. J. and Jungmann, R., Adrenocorticotropic hormone and dibutyryl cyclic AMP-mediated Ca^{2+} uptake by rat adrenal glands, *Biochim. Biophys. Acta*, 329, 196, 1973.
41. Leslie, S. W. and Borowitz, J. L., Evidence for a plasma membrane calcium pump in bovine adrenal medulla but not adrenal cortex, *Biochim. Biophys. Acta*, 394, 227, 1975.
42. Laychock, S. G., Landon, E. J., and Hardman, J. G., The effect of ACTH and nucleotides on Ca^{2+} uptake in adrenal cortical microsomal vesicles, *Endocrinology*, 103, 2198, 1978.
43. Jaanus, S. D. and Rubin, R. P., The effect of ACTH on calcium distributon in the perfused cat adrenal gland, *J. Physiol.*, 213, 581, 1971.
44. Matthews, E. K. and Saffran, M., Effect of ACTH on the electric properties of adrenocortical cells, *Nature (London)*, 219, 1369, 1968.
45. Yanagibashi, K., Kamiya, N., Lin, G., and Matsuba, M., Studies on adrenocorticotropic hormone receptor using isolated rat adrenocortical cells, *Endocrinol. Jpn.*, 25, 545, 1978.
46. Nakamura, M. and Hall, P. F., Inhibition of steroidogenic response to corticotropin in mouse adrenal tumor cells (Y-1) by the ionophore A23187. Role of protein synthesis, *Biochim. Biophys. Acta*, 542, 330, 1978.
47. Lin, T., Murono, E., Osterman, J., and Nankin, H., The effect of calcium ionophore A23187 on interstitial cell steroidogenesis, *Biochim. Biophys. Acta*, 627, 157, 1980.
48. Borle, A. B., Calcium metabolism at the cellular level, *Fed. Proc. Fed. Am. Soc. Exp. Biol.*, 32, 1944, 1973.
49. Bygrave, F. L., Calcium movements in cells, *Trends Biochem. Sci.*, 3, 175, 1978.
50. Lymangrover, J. R. and Martin, R., Effects of ionophore A23187 on in vitro rat adrenal corticosterone production, *Life Sci.*, 23, 1193, 1978.
51. Mackie, C., Warren, R. L., and Simpson, E. R., Investigations into the role of calcium ions in the control of steroid production by isolated adrenal Zona glomerulosa cells of the rat, *J. Endocrinol.*, 77, 119, 1978.
52. Lefkowitz, R. J., Roth, J., and Pastan, I., ACTH-receptor interaction in the adrenal: a model for the initial step in the action of hormones that stimulate adenyl cyclase, *Ann. N.Y. Acad. Sci.*, 185, 195, 1971.
53. Lefkowitz, R. J., Roth, J., and Pastan, I., Effects of calcium on ACTH stimulation of the adrenal: separation of hormone binding from adenyl cyclase activation, *Nature (London)*, 228, 864, 1970.
54. Haksar, A. and Peron, F. G., The role of calcium in the steroidogenic response of rat adrenal cells to ACTH, *Biochim. Biophys. Acta*, 313, 363, 1973.
55. Bär, H. P. and Hechter, O., Adenyl cyclase and hormone action. III. Calcium requirement for ACTH stimulation of adenyl cyclase, *Biochem. Biophys. Res. Commun.*, 35, 681, 1969.
56. Kelly, L. A. and Koritz, S. B., Bovine adrenal cortical adenyl cyclase and its stimulation by ACTH and NaF, *Biochim. Biophys. Acta*, 237, 141, 1971.
57. Dazord, A., Gallet, D., and Saez, J. M., Adenyl cyclase activity in rat, bovine and human adrenal preparations, *Horm. Metab. Res.*, 7, 184, 1975.
58. Farese, R. V., On the requirement for calcium during the steroidogenic effect of ACTH, *Endocrinology*, 89, 1057, 1971.
59. Seelig, S. and Sayers, G., Calcium acts on the plasma membrane of the adrenal cortex cell to amplify the signal generated by the interaction of ACTH and receptor, *Fed. Proc. Fed. Am. Soc. Exp. Biol.*, 33, 205, 1974.
60. Glossmann, H. and Gips, H., Adrenal cortex adenylate cyclase, is Ca^{2+} involved in ACTH stimulation? *Naunyn-Schmiedeberg's Arch. Pharmacol.*, 292, 199, 1976.
61. Rubin, R. P., Carchman, R. A., and Jaanus, S. D., Role of calcium and cyclic AMP in action of ACTH, *Nature New Biol.*, 240, 150, 1972.
62. Sharma, R., Ahmed, N. K., Sutliff, L. S., and Brush, J. S., Metabolic regulation of steroidogenesis in isolated adrenal cells of the rat, *FEBS Lett.*, 45, 107, 1974.
63. Harrington, C. A., Fenimore, D. C., and Farmer, R. W., Regulation of adrenocortical steroidogenesis by cyclic GMP in isolated rat adrenal cells, *Biochem. Biophys. Res. Commun.*, 85, 55, 1978.

64. Hayashi, K., Sala, G., Catt, K., and Dufau, M. L., Regulation of steroidogenesis by ACTH in isolated adrenal cells, the intermediate role of cyclic nucleotides, *J. Biol. Chem.*, 254, 6678, 1979.
65. Mahaffee, D. D. and Ontjes, D. A., The role of calcium in the control of adrenal adenylate cyclase, *J. Biol. Chem.*, 255, 1565, 1980.
66. Means, A. R. and Dedman, J. R., Calmodulin in endocrine cells and its multiple roles in hormone action, *Mol. Cell. Endocrinol.*, 19, 215, 1980.
67. Dedman, J. R., Potter, J. D., Jackson, R. L., Johnson, D. J., and Means, A. R., Physicochemical properties of rat testis Ca^{2+}-dependent regulator protein of cyclic nucleotide phosphodiesterase, *J. Biol. Chem.*, 252, 8415, 1977.
68. Farese, R. V., Calcium as mediator of ACTH action on adrenal protein synthesis, *Science*, 173, 447, 1971.
69. Farese, R. V., Stimulatory effects of Ca^{2+} on protein synthesis in adrenal cell-free systems as related to trophic hormone action, *Endocrinology*, 89, 1064, 1971.
70. Curnow, R. T. and Larner, J., Hormonal and metabolic control of phosphoprotein phosphatases, in *Biochemical Actions of Hormones*, Vol. 6, Litwack, G., Ed., Academic Press, New York, 1979, 77.
71. Krebs, E. G. and Beavo, J. A., Phosphorylation-dephosphorylation of enzymes, *Ann. Rev. Biochem.*, 48, 923, 1979.
72. Greengard, P., Phosphorylated proteins as physiological effectors, *Science*, 199, 146, 1978.
73. Garrison, J. C., Borland, M. K., Floris, V. A., and Twible, D. A., The role of Ca^{2+} as a mediator of the effects of angiotensin. II. Catecholamines and vasopressin on the phosphorylation and activity of enzymes in isolated hepatocytes, *J. Biol. Chem.*, 254, 7147, 1979.
74. Beckett, G. J. and Boyd, G. S., Purification and control of bovine adrenal cortical cholesterol ester hydrolase and evidence for the activation of the enzyme by a phosphorylation, *Eur. J. Biochem.*, 72, 223, 1977.
75. Ullman, B. and Perlman, R. L., Purification and characterisation of a phosphoprotein phosphatase from bovine adrenal cortex, *Biochim. Biophys. Acta*, 403, 393, 1975.
76. Laychock, S. G., Franson, R. C., Weglicki, W. B., and Rubin, R. P., Identification and partial characterization of phospholipases in isolated adrenocortical cells, the effects of synacthen and calcium ions, *Biochem. J.*, 164, 753, 1977.
77. Schrey, M. P. and Rubin, R. P., Characterization of a Ca^{2+}-mediated activation of arachidonic acid turnover in adrenal phospholipids by ACTH, *J. Biol. Chem.*, 254, 11234, 1979.
78. Laychock, S. G., Warner, W., and Rubin, R. P., Further studies on the mechanisms controlling prostaglandin biosynthesis in the cat adrenal cortex: the role of Ca^{2+} and cyclic AMP, *Endocrinology*, 100, 74, 1977.
79. Denton, R. M. and Hughes, W. A., Pyruvatedehydrogenase and the hormonal regulation of fat synthesis in mammalian tissue, *Int. J. Biochem.*, 9, 545, 1978.
80. Pfeiffer, D. R. and Tchen, T. T., The activation of adrenal cortex mitochondrial malic enzyme by Ca^{2+} and Mg^{2+}, *Biochemistry*, 14, 89, 1975.
81. Pfeiffer, D. R., Kuo, T. H., and Tchen, T. T., Some effects of Ca^{2+}, Mg^{2+}, and Mn^{2+} on the ultra structure, light scattering properties, and malic enzyme activity of adrenal cortex mitochondria, *Arch. Biochem. Biophys.*, 176, 556, 1976.
82. Lambeth, J. D., Seybert, D. W., and Kamin, H., Ionic effects on adrenal steroidogenic electron transport, *J. Biol. Chem.*, 254, 7255, 1979.
83. Kimura, T., Wang, H. P., Chu, J. W., Churchill, P. F., and Parcells, J., Membrane association of steroid hydroxylases in adrenal cortex mitochondria, in *The Structural Basis of Membrane Function*, Hatefi, Y. and Djavadi-Ohaniance, L., Eds., Academic Press, New York, 1976, 447.
84. Simpson, E. R., Jefcoate, C. R., McCarthy, J. L., and Boyd, G., Effect of calcium ions on steroid-binding spectra and pregnenolone formation in rat adrenal mitochondria, *Eur. J. Biochem.*, 45, 181, 1974.
85. Simpson, E. R. and Williams-Smith, D. L., Effect of calcium ion uptake by rat adrenal mitochondria on pregnenolone formation and spectral properties of cytochrome P-450, *Biochim. Biophys. Acta*, 404, 309, 1975.
86. Simpson, E. R., Trzeciak, W. H., McCarthy, J. L., Jefcoate, C. R., and Boyd, G. S., Factors affecting cholesterol esterase and cholesterol side chain cleavage activities in rat adrenal, *Biochem. J.*, 129, 10P, 1972.
87. Farese, R. V. and Prudente, W. J., On the role of Ca^{2+} in ACTH-induced changes in mitochondrial pregnenolone synthesis, *Endocrinology*, 103, 1264, 1978.
88. Van der Vusse, G. J., Kalkman, M. L., Van Winsen, M. P. I., and Van der Molen, H. J., Effect of Ca^{2+}, Ruthenium red and ageing on pregnenolone production by mitochondrial fractions from normal and luteinizing hormone treated rat testes, *Biochim. Biophys. Acta*, 428, 420, 1976.
89. Koritz, S. B., On the regulation of pregnenolone synthesis, in *Functions of the Adrenal Cortex*, Vol. 1, McKerns, K. W., Ed., North-Holland, Amsterdam, 1968, 27.

90. Peron, F. G., Guerra, F., and McCarthy, J. L., Further studies on the effect of Ca^{2+} on corticosteroidogenesis. II. Adrenal mitochondrial swelling by Ca^{2+}, *Biochim. Biophys. Acta*, 110, 277, 1965.
91. Peron, F. G., McCarthy, J. L., and Guerra, F., Further studies on corticosteroidogenesis. IV. Inhibition of utilization of biological substrates for corticoid synthesis by high calcium concentrations. Possible role of transhydrogenase in corticosteroidogenesis, *Biochim. Biophys. Acta*, 117, 450, 1966.
92. Van der Vusse, G. J., Kalkman, M. L., Van Winsen, M. P. I., and Van der Molen, H. J., On the regulation of rat testicular steroidogenesis. Short term effect of LH and cycloheximide in vivo and Ca^{2+} in vitro on steroid production in cell-free systems, *Biochim. Biophys. Acta*, 398, 28, 1975.
93. Kimura, T., Chu, J. W., Mukai, K., Ishizuka, I., and Yamakawa, T., Some properties of a glycoprotein isolated from adrenal cortex mitochondria, *Biochem. Biophys. Res. Commun.*, 49, 1678, 1972.
94. Kahnt, F. W., Milani, A., Steffen, H., and Neher, R., The rate-limiting step of adrenal steroidogenesis and cyclic AMP, *Eur. J. Biochem.*, 44, 243, 1974.
95. Arthur, J. R., Mason, J. I., and Boyd, G., The effect of Ca^{2+} on the metabolism of exogenous cholesterol by rat adrenal mitochondria, *FEBS Lett.*, 66, 206, 1976.
96. Mason, J. I., Arthur, J. R., and Boyd, G. S., Regulation of cholesterol metabolism in rat adrenal mitochondria, *Mol. Cell. Endocrinol.*, 10, 209, 1978.
97. Rubin, R. P., Carchman, R. A., and Jaanus, S. D., Role of cyclic AMP on corticosteroid synthesis and release from the intact adrenal gland, *Biochem. Biophys. Res. Commun.*, 47, 1492, 1972.
98. Carchman, R. A., Jaanus, S. D., and Rubin, R. P., The role of ACTH and Ca^{2+} in cyclic AMP production and steroid release from the isolated, perfused rat adrenal gland, *Mol. Pharmacol.*, 7, 491, 1971.
99. Shima, S., Kawashima, Y., Hirai, M., and Asakura, M., Studies on cyclic nucleotides in the adrenal gland. X. Effects of ACTH and prostaglandin on adenylate cyclase activity in the adrenal cortex, *Endocrinology*, 106, 948, 1980.
100. Batta, S. K., Effect of prostaglandins on steroid biosynthesis, *J. Steroid Biochem.*, 6, 1080, 1975.
101. Rolland, P. H. and Chambaz, E. M, Effect of prostaglandins on steroidogenesis by bovine adrenal cortex mitochondria, *Mol. Cell. Endocrinol.*, 7, 325, 1977.
102. Bassett, J. R. and Pollard, I., The involvement of coated vesicles in the secretion of corticosterone by the Zona fasciculata of the rat adrenal cortex, *Tiss. Cell*, 12, 101, 1980.

Chapter 2

CALCIUM AND INSULIN RELEASE

A. Herchuelz and W. J. Malaisse

TABLE OF CONTENTS

I.	Introduction	18
II.	Participation of Calcium In the Process of Insulin Release	18
	A. Effect of Extracellular Calcium Upon Insulin Release	18
	B. Accumulation of Calcium In the β-cell	18
III.	Mode of Action of Calcium Upon Insulin Release	20
IV.	Modalities of Action of Glucose on Calcium Fluxes	21
	A. The Effect of Glucose on Calcium Inflow	21
	B. The Effect of Glucose on Calcium Extrusion	24
	C. The Links Between Glucose Metabolism and Calcium Fluxes	26
	D. Effect of Glucose on Intracellular Calcium Distribution	26
Acknowledgments		27
References		28

I. INTRODUCTION

The participation of calcium in the process of insulin release was first documented in a series of studies conducted by Grodsky and his colleagues[1-3] and by Milner and Hales,[4-6] the suggestion being made that the accumulation of calcium in a critical site of the pancreatic β-cell (possibly the cytosol) triggers the process of insulin release. This review deals with the following topics:

1. The evidences that calcium accumulation in the β-cell indeed participates in the insulin release process
2. The possible mechanism(s) by which calcium may cause the release of insulin
3. The modalities by which glucose, the main physiological stimulus of insulin release, affects calcium movements in the β-cell

II. PARTICIPATION OF CALCIUM IN THE PROCESS OF INSULIN RELEASE

The postulated participation of calcium in the process of insulin release is based mostly on two series of observations dealing, respectively, with the influence of changes in the extracellular Ca^{2+} concentration upon the secretory process and the accumulation of calcium in the stimulated β-cell.

A. Effect of Extracellular Calcium Upon Insulin Release

The absence of extracellular calcium abolishes insulin release evoked in vitro by glucose and various other secretagogues.[1,2,4-6] Under suitable experimental conditions, a very low concentration of extracellular calcium may be sufficient for glucose to stimulate insulin release.[7] The inhibitory effect of extracellular calcium deprivation upon insulin release is immediate and rapidly reversible.[7-8] Various inorganic (e.g., magnesium and cobalt) and organic (e.g., verapamil and suloctidil) calcium antagonists inhibit glucose-induced insulin release by impeding the entry of calcium into the β-cell.[3-5,9-13] On the contrary, the ionophore A23187 stimulates insulin release, an effect which is abolished in the absence of extracellular calcium[14-17] Calcium itself, when used at a sufficiently high extracellular concentration, stimulates the release of insulin from either isolated islets or the perfused pancreas.[18-19] Several studies carried out in vivo have also underlined the important role played by calcium in the regulation of insulin release. Thus, a blunted insulin response to glucose, arginine, or sulfonylureas is observed in pharmacologically provoked or disease-induced hypocalcemic states,[20-25] while hypercalcemic states are associated with enhanced insulin response to glucose, tolbutamide, and glucagon.[23,25-29]

B. Accumulation of Calcium In the β-cell

Several lines of evidence indicate that glucose provokes the accumulation of calcium in the β-cell.

First, when islets are incubated in the presence of ^{45}Ca and then submitted to repeated washes in order to remove radioactive extracellular contamination, more ^{45}Ca is recovered in the islets if the incubation was carried out in the presence as distinct from absence of glucose.[30] When other procedures are used to remove or to correct for radioactive extracellular contamination, (e.g., by use of an extracellular marker and separation of the islets from the incubation medium by rapid centrifugation through a layer of oil, or by displacement of extracellular calcium by lanthanum), glucose was also shown to stimulate the net uptake of ^{45}Ca by the islets.[31-35]

Second, glucose enhances the total ^{40}Ca content of the islets as measured by atomic absorption.[36] The latter observation, however, could not be confirmed.[37]

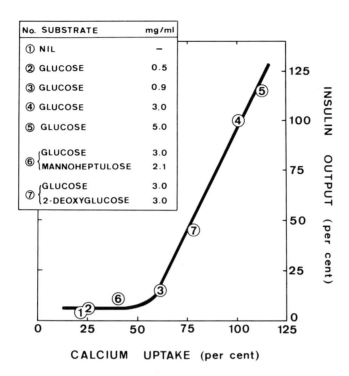

FIGURE 1. Relationship between ^{45}Ca net uptake and insulin output by isolated islets. Both parameters are expressed in percent of the value found in paired groups of islets incubated in the presence of glucose (16.7 mM) at normal calcium concentration.

Last, ultrastructural studies also indicate that glucose favors the accumulation of calcium in the β-cell.[38-40]

The existence of a cause-to-effect link between the accumulation of calcium into the β-cell and the release of insulin is suggested by the dose-action relationships characterizing the stimulant action of glucose upon these two variables.[30] These relationships both display a sigmoidal pattern.[30,41] The relationship between the net uptake of ^{45}Ca and the release of insulin is characterized by a threshold value for the stimulant action of calcium upon insulin release; once the islet calcium content exceeds such a threshold value, insulin output augments above basal value in proportion to the supplemental amount of calcium accumulated in the islets (Figure 1).[42] This relationship is not altered by various metabolic, ionic, or pharmacologic agents susceptible to interfere with insulin secretion.[30] It was also observed with other insulinotropic agents, like mannose and leucine.[43] Some experimental conditions are susceptible, however, to alter the above-mentioned relationship between the cellular calcium content of the islets and the release of insulin. Such is the case, for instance, when the insular content of cyclic AMP is modified. Thus, theophylline, an inhibitor of phosphodiesterase, enhances the insular content of cyclic AMP,[44-45] potentiates the insulinotropic effect of glucose,[46] but fails to affect ^{45}Ca net uptake by the islets.[47] This dissociated response does not invalidate the hypothesis according to which calcium accumulation in a given site of the β-cell provokes insulin release. Indeed, several observations suggest that, in the β-cell, theophylline provokes an intracellular translocation of calcium from an organelle-bound pool into the cytosol.[47-49] By doing so, theophylline may increase the cytoplasmic concentration of calcium and, hence, facilitate the release of insulin.

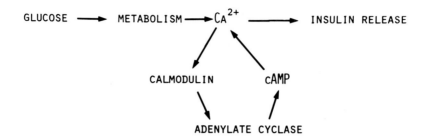

FIGURE 2. Selfamplification of the insulin secretory response to glucose via the Ca^{2+}-dependent activation of adenylate cyclase by endogenous calmodulin.

III. MODE OF ACTION OF CALCIUM UPON INSULIN RELEASE

Various modalities have been considered to account for the stimulant action of calcium upon insulin release. Calcium may trigger the release of insulin by activating a contractile system composed of microtubules and microfilaments and responsible for the migration and extrusion of secretory granules by exocytosis.[50] Initially proposed by Lacy, Howell, Young, and Fink,[51] the participation of this system in the process of insulin release has been documented by a series of ultrastructural and biochemical studies, which have been the subject of several reviews.[52-53] A direct regulatory role of calcium upon the activity of the microtubular-microfilamentous system in the β-cell is suggested by the finding that the ionophore A23187 increases the frequency, the maximal amplitude, and the mean speed of outward expansions observable at the boundary of β-cells in monolayer culture.[54] The latter contractile movements indeed depend on the integrity of the microtubular-microfilamentous system.[54]

Calcium may also cause the release of insulin by neutralizing electronegative charge localized at the external face of the secretory granule membrane and, by doing so, facilitate its fusion with the plasma membrane.[55]

In addition to these biophysical effects, calcium apparently plays a role in the regulation of metabolic events in the β-cell. For instance, glucose is thought to increase the production and/or the concentration of cyclic AMP in the islets.[15,56-57] This effect of glucose is abolished in the absence of extracellular calcium.[15,58] The stimulatory effect of glucose on the accumulation of cyclic AMP in islets incubated in the presence of phosphodiesterase inhibitors is also abolished in the absence of extracellular calcium.[59-60]

The mechanism by which glucose and calcium affect cyclic AMP metabolism in the islets should be understood in the light of recent observations indicating the presence in islets cells of a calcium dependent regulatory protein.[61-63] Like it is the case in brain tissue,[64-65] this calcium-dependent regulatory protein, known as calmodulin, stimulates the activity of adenylate cyclase in islets homogenates.[61,63] However, the activation of adenylate cyclase by calmodulin is dependent on the concentration of ionized calcium.[61,63] It is conceivable, therefore, that glucose, by increasing the cystolic concentration of Ca^{2+} activates adenylate cyclase via calmodulin. The increase in cyclic AMP synthesis may, in turn, amplify the secretory response to glucose (Figure 2). It is indeed well-established that cyclic AMP facilitates insulin release.[45-46] The cyclic nucleotide may act, in part at least, by causing an intracellular redistribution of calcium.[47-49]

Another site of action of calcium in the β-cell deserves due mention. Calcium has been reported to play a role in the periodical oscillation of the β-cell membrane potential. At intermediate glucose concentrations and under steady-state conditions, the membrane potential of the β-cell oscillates between a resting potential and a plateau potential onto which spike activity is superimposed.[66-67] As initially suggested by Matthews[68] and later confirmed by Atwater,[69] the repolarization of the membrane which occurs between bursts of spike activity may result, in part at least, from an increase in K⁺ conductance. It is currently proposed that the accumulation of Ca^{2+} in the β-cell is responsible for the latter increase in potassium conductance.[70-71]

The view that Ca^{2+} may somehow exert inhibitory effects in the β-cell, such as that just mentioned, was emphasized by Hellman and his colleagues, who postulated the existence of (an) inhibitory pool(s) of Ca^{2+} in the β-cell. The latter hypothesis is based on the following findings. First, the insulinotropic potency of high concentrations of calcium is increased when the β-cell has been previously exposed to calcium deprived media.[18] Such a pretreatment indeed induces a calcium depletion of the β-cell.[40]

Second, an increase in the extracellular calcium concentration from 2.6 to 20.5 mM induces a transient secretory response from perifused islets. When the extracellular calcium concentration is reduced to its original value, a transient stimulation of insulin release is noticed.[8]

Third, in islets exposed to glucose or IBMX,* a potent phosphodiesterase inhibitor, high concentrations of extracellular calcium (20 to 40 mM) exert a less potent insulinotropic effect than intermediate calcium concentrations (10 mM).[8]

Fourth, basal release of insulin is inhibited when β-cells are exposed to calcium analogues supposed to displace calcium from its extracellular binding sites and not to penetrate the β-cell. When the latter calcium analogue (lanthanum) is removed from the perifusate, an immediate, sustained, but modest increase in insulin release is observed as if lanthanum had displaced some inhibitory pool of calcium from the plasma membrane.[8]

Fifth, using ^{171}thullium, a radioactive lanthanide to titrate calcium binding sites at the level of the β-cell membrane, Hellman observed that glucose significantly inhibited the islet binding of radioactive thullium.[8]

Last, using chlortetracyline as a probe to investigate the interaction of calcium with the β-cell membrane, Taljedal observed that D-glucose reduced both the intensity and the polarization of the calcium-dependent chlortetracycline fluorescence.[72-73] The author was, however, unable to decide whether these changes reflected a reduction in the amount of calcium bound to the plasma membrane or an increase in the mobility of calcium in the membrane, a change which may reflect a facilitated transport of calcium across the plasma membrane.[74] Further work is required to assess the physiological significance of such hypothetical inhibitory pool(s) of calcium in the process of glucose-induced insulin release.

IV. MODALITIES OF ACTION OF GLUCOSE ON CALCIUM FLUXES

Theoretically, glucose could increase the cytosolic concentration of calcium by three different ways:

1. By increasing the rate of calcium entry into the β-cell
2. By reducing the outflow of the divalent cation across the plasma membrane
3. By provoking an intracellular redistribution of calcium, i.e., by increasing the uptake in and/or reducing the release of calcium from intracellular organelles

A. The Effect of Glucose on Calcium Inflow

Electrophysiological studies suggest that glucose stimulates the entry of calcium into the β-cell. The latter may occur through channels activated by membrane depolarization itself resulting from a decrease in membrane permeability to potassium.[68,76-77] Several studies have indeed shown that glucose reduced K⁺ permeability in islet cells.[78-80]

* 3-isobutyl-1-methylxanthine.

The study of ^{45}Ca efflux from perifused islets has largely confirmed this view. In addition, the radioisotopic studies demonstrated that glucose also reduced calcium efflux from the β-cell. Indeed, when isolated islets, prelabeled with ^{45}Ca, are washed and then perifused, glucose provokes a dual modification of ^{45}Ca efflux. This is best illustrated in Figure 3 which shows the effect of increasing concentrations of glucose on ^{45}Ca efflux from perifused rat pancreatic islets. At noninsulinotropic concentration (2.8, 5.6 mM), glucose only provokes a monophasic and sustained reduction in ^{45}Ca efflux. At a higher or insulinotropic concentration, glucose not only reduces ^{45}Ca efflux but triggers, in addition, a supplementary calcium movement consisting in a secondary rise in ^{45}Ca efflux. The latter movement occurs concomitantly with the release of insulin and masks the initial inhibitory effect of glucose on ^{45}Ca efflux. As illustrated in Figure 4, these two calcium movements display distinct sensitivities towards glucose.[81] The dose-action relationship characterizing the initial inhibitory effect of glucose on ^{45}Ca efflux displays an hyperbolic pattern with a threshold value below 2.8 mM and a half maximal effect at 4.2 mM. This differs vastly from that characterizing the stimulant action of glucose on ^{45}Ca efflux (secondary rise) which is sigmoidal with a threshold value for glucose close to 5.6 mM, and a half maximal effect at 9.1 mM. The latter curve is virtually superimposable to that characterizing the stimulant action of glucose on insulin release.[41] These observations clearly demonstrate that glucose triggers two calcium movements in the pancreatic β-cell.[81]

Although the secondary rise in ^{45}Ca efflux usually occurs concomitantly with insulin release, both phenomena can be dissociated from one another under several experimental conditions, e.g., at low temperature,[82-83] in the presence of somatostatine,[84] epinephrine,[83-84] or verapamil,[85] or in islets removed from fasting animals.[83] Such a dissociation strongly suggests that the secondary rise in ^{45}Ca efflux cannot be viewed as ^{45}Ca included in the secretory granules and released together with insulin at the time and site of exocytosis, as initially proposed by Malaisse et al.[86]

As an alternative explanation, we recently proposed that the secondary rise in ^{45}Ca efflux may reflect the entry of calcium into the β-cell through opened voltage-sensitive calcium channels and that it corresponds to a process of Ca-Ca exchange. Indeed, the glucose-induced secondary rise in ^{45}Ca efflux is impaired under several experimental conditions known to interfere with calcium entry[82,85-88] into the β-cell.[85] It is completely abolished in the absence of extracellular Ca^{2+} or presence of either cobalt[89] or nickel,[90] two specific blockers of the voltage sensitive calcium channels.[91-92]

An increase in ^{45}Ca efflux is observed in response to a depolarization of the β-cell membrane as induced by veratridine or an increase in the extracellular K^+ concentration.[93-94] In the presence of a high concentration of K^+, a close correlation was found between the ability of glucose to stimulate ^{45}Ca efflux and to depolarize the β-cell membrane.[94] Likewise, TEA, a specific blocker of K^+ conductance[95-96] acts synergistically with glucose to depolarize the β-cell membrane, to stimulate ^{45}Ca net uptake, and to increase ^{45}Ca efflux from perifused islets.[97] Quinine and 9-aminoacridine, which are more potent than TEA in reducing K^+ conductance in the β-cell,[98-99] reproduce the effect of glucose to depolarize the membrane, to stimulate ^{45}Ca net uptake, and to increase ^{45}Ca efflux from perifused islets.[99]

Taken as a whole, these data support the hypothesis issued from electrophysiological experiments, i.e., that glucose increases calcium uptake in the β-cell by activating voltage-sensitive calcium channels, the depolarization of the β-cell being the consequence of the inhibitory effect of glucose upon K^+ conductance. In considering the mechanism by which the inflow of ^{40}Ca provokes an outflow of ^{45}Ca, it was proposed that this represented a process of Ca-Ca exchange in which the inflow of ^{40}Ca displaces ^{45}Ca from its intracellular binding sites.[85,89] In agreement with such a view, an increased calcium inflow into the β-cell, as induced by an increase in the extracellular concentration of calcium, stimulates the efflux of ^{45}Ca from perifused islets.[85,87]

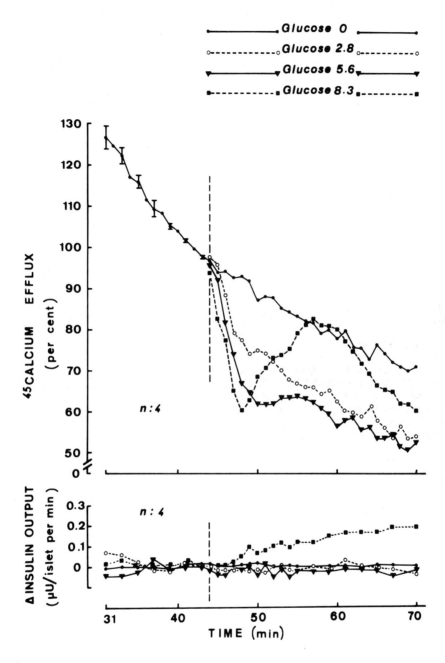

FIGURE 3. Effect of increasing glucose concentrations (2.8 mM ○---○, open circles and dotted line; 5.6 mM ▼—▼, triangles and solid lines; 8.3 mM ■—■, squares and solid lines) on ^{45}Ca efflux (upper panel) and insulin release (lower panel). Also shown are the results of control experiments performed in the absence of glucose (●—●, closed circles and solid lines). Mean values for ^{45}Ca efflux are expressed in percent of the mean value found within the same experiment between the 40th and 44th minute of perifusion and refer to four individual experiments in each case. Mean changes for insulin release are expressed in μU/islets per min and refer to the same experiments. The mean insulin output at the end of the control period averaged 0.21 ± 0.02 μU/islet per in. The SEM are not shown for the clarity of presentation.

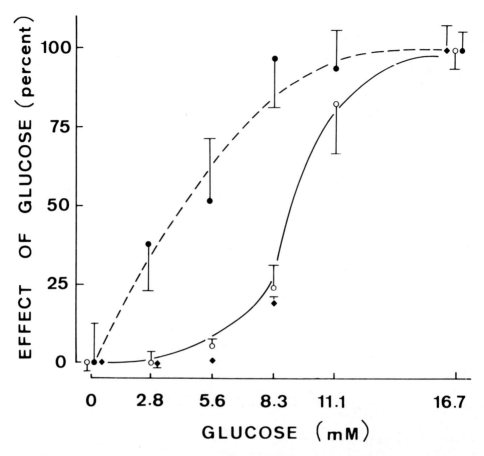

FIGURE 4. Dose-action relationship for the effects of glucose upon ^{45}Ca efflux and insulin release in perifused islets. Mean values (± SEM) are always expressed in percent of the mean reference value seen when the glucose concentration was raised from 0 to 16.7 mM. The *closed circles* refer to the initial inhibitory effect of glucose upon ^{45}Ca efflux: the glucose concentration of the perifusate was raised at the 45th minute from 0 to 2.8, 5.6, 8.3, 11.1, 16.7 mM; the differences between individual values for ^{45}Ca efflux 3 min after the introduction of glucose and the mean control value found at the same time when no glucose was administered are shown as a function of the glucose concentration. The *open circles* refer to the glucose induced secondary rise in ^{45}Ca efflux, when the glucose concentration was raised from 0 to 2.8, 5.6, 8.3, 11.1, or 16.7 mM; the data illustrated refer to the differences in ^{45}Ca efflux between the lowest value seen shortly after introduction of glucose and the highest value observed 3 to 9 min later. The *triangles* refer to the release of insulin seen when the glucose concentration was raised at the 45th minute from 0 to 2.8, 5.6, 8.3, or 16.7 mM; in each experiment, the integrated value for insulin output was calculated for the entire period of exposure to glucose (min 45 to 70).

B. The Effect of Glucose on Calcium Extrusion

It has been shown by Kikuchi, Wollheim, Cuendet, Renold, and Sharp[82] that the glucose-induced decrease in ^{45}Ca efflux reflected a true decrease of ^{45}Ca efflux from the islet cells.[82] It may, therefore, reflect the inhibition by glucose of calcium extrusion from the β-cell.

The existence of a calcium activated ATPase has been reported in insular homogenate of mice and rat.[100-101] No evidence is, however, available to suggest that such an ATPase participates in the active outward transport of calcium across the β-cell plasma membrane. Another mechanism susceptible to extrude calcium from cells against its electrochemical gradient is the Na-Ca countertransport. In this system, the entry of Na$^+$ into the cell down its electrochemical gradient provides the energy for the extrusion of calcium against its electrochemical gradient.[102-103]

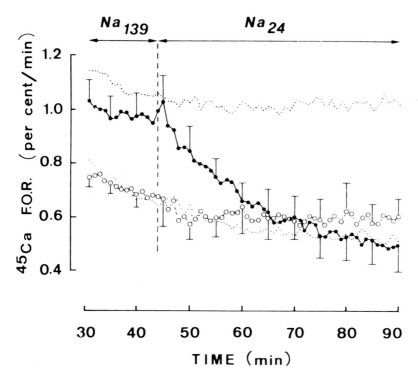

FIGURE 5. Effect of partial replacement of NaCl by sucrose on ^{45}Ca efflux from rat pancreatic islets perifused with a medium deprived of calcium and enriched with 0.5 mM EGTA (ethylene glycol-bis-(aminoethylether) N,N-tetraacetic acid); (●) refer to experiments made in the absence of glucose; (O) refer to experiments made in the presence of glucose 16.7 mM. The stippled lines represent control experiments performed throughout at normal extracellular sodium concentration (139 mM) either in the absence (upper line) or presence (lower line) of glucose 16.7 mM. Mean values for ^{45}Ca efflux are expressed as a fractional outflow rate.

Recently, we have obtained evidence that such an ionophoretic system represents the mechanism by which calcium is actively extruded from the β-cell.[104] An inhibition of the process of Na-Ca countertransport may represent the mechanism by which glucose inhibits ^{45}Ca efflux from perifused islets.[104] The data illustrated in Figure 5 show that a reduction in the extracellular Na$^+$ concentration provokes an immediate and sustained decrease in ^{45}Ca efflux when the islets are perifused in the absence of both extracellular calcium and glucose. Such a decrease fails to occur when the islets are perifused in the presence of glucose. When the extracellular Na$^+$ concentration was decreased, the ^{45}Ca fractional outflow rate from islets perifused in the absence of glucose reaches values similar to those of islets perifused in the presence of glucose. Similar experiments carried out in the presence of extracellular calcium confirmed (1) the existence in islet cells of a process of Na-Ca countertransport apparently responsible for calcium extrusion and (2) its inhibition by glucose.[104]

It has been suggested that a glucose-induced increase in intracellular Na$^+$ concentration may mediate the inhibitory effect of glucose on ^{45}Ca efflux.[105] In our system, however, an increase in intracellular Na$^+$ concentration failed to reduce ^{45}Ca efflux from perifused islets (whether the increase in intracellular Na$^+$ concentration was due to veratridine or to a reduction in the extracellular K$^+$ concentration) and, instead, provoked a dramatic increase in ^{45}Ca efflux.[93] The latter finding is compatible with the existence, in islet cells, of a process of sodium-calcium exchange localized at the level of intracellular organelles.[93] In nerve tissue, the process of Na-Ca countertransport appears dependent on the membrane potential, the ionophoretic process being inhibited by membrane depolarization.[106] This, however, does not appear to be the case in islet cells. Thus, no reduction in ^{45}Ca efflux is observed when the islets are exposed to veratridine or to a high concentration of extracellular K^+.[93-94] Furthermore, TEA and 9-aminoacridine fail to reduce ^{45}Ca efflux from perifused islets.[97,99]

In conclusion, the inhibition by glucose of K^+ conductance and subsequent membrane depolarization does not represent the mechanism by which glucose inhibits Na-Ca countertransport and, by doing so, reduces ^{45}Ca efflux from perifused islets.[97,99] Incidentally, quinine reduces ^{45}Ca efflux from perifused islets. However, the latter effect was shown to result from a direct effect of the drug upon Na-Ca countertransport and was not related with the ability of quinine to reduce K^+ conductance.[99]

C. The Links Between Glucose Metabolism and Calcium Fluxes

Like the release of insulin, the two above-mentioned glucose-induced calcium movements are dependent on the integrity of glucose metabolism in the β-cell.[81] Both movements are indeed impaired in the presence of metabolic inhibitors such as mannoheptulose or iodoacetate,[81] and may be evoked by other nutrient secretagogues, such as glyceraldehyde[81] and α-ketoisocaproic acid.[107]

The glucose-induced stimulation of calcium inflow and subsequent increase in ^{45}Ca efflux are inhibited in an exquisitely sensitive manner, by menadione[108] or NH_4^+.[109] These agents decrease the islet content of reduced pyridine nucleotides.[109-110] The latter findings suggest, therefore, that a glucose-induced increase in the islet content of NAD(P)H plays a role in the mechanism by which glucose increases calcium inflow into the β-cell. Incidentally, the inhibitory effect of glucose on K^+ conductance also appears dependent on the ability of glucose to increase the endogenous production of reduced pyridine nucleotides.[108,111-112] By modifying the β-cell redox state, glucose may thus affect the membrane permeability of potassium, depolarize the membrane and, hence, activate voltage-sensitive calcium channels.[113]

The inhibitory effect of glucose on ^{45}Ca efflux appears to be less sensitive to experimental conditions interfering with the islet content of reduced pyridine nucleotides,[108-109] and may be dependent on the generation of protons through the metabolism of glucose.[114-115] Intracellular acidification, as induced by an increase in the extracellular partial pressure of CO_2 mimics the inhibitory effect of glucose on ^{45}Ca efflux.[114] It is conceivable that the inhibitory effect of glucose on ^{45}Ca efflux may result from a competition of H^+ with Ca^{2+} for exit by the Na-Ca countertransport process.[114,115] Using native ionophores extracted from the islets, Anjaneyulu et al.[116] were recently able to reproduce, in an artificial system, the process of Na-Ca countertransport and its inhibition in response to a change in pH.

Incidentally, intracellular acidification as provoked by the increase in pCO_2, does not reproduce the inhibitory effect of glucose on K^+ conductance.[117] The latter observation again underlines the dissociation existing between the inhibitory effect of glucose on potassium conductance and calcium outflow, respectively.

D. Effect of Glucose on Intracellular Calcium Distribution

The study by electron microscopy of the intracellular localization of calcium by the pyroantimonate precipitation technique, reveals that glucose provokes the accumulation of calcium at the level of the plasma membrane and in the halo surrounding the granule core of secretory granules.[38-40]

When the β-cell is exposed to a stimulatory glucose concentration, the calcium deposits initially appear at the level of the plasma membrane and later into the secretory granules.[118] Significant amounts of calcium have also been found in β-cell mitochondria either by electron microscopy, autoradiography, or by X-ray microanalysis of frozen sections of unfixed islet tissue prepared by cryoultramicrotomy.[49,119] In agreement with the latter observations, glucose was found to promote the accumulation of barium in β-cell mitochondria, the divalent cation being used as an electron opaque substitute for calcium.[120]

Tracer techniques have been another mode of approach to study the uptake of Ca^{2+} by β-cell organelles. For this purpose, the uptake was either studied by incubating islet subcellular fractions in the presence of ^{45}Ca (and other test substances) or by measuring the amount of ^{45}Ca accumulated in subcellular fractions of intact islets previously incubated in the presence of the tracer.

With the first technique, it was observed that ATP promoted the accumulation of calcium in subcellular fractions enriched either in mitochondria,[49,121] secretory granules,[49-122] or endoplasmic reticulum.[122] In each of these fractions, cyclic-AMP was reported to significantly inhibit calcium accumulation.[49,122] These data are compatible with the view that most of the calcium taken up by the β-cell in response to glucose is rapidly trapped by intracellular organelles and that cyclic-AMP potentiates insulin release, at least in part, by decreasing this sequestering phenomenon.

With the second technique, it was observed that glucose enhanced ^{45}Ca accumulation in secretory granules,[123] mitochondria and nuclei-cell debris fractions.[124] In the latter study, theophylline and 3-isobutyl-1-methylxanthine, a potent phosphodiesterase inhibitor, was found to reduce glucose-induced accumulation of ^{45}Ca in the mitochondria-reach subcellular fraction.

In considering the mechanism by which Ca^{2+} accumulates into secretory granules, Hellman postulated, by analogy with the situation found in chromaffin secretory granules,[125-126] that the uptake occurred at the intervention of a "proton translocating ATPase" located at the granule membrane. Direct evidence for the existence of such a process is, however, still lacking.

The stimulation of insulin release as induced by a square wave increase in glucose concentration is characterized by a biphasic pattern. It consists in an initial and rapid surge in insulin release followed by a slow increase of the secretory rate.[127] Recently, Wollheim suggested that intracellular calcium stores may be of significance in the latter release pattern. More precisely, it was proposed that the first phase of insulin release does not depend on the uptake of extracellular calcium, but rather, results from an intracellular calcium redistribution. This hypothesis is essentially based on two main findings. First, in shortterm experiments (5 min), verapamil blocks glucose-induced ^{45}Ca uptake without affecting insulin release.[128] Second, when islets are perifused in the absence of extracellular calcium, the first phase of insulin release which is abolished in normal islets, is preserved when the islets contain increased endogenous calcium stores.[129] However, these data should be weighed against the evidence provided by other authors who suggested either that Ca^{2+} influx into the β-cells is required for glucose to trigger the initial phase of insulin release[130] or that both phases of insulin release are dependent on the availability of minute amounts of extracellular calcium even when extracellular Mg^{2+} is removed from the medium to facilitate Ca^{2+} entry into the islet cells.[7]

ACKNOWLEDGMENTS

The work presented in this review was supported in parts by grants from the Belgian Foundation for Scientific Medical Research. The authors are indebted to J. L. Servranckx and J. De Ligne for secretarial help.

REFERENCES

1. Grodsky, G. M. and Bennett, L. L., Cation requirements for insulin secretion in the isolated perfused pancreas, *Diabetes,* 15, 910, 1966.
2. Curry, D. L., Bennett, L. L., and Grodsky, G. M., Requirement for calcium ion in insulin secretion by the perfused rat pancreas, *Am. J. Physiol.,* 214, 174, 1967.
3. Bennett, L. L., Curry, D. L., and Grodsky, G. M., Calcium-magnesium antagonism in insulin secretion by the perfused rat pancreas, *Endocrinology,* 85, 594, 1969.
4. Milner, R. D. G. and Hales, C. N., The role of calcium and magnesium in insulin secretion from rabbit pancreas studied in vitro, *Diabetologia,* 3, 47, 1967.
5. Hales, C. N. and Milner, R. D. G., Cations and the secretion of insulin from rabbit pancreas in vitro, *J. Physiol. (London),* 199, 177, 1968.
6. Milner, R. D. G. and Hales, C. N., Cations and the secretion of insulin, *Biochim. Biophys. Acta,* 150, 165, 1968.
7. Somers, G., Devis, G., and Malaisse, W. J., Calcium antagonists and islet function. IX. Is extracellular calcium required for insulin release?, *Acta Diabetol. Lat.,* 16, 9, 1979.
8. Hellman, B., Andersson, T., Berggren, P. O., Flatt, P., Gyfle, E., and Kohnert, K. D., The role of calcium in insulin secretion, in *Hormones and Cell Regulation,* Vol. 3, Dumont J., and Nunez, J. Eds., Elsevier/North-Holland, 1979, 69.
9. Malaisse, W. J., Brisson, G., and Malaisse-lagae, F., The stimulus-secretion coupling of glucose-induced insulin release. I. Interaction of epinephrine and alcaline earth cations, *J. Lab. Clin. Med.,* 76, 895, 1970.
10. Henquin, J. C. and Lambert, A. E., Cobalt inhibition of insulin secretion and calcium uptake by isolated rat islets, *Am. J. Physiol.,* 228, 1669, 1975.
11. Malaisse, W. J., Devis, G., Herchuelz, A., Sener, A., and Somers, G., Calcium antagonists and islet function. VIII. The effect of magnesium, *Diabete Metabol.,* 2, 1, 1976.
12. Malaisse, W. J., Herchuelz, A., Levy, J., and Sener, A., Calcium antagonists and islet function, III. The possible site of action of verapamil, *Biochem. Pharmacol.,* 26, 735, 1977.
13. Malaisse, W. J., Calcium-antagonists and islet function. X. Effect of suloctidil, *Arch. Intern. Pharmacodyn. Ther.,* 228, 339, 1977.
14. Wollheim, C. B., Blondel, B., Trueheart, P. A., Renold, A. E., and Sharp, G. W. G., Calcium-induced insulin release in monolayer culture of the endocrine pancreas. Studies with ionophore A23187, *J. Biol. Chem.,* 250(4), 1354, 1975.
15. Charles, M. A., Lawecki, J., Pictet, R., and Grodsky, G. M., Insulin secretion. Interrelationships of glucose, cyclic adenosine 3′ 5′-monophosphate, and calcium, *J. Biol. Chem.,* 250, 6134, 1975.
16. Somers, G., Devis, G., and Malaisse, W. J., Analogy between native and exogenous ionophores in the pancreatic β-cell, *FEBS Lett.,* 66, 20, 1976.
17. Conaway, H. H., Griffey, M. A., Marks, S. R., and Whitney, J. E., Ionophore A23187-induced insulin secretion in the isolated perfused dog pancreas, *Horm. Metab. Res.,* 8, 351, 1976.
18. Devis, G., Somers, G., and Malaisse, W. J., Stimulation of insulin release by calcium, *Biochem. Biophys. Res. Commun.,* 67, 525, 1975.
19. Hellman, B., Stimulation of insulin release after raising extracellular calcium, *FEBS Lett.,* 63, 125, 1976.
20. Littledike, E. T., Witzel, D. A., and Whipp, S. C., Insulin: evidence for inhibition of release in spontaneous hypocalcemia, *Proc. Soc. Exp. Biol.,* 129, 135, 1968.
21. Laron, Z. and Rosenberg, T., Inhibition of insulin release and stimulation of growth hormone release by hypocalcemia in a boy, *Horm. Metab. Res.,* 2, 121, 1970.
22. Witzel, D. A. and Littledike, E. T., Suppression of insulin secretion during induced hypocalcemia, *Endocrinology,* 93, 761, 1973.
23. Pento, J. T., Kagan, A., and Glick, S. M., Influence of altered states of calcium homeostasis on insulin secretion in rats and rabbits, *Horm. Metab. Res.,* 6, 177, 1974.
24. Amend, W. J. C., Steinberg, S. M., Lowrie, E. G., Lazarus, J. M., Soeldner, J. S., Hampers, C. L., and Merril, J. P., The influence of serum calcium and parathyroid hormone upon glucose metabolism in uremia, *J. Lab. Clin. Med.,* 86, 435, 1975.
25. Yasuda, K., Hurukawa, Y., Okuyama, M., Kikuchi, M., and Yoshinaga, K., Glucose tolerance and insulin secretion in patients with parathyroid disorders, *N. Engl. J. Med.,* 292, 501, 1975.
26. Kim, H., Kalkhoff, R. K., Castrini, N. V., Cerletty, J. M., and Jacobson, M., Plasma insulin disturbances in primary hyperparathyroidism, *J. Clin. Invest.,* 50, 2596, 1971.
27. Ziegler, R., Bellwinkel, S., Schmidtchen, D., and Minne, H., Effects of hypercalcemia, hypercalcemia and calcitonine on glucose-stimulated insulin secretion in man, *Horm. Metab. Res.,* 4, 60, 1972.

28. Harter, H. R., Santiago, J. V., Rutherford, W. E., Slatopolsky, E., and Klahr, S., The relative roles of calcium, phosphorus and parathyroid hormone in glucose- and tolbutamide-mediated insulin release, *J. Clin. Invest.*, 58, 359, 1976.
29. Hague, R. V., Cullen, D. R., and Martin, T. J., Glucose-induced insulin secretion in patients with parathyroid disorders, *Clin. Endocrinol.*, 9, 283, 1978.
30. Malaisse-lagae, F. and Malaisse, W. J., Stimulus-secretion coupling of glucose-induced insulin release. III. Uptake of ^{45}calcium by isolated islets of Langerhans, *Endocrinology*, 88, 72, 1971.
31. Wollheim, C. B., Kikuchi, M., Renold, A. E., and Sharp, G. W. G., Somatostatin- and epinephrine-induced modifications of $^{45}Ca^{++}$ fluxes and insulin release in rat pancreatic islets maintained in tissue culture, *J. Clin. Invest.*, 60, 1165, 1977.
32. Naber, S. P., McDaniel, M. L., and Lacy, P. E., The effect of glucose on the acute uptake and efflux of calcium-45 in isolated rat islets, *Endocrinology*, 101, 686, 1977.
33. Malaisse, W. J., Hutton, J. C., Sener, A., Levy, J., Herchuelz, A., Devis, G., and Somers, G., Calcium antagonists and islet function. VII. Effect of calcium deprivation, *J. Membr. Biol.*, 38, 193, 1978.
34. Frankel, B. J., Kromhout, J. A., Imagawa, W., Landahl, H. D, and Grodsky, G. M., Glucose-stimulated ^{45}Ca uptake in isolated rat islets, *Diabetes*, 27, 365, 1978.
35. Hellman, B., Sehlin, J., and Taljedal, I.-B., Effects of glucose on $^{45}Ca^{2+}$ uptake by pancreatic islets as studied with the lanthanum method, *J. Physiol. (London)*, 254, 639, 1976.
36. Malaisse, W. J., Herchuelz, A., Devis, G., Somers, G., Boschero, A. C., Hutton, J. C., Kawazu, S., Sener, A., Atwater, I. J., Duncan, G., Ribalet, B., and Rojas, E., Regulation of calcium fluxes and their regulatory roles in pancreatic islets, *Ann. N.Y. Acad. Sci.*, 307, 562, 1978.
37. Hellman, B., Abrahamson, H., Andersson, T., Berggren, P. O., Flatt, P., Gylfe, E., and Hahn, H. J., Calcium movements in relation to glucose-stimulated insulin secretion, *Horm. Metab. Res.*, Suppl. 10, 122, 1980.
38. Herman, L., Sato, T., and Hales, C. N., The electron microscopic localization of cations to pancreatic islets of Langerhans and their possible role in insulin secretion, *J. Ultrastruct. Res.*, 42, 298, 1973.
39. Schäfer, H. J. and Klöppel, G., The significance of calcium in insulin secretion. Ultrastructural studies on identification and localization of calcium in activated and inactivated β-cells of mice, *Virchows Arch. A.*, 362, 231, 1974.
40. Ravazzola, M. F., Malaisse-lagae, F., Amherdt, M., Perrelet, A., Malaisse, W. J., and Orci, L., Patterns of calcium localization in pancreatic endocrine cells, *J. Cell. Sci.*, 27, 107, 1976.
41. Malaisse, W. J., Malaisse-lagae, F., and Wright, P. H., A new method for the measurement in vitro of pancreatic insulin secretion, *Endocrinology*, 80, 99, 1967.
42. Malaisse, W. J., Role of cations, in *Handbook of Experimental Pharmacology*, Hasselblatt, A., and Bruchhausen, F. V., Eds., Springer-verlag Berlin, 1975, 145.
43. Malaisse-lagae, F., Brisson, G. R., and Malaisse, W. J., The stimulus-secretion coupling of glucose-induced insulin release. VI. Analogy between the insulinotropic mechanisms of sugars and amino acids, *Horm. Metab. Res.*, 3, 374, 1971.
44. Turtle, J. R. and Kipnis, D. M., An adrenergic receptor mechanism for the control of cyclic 3′ 5′ adenosine monophosphate synthesis in tissues, *Biochem. Biophys. Res. Commun.*, 28, 797, 1967.
45. Montague, W. and Cook, J. R., The role of adenosine 3′ 5′ cyclic monophosphate in the regulation of insulin release by isolated rat islets of Langerhans, *Biochem. J.*, 122, 115, 1971.
46. Malaisse, W. J., Malaisse-Lagae, F., and Mayhew, D., A possible role for the adenylcyclase system in insulin secretion, *J. Clin. Invest.*, 46, 1724, 1967.
47. Brisson, G. R., Malaisse-Lagae, F., and Malaisse, W. J., The stimulus-secretion coupling of glucose-induced insulin release. VII. A proposed site of action for adenosine — 3′ 5′-cyclic monophosphate, *J. Clin. Invest.*, 51, 232, 1972.
48. Brisson, G. R. and Malaisse, W. J., The stimulus-secretion coupling of glucose-induced insulin release. XI. Effect of theophylline and epinephrine on ^{45}Ca efflux from perifused islets, *Metabolism*, 22, 455, 1973.
49. Howell, S. L., Montague, W., and Tyhurst, M., Calcium distribution in islets of Langerhans: a study of calcium condentration and of calcium accumulation in B-cell organelles, *J. Cell Sci.*, 19, 395, 1975.
50. Malaisse, W. J. and Malaisse-Lagae, F., A possible role for calcium in the stimulus-secretion coupling for glucose-induced insulin secretion, *Acta Diabet. Lat.*, 7, 264, 1970.
51. Lacy, P. E., Howell, S. L., Young, D. A., and Fink, C. J., New hypothesis of insulin secretion, *Nature (London)*, 219, 1177, 1968.
52. Lacy, P. E. and Malaisse, W. J., Microtubules and beta cell secretion, *Recent Progr. Horm. Res.*, 29, 199, 1973.

53. Malaisse, W. J., Devis, G., Somers, G., Van Obberghen, E., Malaisse-Lagae, F., Ravazzola, M., Blondel, B., and Orci, L., The role of microtubules and microfilaments in secretion, in *Stimulus-Secretion Coupling in the Gastrointestinal Tract*, Case, R. M. and Goebell, H., Eds., MTP, Lancaster, 1975, 65.
54. Somers, G., Blondel, B., Orci, L., and Malaisse, W. J., Motile events in pancreatic endocrine cells, *Endocrinology*, 104, 255, 1979.
55. Dean, P. M., Surface electrostatic-charge measurement on islet and zymogen granules: effect of calcium ions, *Diabetologia*, 10, 427, 1974.
56. Grill, V. and Cerasi, E., Activation by glucose of adenylcyclase in pancreatic islets of the rat, *Febs Lett.*, 33, 311, 1973.
57. Grill, V. and Cerasi, E., Stimulation by D-glucose of cyclic adenosine 3'5'-monophosphate accumulation and insulin release in isolated pancreatic islets of the rat, *J. Biol. Chem.*, 249, 4196, 1974.
58. Zawalich, W. S., Karl, R. C., Ferrendelli, J. A., and Matschinsky, F. M., Factors governing glucose-induced elevation of cyclic 3'5' AMP levels in pancreatic islets, *Diabetologia*, 11, 231, 1975.
59. Hellman, B., Effect of starvation and Ca^{++} on glucose-induced accumulation of 3'5'-AMP in pancreatic islets, *Experientia*, 32, 155, 1976.
60. Katada, T. and Ui, M., Islet-activating protein. Enhanced insulin secretion and cyclic AMP accumulation in pancreatic islets due to activation of native calcium ionophores, *J. Biol. Chem.*, 254, 469, 1979.
61. Valverde, I., Vandermeers, A., Anjaneyulu, R., and Malaisse, W. J., Calmodulin activation of adenyl-cyclase in pancreatic islets *Science*, 206, 225, 1979.
62. Sugden, M. C., Christie, M. R., and Ashcroft, S. J. H., Presence and possible role of calcium-dependent regulator (calmodulin) in rat islets of Langerhans, *Febs Lett.*, 105, 95, 1979.
63. Sharp, G. W. G., Wiedenkeller, D. E., Kaelin, D., Siegel, E. G., and Wollheim, C. B., Stimulation of adenylate cyclase by Ca^{2+} and calmodulin in rat islets of Langerhans explanation for the glucose-induced increase in cyclic AMP levels, *Diabetes*, 29, 74, 1980.
64. Brostrom, C. D., Huang Y. C., Breckenridge, B., and Wolff, D. J., Identification of a calcium-binding protein as a calcium-dependent regulator of brain adenylate cyclase, *Proc. Natl. Acad. Sci. USA*, 72, 64, 1975.
65. Brostrom, C. O., Brostrom, M. A., and Wolff, D. J., Calcium-dependent adenylate cyclase from rat cerebral cortex, *J. Biol. Chem.*, 252, 5677, 1977.
66. Meissner, H. P. and Schmelz, H., Membrane potential of beta-cells in pancreatic islets, *Pfluegers Arch.*, 351, 195, 1974.
67. Meissner, H. P., Electrical characteristics of the beta-cells in pancreatic islets, *J. Physiol. (Paris)*, 72, 757, 1976.
68. Matthews, E. K., Calcium and stimulus-secretion coupling in pancreatic islet cells, in *Calcium Transport in Contraction and Secretion*, Carafoli, E., Clementi, F., Drabikowsky, W., and Margreth, A., Eds., North-Holland, Amsterdam, 1975, 203.
69. Atwater, I. and Beigelman, P. M., Dynamic characteristics of electrical activity in pancreatic β-cells. I. Effects of calcium and magnesium removal, *J. Physiol. (Paris)*, 72, 769, 1976.
70. Atwater, I., Dawson, C. M., Ribalet, B., and Rojas, E., Potassium permeability activated by intracellular calcium ion concentration in the pancreatic β-cell, *J. Physiol. (London)*, 288, 575, 1979.
71. Henquin, J. C., Meissner, H. P., and Preissler, M., 9-aminoacridine- and tetraethylammonium induced reduction of the potassium permeability in pancreatic β-cells. Effects on insulin release and electrical properties, *Biochim. Biophys. Acta*, 587, 579, 1979.
72. Täljedal, I.-B., Chlorotetracycline as a fluorescent Ca^{2+} probe in pancreatic islet cells. Methodological aspects and effects of alloxan, sugars, methylxanthines, and Mg^{2+}, *J. Cell. Biol.*, 76, 652, 1978.
73. Täljedal, I.-B., Polarization of chlorotetracycline fluorescence in pancreatic islet cell and its response to calcium ions and D-glucose, *Biochim. J.*, 178, 187, 1979.
74. Täljedal, I.-B., Fluorescent probing of calcium ions in islet cells, *Horm. Metabol. Res.*, Suppl. 10, 130, 1980.
75. Dean, P. M. and Matthews, E. K., Electrical activity in pancreatic islet cells: effect of ions, *J. Physiol. (London)*, 210, 265, 1970.
76. Atwater, I., Ribalet, R., and Rojas, E., Cyclic changes in potential and resistance of the β-cell membrane induced by glucose in islets of Langerhans from mouse, *J. Physiol. (London)*, 278, 117, 1978.
77. Meissner, H. P. and Preissler, M., Glucose-induced changes of the membrane potential of pancreatic β-cells: their significance for the regulation of insulin release, in *Treatment of Early Diabetes*, Camerini-Davalos, R. A. and Hanover, B., Eds., Plenum Press, New York, 1979, 97.
78. Sehlin, J. and Täljedal, I. B., Glucose-induced decrease in Rb^+ permeability in pancreatic β-cells, *Nature (London)*, 253, 635, 1975.
79. Boschero, A. C., Kawazu, S., Duncan, G., and Malaisse, W. J., Effect of glucose on K^+ handling by pancreatic islets, *FEBS Lett.*, 83, 151, 1977.

80. Henquin, J. C., D-glucose inhibits potassium efflux from pancreatic islet cells, *Nature (London),* 271, 271, 1978.
81. Herchuelz, A. and Malaisse, W. J., Regulation of calcium fluxes in pancreatic islets: two calcium movements' dissociated response to glucose, *Am. J. Physiol.,* 238, E87, 1980.
82. Kikuchi, M., Wollheim, C. B., Cuendet, G. S., Renold, A. E., and Sharp, G. W. G., Studies on the dual effects of glucose in ^{45}Ca efflux from isolated rat islets, *Endocrinology,* 102, 1339, 1978.
83. Gylfe, E. and Hellman, B., Calcium and pancreatic β-cell function. II. Mobilization of glucose-sensitive ^{45}Ca from perifused islets rich in β-cells, *Biochim. Biophys. Acta,* 538, 249, 1978.
84. Wollheim, C. B., Kikuchi, M., Renold, A. E., and Sharp, G. W. G., Somatostatin and epinephrine induced modifications of ^{45}Ca^{++} fluxes and insulin release in rat pancreatic islets maintained in tissue culture, *J. Clin. Invest.,* 60, 1165, 1977.
85. Herchuelz, A. and Malaisse, W. J., Regulation of calcium fluxes in pancreatic islets: dissociation between calcium and insulin release, *J. Physiol. (London),* 283, 409, 1978.
86. Malaisse, W. J., Brisson, G. R., and Baird, L. E., Stimulus-secretion coupling of glucose-induced insulin release. X. Effect of glucose on ^{45}Ca efflux from perifused islets, *Am. J. Physiol.,* 224, 389, 1973.
87. Frankel, B. J., Imagawa, W. T., O'Connor, M. D. L., Lundquist, I., Kromhout, J. A., Fanska, R. E., and Grodsky, G. M., Glucose-stimulated ^{45}Calcium efflux from isolated rat pancreatic islets, *J. Clin. Invest.,* 62, 525, 1978.
88. Gylfe, E., Buitrago, A., Berggren, P. O., Hammarström, K., and Hellman, B., Glucose inibition of ^{45}Ca efflux from pancreatic islets, *Am. J. Physiol.,* 235, E191, 1978.
89. Herchuelz, A., Couturier, E., and Malaisse, W. J., Regulation of calcium fluxes in pancreatic islets. Glucose-induced calcium-calcium exchange, *Am. J. Physiol.,* 238, E96, 1980.
90. Bukowiecki, L. and Freinkel, N., Relationship between efflux of ionic calcium and phosphorus during excitation of pancreatic islets with glucose, *Biochim. Biophys. Acta,* 436, 190, 1976.
91. Baker, P. F., Meves, H., and Ridgway, F. B., Effect of manganese and other agents on the calcium uptake that follows depolarization of squid axons, *J. Physiol.,* 231, 511, 1973.
92. Kohlhard, M., Bauer, B., Krause, H., and Fleckenstein, A., Selective inhibition of the transmembrane Ca conductivity of mammalian myocardial fibers by Ni, Co and Mn ions, *Pfleugers Arch.,* 338, 115, 1973.
93. Herchuelz, A. and Malaisse, W. J., Regulation of calcium fluxes in rat pancreatic islets: dissimilar effects of glucose and of sodium ion accumulation, *J. Physiol. (London),* 302, 263, 1980.
94. Herchuelz, A., Thonnart, N., Sener, A., and Malaisse, W. J., Regulation of calcium fluxes in pancreatic islets. The role of membrane depolarization, *Endocrinology,* 107, 491, 1980.
95. Hille, B., Ionic channels in nerve membranes, *Progr. Biophys. Mol. Biol.,* 21, 3, 1970.
96. Narahashi, T., Chemicals as tools in the study of exitable membranes, *Physiol. Rev.,* 54, 813, 1974.
97. Herchuelz, A., Thonnart, N., Carpinelli, A., Sener, A., and Malaisse, W. J., Regulation of calcium fluxes in rat pancreatic islets. The role of K$^+$ conductance, *J. Pharmacol. Exp. Ther.,* 215, 213, 1980.
98. Carpinelli, A. and Malaisse, W. J., Regulation of ^{86}Rb outflow from pancreaic islets. I. Reciprocal changes in the response to glucose, tetraethylammonium and quinine, *Mol. Cell. Endocrinol.,* 17, 103, 1980.
99. Herchuelz, A., Lebrun, P., Carpinelli, A., Thonnart, N., Sener, A., and Malaisse, W. J., Regulation of calcium fluxes in rat pancreatic islets. Quinine mimics the dual effect of glucose on calcium movements, *Biochim. Biophys. Acta,* 640, 10, 1981.
100. Formby, B., Capito, K., Egeberg, J., and Hedeskow, C. J., Ca-activated ATPase activity in subcellular fractions of mouse pancreatic islets, *Am. J. Physiol.,* 230, 441, 1978.
101. Levin, S. R., Kasson, B. G., and Driessen, J. F., Adenosine triphosphatase of rat pancreatic islets, *J. Clin. Invest.,* 62, 692, 1978.
102. Baker, P. F., Transport and metabolism of calcium ions in nerve, *Progr. Biophys. Mol. Biol.,* 24, 177, 1972.
103. Blaustein, M. P., The interrelationship between sodium and calcium fluxes across cell membranes, *Rev. Physiol. Biochem. Pharmacol.,* 70, 33, 1974.
104. Herchuelz, A., Sener, A., and Malaisse, W. J., Regulation of calcium fluxes in rat pancreatic islets. IV. Calcium intrusion by sodium- calcium counter transport, *J. Membr. Biol.,* 57, 1, 1980.
105. Donatsch, P., Lowe, D. A., Richardson, B. P., and Taylor, P., The functional significance of sodium channels in pancreatic beta-cells membranes, *J. Physiol.,* 267, 357, 1977.
106. Mullins, L. J. and Brinley, J. R., Sensitivity of calcium efflux from squid axon to changes in membrane potential, *J. Gen. Physiol.,* 65, 135, 1975.
107. Hutton, J. C., Sener, A., Herchuelz, A., Atwater, I., Kawazu, S., Boschero, A. C., Somers, G., Devis, G., and Malaisse, W. J., Similarities in the stimulus-secretion coupling mechanisms of glucose and 2-keto acid-induced insulin release, *Endocrinology,* 106, 203, 1980.

108. Malaisse, W. J., Sener, A., Boschero, A. C., Kawazu, S., Devis, G., and Somers, G., The stimulus-secretion coupling of glucose-induced insulin release. Cationic and secretory effects of menadione in the endocrine pancreas, *Eur. J. Biochem.*, 87, 111, 1978.
109. Sener, A., Hutton, J. C., Kawazu, S., Boschero, A. C., Somers, G., and Devis, G., The stimulus-secretion coupling of glucose-induced insulin release. Metabolic and functional effects of NH_4^+ in rat islets, *J. Clin. Invest.*, 62, 868, 1978.
110. Malaisse, W. J., Hutton, J. C., Kawazu, S., and Sener, A., The stimulus-secretion coupling of glucose-induced insulin release. Metabolic effects of Menadione in isolated islets, *Eur. J. Biochem.*, 87, 111, 1978.
111. Boschero, A. C. and Malaisse, W. J., Stimulus-secretion coupling of glucose-induced insulin release. XXIX. Regulation of ^{86}Rb efflux from perifused islets, *Am. J. Physiol.*, 236, E139, 1979.
112. Henquin, J. C., The potassium permeability of pancreatic islet cells: mechanisms of control and influence on insulin release, *Horm. Metab. Res.*, Suppl. 10, 66, 1980.
113. Malaisse, W. J., Sener, A., Herchuelz, A., and Hutton, J. C., Insulin release: the fuel hypothesis, *Metabolism*, 28, 373, 1979.
114. Carpinelli, A. R., Sener, A., Herchuelz, A., and Malaisse, W. J., The stimulus-secretion coupling of glucose-induced insulin release. XXXX. Effect of intracellular acidification upon calcium efflux from islet cells, *Metabolism*, 29, 540, 1980.
115. Malaisse, W. J., Herchuelz, A., and Sener, A., The possible significance of intracellular pH in insulin release, *Life Sci.*, 26, 1367, 1980.
116. Anjaneyulu, K., Anjaneyulu, R., and Malaisse, W. J., The stimulus-secretion coupling of glucose-induced insulin release. XLIII. Na-Ca counter transport mediated by pancreatic islet native ionophores, *J. Inorg. Biochem.*, 1980, in press.
117. Carpinelli, A. and Malaisse, W. J., Regulation of ^{86}Rb outflow from pancreatic islets. II. Effect of changes in extracellular and intracellular pH, *Diabete Metab.*, 6, 193, 1980.
118. Klöppel, G. and Bommer, G., Ultracytochemical calcium distribution in β-cells in relation to biphasic glucose-stimulated insulin release by the perfused rat pancreas, *Diabetes*, 28, 585, 1979.
119. Howell, S. L. and Tyhurst, M., ^{45}Calcium localization in islets of Langerhans, a study by electron-microscopy autoradiography, *J. Cell. Sci.*, 21, 415, 1976.
120. Howell, S. L. and Tyhurst, M., Barium accumulation in rat pancreatic β-cells, *J. Cell. Sci.*, 22, 455, 1976.
121. Sugden, M. C. and Ashcroft, S. J. H., Effects of phosphoenolpyruvate, other glycolytic intermediates and methylxanthines on calcium uptake by a mitochondrial fraction from rat pancreatic islets, *Diabetologia*, 15, 173, 1980.
122. Sehlin, J., Calcium uptake by subcellular fractions of pancreatic islets. Effects of nucleotides and theophylline, *Biochem. J.*, 156, 63, 1976.
123. Bloom, G. D., Hellman, B., Sehlin, J., and Täljedal, I. B., Glucose stimulated and La^{3+} nondisplaceable Ca^{2+} pool in pancreatic islets, *Am. J. Physiol.*, 232, E114, 1977.
124. Hahn, H. J., Gylfe, E, and Hellman, B., Glucose-dependent effect of methylxanthines on the ^{45}Ca distribution in pancreatic β-cells, *FEBS Lett.*, 103, 348, 1979.
125. Njus, D. and Radda, G. K., Bioenergetic processes in chromaffin granules a new perspective on some old problems, *Biochim. Biophys. Acta*, 463, 219, 1978.
126. Aberer, W., Kostron, H., Huber, E., and Winkler, H., A characterization of the nucleotide uptake by chromaffin granules of bovine adrenal medulla, *Biochem. J.*, 172, 353, 1978.
127. Curry, D. L., Bennet, L. L., and Grodsky, G. M., Dynamics of insulin secretion by the perfused rat pancreas, *Endocrinology*, 83, 572, 1968.
128. Wollheim, C. B., Kikuchi, M., Renold, A. E., and Sharp, G. W. G., The roles of intracellular and extracellular Ca^{++} in glucose-stimulated biphasic insulin release by rat islets, *J. Clin. Invest.*, 62, 451, 1978.
129. Kikuchi, M., Wollheim, C. B., Siegel, E. G., Renold, A. E., and Sharp, G. W. G., Biphasic insulin release in rat islets of Langerhans and the role of intracellular Ca^{++} stores, *Endocrinology*, 105, 1013, 1979.
130. Henquin, J. C., Relative importance of extracellular and intracellular calcium for the two phases of glucose stimulated insulin release: studies with theophylline, *Endocrinology*, 102, 723, 1978.

Chapter 3

CALCIUM IN TASTE AND OLFACTION MECHANISMS

Kenzo Kurihara and Kiyonori Yoshii

TABLE OF CONTENTS

I.	Introduction	34
II.	Role of Ca^{2+} Bound to Taste Receptor Membranes	34
	A. Ca^{2+} Removal by the ANS and the Alkaline Treatments	34
	B. Effect of Ca^{2+} Removal on Responses to Salts	36
	C. Effect of Ca^{2+} Removal on Responses to Various Stimuli	36
	D. Suppression of the Enhanced Responses by Ca^{2+}	38
	E. Effect of Binding of Extra Ca^{2+} on the Gustatory Responses	39
	F. Binding and Release of ^{45}Ca	39
III.	Role of Ca^{2+} in Taste Transduction	40
IV.	Role of Ca^{2+} in Olfactory Reception	43
References		44

I. INTRODUCTION

In higher vertebrates, chemical stimuli in the external environments are received at the gustatory and the olfactory cells. The initial event of chemoreception in these cells is adsorption of chemical stimuli on the receptor membranes,[1,2] which induces the receptor potential in the cells. In taste cells of vertebrates which are the secondary sensory cell, the receptor potential releases a chemical transmitter from taste cells and induces the gustatory nerve responses. In olfactory cells which are the primary sensory cells, the receptor potential induces directly the olfactory nerve responses.

The first section of this chapter describes the role of Ca^{2+} bound to the taste receptor membranes in the frog showing that the membrane-bound Ca^{2+} regulates the magnitude of the gustatory responses; in the second section, the role of Ca^{2+} in the taste transduction process is dealt with where it is shown that Ca^{2+} is needed to release a chemical transmitter from taste cells; and the third section describes that Ca^{2+} in the external medium is needed to induce the olfactory responses.

II. ROLE OF Ca^{2+} BOUND TO TASTE RECEPTOR MEMBRANES

In some membranes, reducing Ca^{2+} concentration in the external medium leads easily to a release of Ca^{2+} from the membranes. On the other hand application of Ca^{2+}-free solution to the tongue does not bring about any particular effect on the function of the gustatory receptor membrane because Ca^{2+} is so tightly bound to the receptor membranes that reducing Ca^{2+} concentration in the external medium does not lead to a release of Ca^{2+} from the membranes. The treatment of the tongue with EDTA leads to a small enhancement in the gustatory responses in some animals, but often does an irreversible damage to the gustatory receptors.[3]

In 1977, Kashiwagura et al.[3] showed that the treatment of the frog tongue with 1-anilinonaphthalene-8-sulfonate (ANS) led to a large enhancement of the gustatory responses to certain stimuli and that the enhancement was attributed to a release of Ca^{2+} from the gustatory receptor membranes. Later, Kamo et al.[4] found that the treatment of the frog tongue with an alkaline solution released Ca^{2+} from the receptor membranes. The effect of the alkaline treatment on the gustatory responses was not identical to that of the ANS treatment but was similar to that in most cases. In the following, the results obtained with both treatments are reviewed together.

A. Ca^{2+} Removal by the ANS and the Alkaline Treatments

The enhancing effect of the ANS treatment on the salt responses depended on the temperature of an ANS solution to be applied to the tongue. When the tongue was incubated in ANS solution below 10°C, the enhancing effect was much larger than that obtained with ANS solution at 20°C. Then the tongue was treated with 1 mM ANS solution at 5°C for 2 min and the ANS solution was washed away. A mechanism on the removal of Ca^{2+} from the frog gustatory receptors by the ANS treatment is unknown, but it was speculated as follows: when ANS is applied to the frog tongue, ANS is adsorbed in the hydrophobic region of the gustatory membrane, which will induce the expansion of the receptor membrane. This expansion of the membrane will lead to cleavage of the salt bridge of Ca^{2+} between two negative sites on the membrane and then Ca^{2+} will be removed from the receptor membrane. A conformational change of the receptor membrane induced by cooling the tongue will be helpful for the removal of Ca^{2+}.

The condition for the alkaline treatment was examined as follows. The records in Figure 1A are the integrated responses of the frog glossopharyngeal nerves to various salt stimuli before and after the tongue was treated with 2.5 mM carbonate buffer of

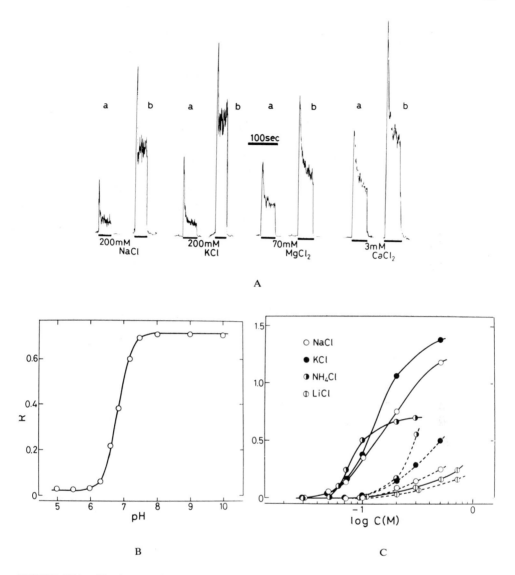

FIGURE 1(A). The integrated responses of the frog glossopharyngeal nerve to various inorganic salts before (a) and after (b) the tongue was treated with 2.5 mM carbonate buffer of pH 10.0 for 1 min. The ordinate scale of the response to CaCl$_2$ is reduced to one half. (B) Relative magnitude of the response (R) to 200 mM NaCl as a function of pH of solutions with which the tongue was treated. The tongue was treated with 2.5 mM phosphate buffer of varying pH[4]. (C) Relative magnitude of responses (R) to various salts as a function of salt concentration. The interrupted and continuous lines represent responses before and after the alkaline treatment respectively.[4] (From Kamo, N., Kashiwagura, T., Kobatake, Y., and Kurihara, K., *J. Physiol. (London)*, 282, 115, 1978. With permission.)

pH 10.0. The alkaline treatment does not affect the spontaneous activities of the nerves but leads to a large enhancement of the responses. In certain experiments, the tongue was treated with 2.5 mM phosphate buffer of varying pH and stimulating solution (200 mM NaCl solution) was applied to the tongue. Figure 1B shows relative magnitude of the tonic response (R) to 200 mM NaCl as a function of pH of solutions with which the tongue was treated. The enhanced effect of the treatment appears above pH 6.3 and reaches a saturation level around pH 7.5. The pH to give half the maximum response is 6.8. For some frogs, especially frogs in winter season, the pH to give half the maximum is shifted to an alkaline side.

The enhanced effect of the alkaline treatment greatly depends on the salt concentration of the treatment solutions. The tongue was treated with solutions containing various concentrations of NaCl where the pH of the solutions was fixed at pH 10.0 and stimulating solution was applied to the tongue. The largest enhancement was observed when the tongue was treated with an alkaline solution containing NaCl below 30 mM and the enhanced effect was decreased by an increase in NaCl concentration. Other salts added to the alkaline solution also exhibited a suppressive action similar to that of NaCl.

A possible explanation for Ca^{2+} release by the alkaline treatment is as follows: application of an alkaline solution to the tongue leads to an increase of negative charges on the receptor membrane. Electrostatic repulsion between negative charges will cause a conformational change of the membrane which brings about a release of Ca^{2+} from the membrane. The presence of salts in high concentration reduces electrostatic repulsion between negative charges and, hence, leads to suppression of the conformational change.

Various chemicals such as inorganic salts, distilled water, sugars, amino acids, bitter substances, and acids stimulate the frog gustatory receptors. Typical chemicals used usually as stimuli are NaCl, $CaCl_2$, distilled water, D-galactose, L-threonine, L-leucine, quinine, and HCl. The effect of removal of Ca^{2+} from the gustatory receptor membranes varies with species of chemical stimuli.

B. Effect of Ca^{2+} Removal on Responses to Salts

In Figure 1C, the magnitude of the tonic responses to salts of various monovalent cations is plotted against salt concentration. Interrupted and continuous lines represent the responses before and after the alkaline treatment, respectively. The magnitude of the enhanced response to 200 mM NaCl, for example, is about nine times that before the treatment and that to 500 mM NaCl is about five times. The response to LiCl is still small after the treatment; thus, the degree of the enhancement varies with species of cations. These results imply that the specificity of the receptor membrane to cations is changed by Ca^{2+} removal, although the chemical composition of the membrane is unchanged by the treatment. Probably, an assembly of chemical components of the receptor membranes determines the specificity to cations.

C. Effect of Ca^{2+} Removal on Responses to Various Stimuli

Figure 2A shows records of the integrated response to 1 mM quinine, 0.1 mM HCl, 1 M D-galactose, and distilled water. The responses to quinine, HCl, and galactose are unaffected by the treatment, while the water response is appreciably enhanced. Sometimes we encountered frogs which hardly responded to distilled water (records at the right most in Figure 2A). Such frogs, however, gave a large water response after the alkaline treatment.

In the experiments for Figure 2A, the frog tongue was treated with an alkaline solution for a short time (1 min). Longer treatment (e.g., 15 min) of the tongue enhanced the responses to D-galactose, while responses to quinine and HCl were only slightly enhanced or unchanged. Longer treatment of the tongue also enhanced the responses to amino acids.[5] Under the natural condition, responses to amino acids vary greatly with individual frogs. Hydrophilic amino acids such as L-threonine, glycine, L-alanine, L-serine, L-proline, and L-cysteine induced the responses of a tonic type. The long treatment of the tongue led to enhancement of the responses to hydrophilic amino acids. The records at the left most in Figure 2B show the integrated responses to 50 mM L-threonine before and after the long alkaline treatment. On the other hand, hydrophobic amino acids such as L-leucine, L-valine, L-isoleucine, L-phenylalanine, L-tryptophan, and L-methionine elicited the response patterns which usually consist of

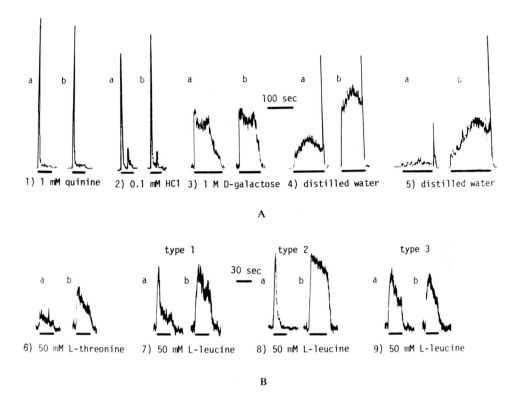

FIGURE 2. The integrated responses of the frog glossopharyngeal nerve to various chemical stimuli before (a) and after (b) the tongue was treated with 2.5 mM carbonate buffer of pH 10.0 for 1 min (A) and 15 min (B)[5]. The responses to distilled water and amino acids vary with individual frogs. The records 4 and 5 are the responses of frogs which respond well and hardly to distilled water, respectively. The records 7, 8, and 9 are the responses of frogs to 50 mM L-leucine which give typical responses.[5] (Figure 2A from Kamo, N., Kashiwagura, T., Kobatake, Y., and Kurihara, K., *J. Physiol. (London)*, 282, 115, 1978. With permission.)

an initial phasic component and a following tonic component. In Figure 2B, the response patterns for 50 mM L-leucine are represented. Most frogs exhibited the response pattern of type 1. Some frogs exhibited the pattern of type 2 which consists of only a phasic component, and other frogs exhibited the pattern of type 3 which has a large tonic component. The frog tongues which exhibited any type of the response pattern came to exhibit the responses having a large tonic component after the alkaline treatment (Figure 2B).

Most bitter stimuli are hydrophobic substances[6] and hydrophobic amino acids actually elicit bitter taste in humans. In addition, the response of the frog to quinine was not affected by removal of membrane-bound Ca^{2+} as shown in Figure 2A. Therefore, it was suggested that hydrophobic amino acids stimulate the same receptor site as that for bitter stimuli. The tonic responses to both hydrophobic and hydrophilic amino acids were diminished after the tongue was treated with pronase E, which suggested that the receptor molecule responsible for the tonic component is protein(s). Based on these results, the effect of Ca^{2+} removal on the responses to amino acids is summarized as follows. The Ca^{2+} removal enhances responses to hydrophilic amino acids and tonic component of responses to hydrophobic amino acids by affecting the receptor sites containing receptor protein(s). On the other hand, the Ca^{2+} removal does not affect the phasic component of responses to hydrophobic amino acids which are induced by stimulating the same receptor site as that for bitter stimuli.

In general, biological membranes have heterogeneous structure and the binding

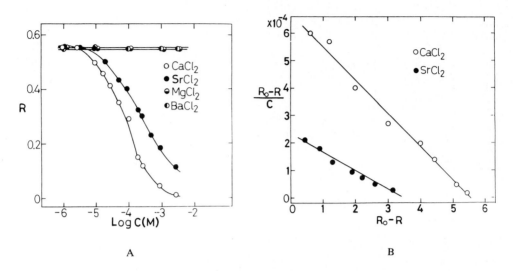

FIGURE 3(A). Effect of divalent cations on the enhanced response. After the tongue was treated with ANS, 100 mM NaCl solutions containing divalent cations of various concentrations were applied to the tongue.[3] (B) Plots of $(R_o - R)/C$ against $(R_o - R)$ where the data were taken from Figure 3A. Details are described in the text. (From Kashiwagura, T., Kamo, N., Kurihara, K., and Kobatake, Y., *J. Membr. Biol.,* 35, 205, 1977. With permission.)

properties of the membrane-bound Ca^{2+} also seem to be heterogeneous. The results obtained by the ANS and the alkaline treatment indicated that the effect of Ca^{2+} removal on the responses to various species of stimuli is decreased in the following order: NaCl, KCl, NH_4Cl > $CaCl_2$, distilled water > D-galactose, amino acids > HCl, quinine. The responses to the first group are easily enhanced by both the ANS and the alkaline treatment. Those to the second group are not enhanced by the ANS treatment, but enhanced by the alkaline treatment. Those to the third group are not or only slightly enhanced by the short alkaline treatment, but enhanced greatly by the long treatment. Those to the fourth group are practically not enhanced even by the long treatment.

D. Suppression of the Enhanced Responses by Ca^{2+}

The enhanced responses to chemical stimuli after the ANS and the alkaline treatment are suppressed by addition of $CaCl_2$ or $SrCl_2$ to stimulating solutions. Figure 3A shows the effect of divalent cations of varying concentrations on the enhanced response to 100 mM NaCl after the ANS treatment. Note that 1 mM $CaCl_2$ or $SrCl_2$ alone induces a large response in the frog gustatory nerve, but the response is suppressed in the presence of 100 mM NaCl. As seen from the figure, addition of $CaCl_2$ or $SrCl_2$ returned the response to the original level before the ANS treatment. The enhanced response to other stimuli also returned to the original level by addition of $CaCl_2$ or $SrCl_2$. On the other hand, $MgCl_2$ and $BaCl_2$ exhibited practically no effect on the enhanced responses.

The above results suggest that the binding of Ca^{2+} or Sr^{2+} to the sites in the gustatory receptor membrane treated with ANS leads to the suppression. In Figure 3B, $(R_o - R)/C$ is plotted against $(R_o - R)$ where R_o stands for magnitude of the response to 100 mM NaCl in the absence of Ca^{2+} or Sr^{2+}, R, magnitude of the response to 100 mM NaCl in the presence of Ca^{2+} or Sr^{2+} of varying concentrations and C, molar concentration of Ca^{2+} or Sr^{2+}. The figure indicates that the enhanced response to NaCl after the ANS treatment is suppressed by the Langmuir type binding of Ca^{2+} or Sr^{2+} to the sites in the gustatory receptor membrane. From the slope of straight line shown in Figure 3B, the apparent equilibrium constants for Ca^{2+} and Sr^{2+} were obtained to be 1.2×10^4 M^{-1} and 6.7×10^3 M^{-1}, respectively.

As shown above, addition of $CaCl_2$ to stimulating solutions brought about the suppression of the enhanced response. However, responses to stimulating solution free of Ca^{2+} were still in the enhanced level even after the alkali-treated tongue was bathed in Ringer solution of pH 7.0 containing 2.5 mM $CaCl_2$ for 15 min. This was because Ca^{2+} was not tightly rebound to the membrane under this condition. The incubation of the alkali-treated tongue in Ringer solution of pH 6.0 restored reversibly the function of the gustatory receptor to that before the treatment: application of stimulating solution free of Ca^{2+} to the tongue thus incubated induced as small a response as had been obtained before the alkaline treatment. Repeated application of stimulating solution free of Ca^{2+} to the tongue thus incubated did not bring about any enhancement of salt responses unless the tongue was subjected to further alkaline treatment. The above results indicated that incubation of the alkali-treated tongue in a solution containing $CaCl_2$ of pH 6.0 let Ca^{2+} rebind tightly to the receptor membrane.

E. Effect of Binding of Extra Ca^{2+} on the Gustatory Responses

Figure 1B shows that the membrane-bound Ca^{2+} is easily removed at alkaline pH. On the other hand, the incubation of the frog tongue in an acidic solution containing $CaCl_2$ brought about the binding of extra Ca^{2+} to the receptor membrane. The records a and b in Figure 4 are the integrated responses before and after the tongue is incubated in Ringer solution containing 2.5 mM $CaCl_2$ of pH 5.3, respectively. The responses to 400 mM NaCl are reduced to about one fourth of that before the incubation. The reduction of the water response is more pronounced; the tongue, thus incubated, comes to exhibit practically no water response. The reduced responses to NaCl and distilled water are fully restored to those before the incubation in the acidic Ringer when the tongue is incubated in Ringer solution of pH 7.0 for about 1 hr (see records c). The response to 50 mM L-threonine is also greatly reduced by the acidic Ringer treatment. As mentioned before, the response pattern for L-leucine varies with individual frogs. With any frog, the acidic Ringer treatment suppresses only the tonic component of the response to L-leucine and unaffects the phasic component. The response to quinine is not affected, as is similar to the phasic response to L-leucine.

F. Binding and Release of ^{45}Ca

In order to examine more directly the binding properties of Ca^{2+} to the frog tongue, the experiments using ^{45}Ca were carried out. The tongue was first treated with alkaline solution (to release Ca^{2+}) and then half a piece of the frog tongue was incubated in solutions of pH 5.3 and 7.0 containing ^{45}Ca. After each piece was washed thoroughly, 5 mM KCl solution was passed over the tongue surface and the radioactivity of the effluent was measured. After the radioactivity of the effluent became sufficiently small, a carbonate buffer of pH 10.0 was flowed on the tongue surface. Application of the alkaline solution to the tongue incubated in pH 5.3 led to a large increase in radioactivity of the effluent compared with that of the previous effluent, while that to the tongue incubated in pH 7.0 brought about a small increase. The above tracer experiments indicated that the alkaline treatment led to removal of Ca^{2+} from the tongue and that incubation in $CaCl_2$ solution in acidic pH led to binding of a larger amount of Ca^{2+} than incubation in neutral pH, which was consistent with the conclusion obtained from electrophysiological experiments.

Much evidence has accumulated to indicate that Ca^{2+} stabilizes the structure of the biological membranes. Hence, the removal of Ca^{2+} from the gustatory receptor membrane may unstabilize the receptor domains so that a conformational change of the domains is easily induced by adsorption or desorption (water response) of chemicals. On the other hand, binding of extra Ca^{2+} to the receptor membrane will stabilize the receptor domains so that adsorption or desorption of chemicals does not easily induce a conformational change.

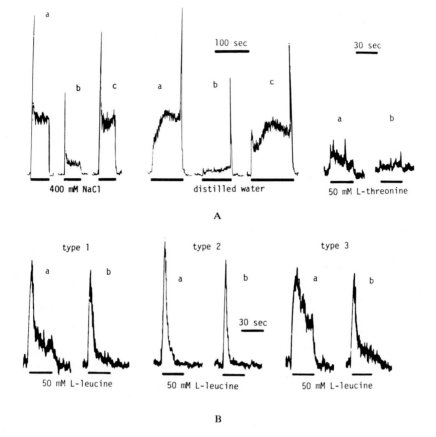

FIGURE 4. Effect of binding of extra Ca^{2+} on the gustatory responses to various chemical stimuli. The records a and b show the integrated responses before (a) and after (b) the tongue was incubated in Ringer solution of pH 5.3 for 15 min.[4] Since the responses to amino acids vary with individual frogs, typical responses to 50 mM L-leucine are represented.[5] The records c show the responses after the tongue incubated in the acidic Ringer solution was incubated in Ringer solution of pH 7.0 for 1 hr. (From Kamo, N., Kashiwagura, T., Kobatake, Y., and Kurihara, K., *J. Physiol. (London)*, 282, 115, 1978.)

III. ROLE OF Ca^{2+} IN TASTE TRANSDUCTION

Chemical stimuli applied to the tongue are adsorbed on the microvillus membrane of taste cells, which induces a depolarization of taste cells. It is generally considered that the depolarization induces a release of a chemical transmitter from taste cells and elicits the gustatory nerve impulses, but the detailed transduction mechanism is still unknown.

Morimoto[7] and Nagahama and Kurihara[8] perfused the frog lingual artery with Ringer solution and examined the role of Ca^{2+} in the taste transduction mechanism. Adult bullfrogs were anaesthetized with urethan. A polyethylene tube was cannulated into the lingual artery and Ringer solution (112 mM NaCl, 3.4 mM KCl, 3.6 mM MgSO$_4$, 2.5 mM NaHCO$_3$, pH 7.2) containing CaCl$_2$ of varying concentrations was perfused through the tube by using a peristaltic pump. Under this condition, the gustatory responses in the glossopharyngeal nerve were measured.

Figure 5A shows the responses to 1 mM CaCl$_2$ (CaCl$_2$ is potent stimulus for the frog gustatory receptor) when CaCl$_2$ concentration in the perfusing solution is changed

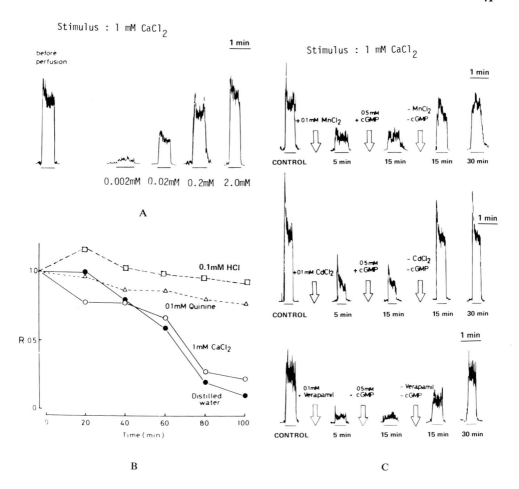

FIGURE 5 (A and B). The integrated responses of the frog gustatory nerves to 1 mM CaCl₂ under the conditions that the lingual artery was perfused with Ringer solutions containing various concentrations of CaCl₂ (A) and with Ringer solutions containing 0.2 mM CaCl₂ and 0.1 mM Ca-channel blockers (B).[8] (C). The magnitude of the responses to various stimuli as a function of time after the lingual artery was perfused with Ringer solution containing low concentration of CaCl₂. (0.002 mM).

from 0.002 to 2.0 mM. The response is greatly diminished by decreasing CaCl₂ concentration in the perfusing solution to 0.002 mM and increased with increasing CaCl₂ concentration in the solution. The response is also diminished by addition of Ca channel blockers to the perfusing solution containing sufficient amount of CaCl₂; the response to 1 mM CaCl₂ is greatly decreased by addition of 0.1 mM CaCl₂, 0.1 mM verapamil, or 0.1 mM MnCl₂ to the perfusing solution containing 0.2 mM CaCl₂ (Figure 5B). The decreased response is reversibly recovered by removal of the blockers from the perfusing solution. These results suggested that Ca influx from intercellular medium into taste cells is needed for a release of a chemical transmitter from taste cells.

In the experiments for Figure 5A, the response to 1 mM CaCl₂ reached respective constant levels within, at most, 5 min when CaCl₂ concentration in perfusing solution was increased from 0.002 to 0.02, 0.2, and 2.0 mM. However, it took a long time until the response to 1 mM CaCl₂ was completely diminished by switching the perfusing solution from high Ca Ringer to low Ca Ringer. This implies that perfusion with low Ca Ringer for a long time is needed to eliminate completely Ca²⁺ contaminated from high Ca Ringer. Figure 5C shows the time course of the responses to CaCl₂, distilled

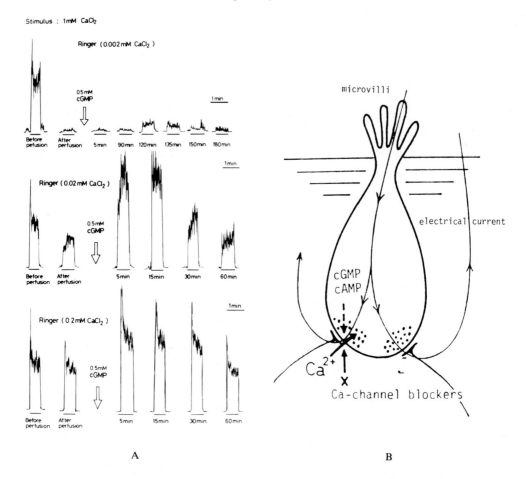

FIGURE 6(A). Effect of cGMP on the responses of the frog gustatory nerve to 1 mM CaCl$_2$ where the lingual artery was perfused with Ringer solution containing various concentrations of CaCl$_2$. cGMP was added to the perfusing solution. (B). A schematic diagram illustrating taste transduction mechanism.

water, quinine, and HCl, after CaCl$_2$ concentration in the perfusing solution is decreased from 2.0 to 0.002 mM. The responses to CaCl$_2$, distilled water, D-galactose, and L-threonine (data not shown group 1) are decreased rapidly during perfusion, while the responses to quinine, HCl and ethanol (group 2) are not decreased within 100 min in this preparation.

Recently, Nagahama et al.[9] found that addition of cGMP to the perfusing solution greatly enhanced the frog gustatory responses to chemical stimuli of group 1 and that cAMP suppressed the responses. Figure 6A shows the effect of cGMP on the responses to 1 mM CaCl$_2$ when the lingual artery is perfused with Ringer solutions containing various CaCl$_2$ concentrations. Addition of cGMP brought about only a small enhancement of the response about 120 min after addition of cGMP when the artery is perfused with low Ca (0.002 mM) Ringer. The response is greatly enhanced within, at most, 5 min after addition of cGMP when the artery is perfused with Ringer solution containing CaCl$_2$ above 0.02 mM. The responses to chemical stimuli of group 2 are affected neither by cGMP nor cAMP.

Figure 6B represents a schematic diagram illustrating the taste transduction mechanism of the responses to chemical stimuli of group 1. Application of the chemical stimuli to the tongue induces a depolarization at the microvillus membrane of a taste cell. The depolarization spreads electrotonically to the synaptic area of the taste cell,

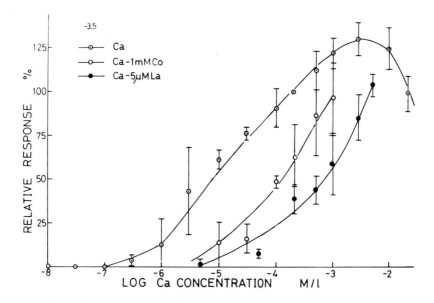

FIGURE 7. Relative magnitude of the integrated response of the olfactory bulb in the rainbow trout to 0.3 mM L-glutamine as a function of $CaCl_2$ concentration in a perfusing solution for the olfactory epithelium.[11]

which opens the voltage-dependent Ca-channel and induces Ca-influx from the intercellular medium into the taste cell. The Ca-influx will lead to a release of a chemical transmitter. Since the cyclic nucleotides did not affect the spontaneous activities of the gustatory nerve, the nucleotides themselves seem not to contribute to changes in the membrane potential of a taste cell. cGMP did not exhibit appreciable effect on the gustatory response under the conditions that $CaCl_2$ concentration in the perfusing solution is low (Figure 6A) or Ca channel blockers are present in the solution (Figure 5B). Probably, the cyclic nucleotides modulate Ca influx triggered by depolarization. The responses to chemical stimuli of group 2 are induced without accompanying Ca influx.

IV. ROLE OF Ca^{2+} IN OLFACTORY RECEPTION

The olfactory epithelium of aquatic animals contacts directly with water in external environments and then chemical stimuli dissolved in the water stimulate the olfactory receptors of the animals. The olfactory receptors of certain species of aquatic animals respond sensitively to amino acids. Suzuki examined the effect of external Ca^{2+} concentration on the olfactory responses of the lamprey[10] and the rainbow trout[11] to amino acids. The olfactory responses were measured by recording spike activity of a single olfactory receptor (lamprey) or the integrated responses of olfactory bulb (rainbow trout).

Perfusion of the olfactory epithelium with Ca^{2+}-free soluton did not affect or only partly decreased the olfactory responses to amino acids. However, perfusion with a solution containing EGTA led to a marked suppressive effect on the olfactory responses in the lamprey and the rainbow trout. Figure 7A shows relative magnitude of the olfactory response in the rainbow trout to 0.3 mM L-glutamine as a function of Ca^{2+} concentration in the external medium, where Ca^{2+} concentration is varied using Ca^{2+}-EGTA buffer. The response completely disappears at low Ca^{2+} concentration. The threshold Ca^{2+} concentration is in the range between 3×10^{-7} and $1 \times 10^{-6} M$. The response increases with increasing Ca^{2+} concentration up to $3 \times 10^{-3} M$ and decreases with further increase in Ca^{2+} concentration. The reduced response by perfusion with low Ca solution restored its original response by addition of $SrCl_2$. Ca-channel blockers such as $LaCl_3$, $CoCl_2$, or verapamil showed suppressive effect on the response.

Actual role of Ca^{2+} in the olfactory reception is unknown. Suzuki[10] pointed out the following two possibilities: (1) Ca^{2+} acts as an inward current carrier through the apical membrane of the olfactory receptor, which induces impulses in the olfactory nerve (2) Ca^{2+} affects the affinity between amino acids and the receptor site at the apical membrane.

REFERENCES

1. Beidler, L. M., Taste receptor stimulation with salts and acids, in *Handbook of Sensory Physiology*, Beidler, L. M., Ed., Springer-Verlag, Basel, 1971, 200.
2. Kurihara, K., Kamo, N., and Kobatake, Y., Transduction mechanism in chemoreception, in *Advances in Biophysics*, Vol. 10, Kotani, M., Ed., Japan Scientific Societies Press, Tokyo, 1978, 27.
3. Kashiwagura, T., Kamo, N., Kurihara, K., and Kobatake, Y., Enhancement of salt responses in frog gustatory nerve by removal of Ca^{2+} from the receptor membrane treated with ANS, *J. Membr. Biol.*, 35, 205, 1977.
4. Kamo, N., Kashiwagura, T., Kobatake, Y., and Kurihara, K., Role of membrane-bound Ca^{2+} in frog gustatory receptors, *J. Physiol. (London)*, 282, 115, 1978.
5. Yoshii, K., Kobatake, Y., and Kurihara, K., Selective enhancement and suppression of frog gustatory responses to hydrophilic and hydrophobic amino acids, *J. Gen. Physiol.* 77, 373, 1981.
6. Koyama, N. and Kurihara, K., Mechanism of bitter taste reception: interaction of lipid monolayers from bovine circumvallate papillae with bitter compounds, *Biochim. Biophys. Acta*, 288, 22, 1972.
7. Morimoto, K., unpublished data.
8. Kurihara, K., Nagahama, S., and Kashiwayanagi, M. A molecular mechanism of taste transduction in the frog, in *Olfaction and Taste*, Vol. 7, van der Starre, H., Ed., IRL Press, London, 1980, 69.
9. Nagahama, S., Kobatake, Y., and Kurihara, K., Effects of Ca^{2+}, cyclic GMP and cyclic AMP added to artificial solution perfusing lingual artery on frog gustatory nerve responses, *J. Gen. Physiol.* in press, 1982.
10. Suzuki, N., Effects of different ionic environments on the responses of single olfactory receptors in the lamprey, *Comp. Biochem. Physiol.*, 61A, 461, 1978.
11. Suzuki, N., The dependence of fish olfactory receptor responses to amino acids on the external concentration of calcium ions, *Proc. 12th Jpn. Symp. Taste and Smell*, 12, 9, 1978.

Chapter 4

CALCIUM IN NUTRITION

Margaretha Jägerstad

TABLE OF CONTENTS

I. Introduction 46

II. Calcium in Foods 46

III. Availability of Dietary Calcium 46
    A.   Absorption 46
    B.   Adaption 48

IV. Influence of Calcium Retention from Other Nutrients 48
    A.   Protein 48
    B.   Calcium/Phosphorus Ratio 49
    C.   Minerals and Trace Elements 49

V. Calcium Intake 50

VI. Dietary Calcium Requirements 50

VII. Recommendations 52

References 53

I. INTRODUCTION

Calcium is an essential nutrient with several important functions within the human body. About 99% of the calcium content of the body is deposited in the skeleton and teeth for structural purposes. One percent is present in tissues and extracellular fluids for normal function of cells and for many enzyme-controlled reactions.

An adult man contains between 950 and 1300 g of calcium in his skeleton depending on his height. Corresponding figures for women of varying heights are between 770 and 920 g.[1,2] In an adult man, about 700 mg of calcium enters and leaves the bones daily.[3]

The calcium levels in serum are carefully regulated and under homeostatic control mediated by the parathyroid glands and vitamin D. Changes in the concentration of serum calcium lead to disturbances in the function of the nervous system, the heart, and other muscles. In the case of decreasing serum calcium concentrations, the absorption from the gastrointestinal tract increases, or calcium bound to the skeleton is mobilized into the circulation.

The dietary calcium intake varies greatly throughout the world. Although many populations have intake as low as 150 to 200 mg/day, they have no problem with their bones or teeth. In the western countries, the dietary habits include several milk-based food items and, therefore, the calcium intake usually exceeds 500 mg daily. In spite of this high consumption of calcium, disturbances in the calcium metabolism causing increasing fragility of bone are common in elderly individuals. Femoral-neck fractures due to falls at home are the most common accidents in the elderly, taking a profound economic, social, and human toll.[4] Whether dietary calcium in any way contributes as causal agent or therapeutic tool for disorders such as osteoporosis is still the subject of much research.

II. CALCIUM IN FOODS

Cows' milk is a good source of calcium. It contains on average 120 mg/100 g, which is almost four times as high as human milk, which contains only about 30 mg/100 g. All kinds of cheeses are rich in calcium, averaging from 400 mg up to 1200 mg/100 g.

In Table 1, a comparison of the calcium content of some important groups of food items is presented. These values constitute ranges of average values obtained from different food composition tables.[5-8] In foods of animal origin except milk and dairy products, e.g., meat, eggs, and fish, the calcium content varies between 10 and 80 mg/100 g.

Vegetables like roots, pulses, and greens contain varying amounts from 5 mg calcium per 100 g up to almost 100 mg/100 g. Fruits and berries seldom contain more than 40 mg, but usually around 10 to 20 mg/100 g. Cereals, especially after grinding into flour, lose most of their mineral and vitamin content which is present in the hull. Such refined products contain on average 10 to 30 mg/100 g. Several food items of vegetable origin may contain high amounts of calcium, e.g., nuts (100 to 200 mg/100 g) and nettles (200 mg/100 g).[5]

III. AVAILABILITY OF DIETARY CALCIUM

A. Absorption

The calcium content of a diet is no direct index of the amount actually utilized by the human body, since normally a large part of the dietary calcium is excreted in the feces. The amount of calcium absorbed from the small intestine is far from complete and depends on several factors. Most important are the requirements of the body. This regulation seems now to have been resolved as discussed in detail elsewhere in this

Table 1
CALCIUM CONTENT OF SOME COMMON FOODS (SELECTED VALUES FROM VARIOUS FOOD TABLES)[5-8]

Food	Description	Range (mg/100 g or 100 ml)
Milk	Cow's — fresh whole	80—130
Cheese	Hard — from whole or skimmed milk	700—1200
Cheese	Soft — from whole or skimmed milk	400—900
Eggs	Fresh, whole	50
Beef	Raw — edible portions of various animals	7—20
Fish	Raw, fresh, various species	20—80
Nuts	Various — shelled	130—190
Roots	Various except potatoes, raw	6—66
Potatoes	Raw	5—10
Pulses	Various — raw	20—60
Vegetables	Various, raw, green	20—90
Fruits	Fresh	5—40
Berries	Fresh	15—30
Wheat flour	Refined, not fortified	20—30
Rice	Raw, polished	10—25

volume. An active metabolite of vitamin D_3 plays a key role. After two hydroxylations of vitamin D_3, the first one occurring in the liver, and the second in the kidney, 1,25-dihydroxy vitamin D_3 is formed. This compound exerts a stimulating effect on the calcium absorption in the small intestine. The amounts of 1,25-$(OH)_2$ D_3 synthesized in the kidney are regulated by the parathyroid hormone.[9]

Dietary components interfering with calcium absorption are shown in Table 2. The role of vitamin D has already been discussed. The enhancing effect of lactose on the calcium absorption has most often been explained that lactose is relatively slowly absorbed and the free sugar in the lower gut may modify the flora and produce a lower pH in the gut which is favorable for calcium absorption.[10] There is, however, evidence that lactose may form a chelate of calcium which may protect it from precipitation or, otherwise, assist in absorption.[11] Proteins exert their enhancing effect via amino acids which also form soluble chelates.

Other dietary factors impair the calcium absorption by forming insoluble compounds not easily absorbed. Several of these inhibitory dietary factors, e.g., phytic acid, oxalate, and dietary fiber are present in diet in amounts sufficient to precipitate all of the calcium in the diet, at least theoretically. It seems, however, that the inhibitory effects occur only temporarily. Adaptive mechanisms rapidly ameliorate or abolish these effects unless the amounts involved are excessive. It is unlikely, therefore, that calcium needs are influenced to any serious degree by the amounts of such substances ordinarily consumed.

The fraction of dietary calcium absorbed is inversely proportional to the amount ingested. In longterm studies, Malm found that adult men absorbed 45% of the dietary calcium at an intake of 450 mg/day and 31% at an intake of 940 mg/day.[12] Heaney et al.[13] studied absorption of radioactive calcium in adults; a graph in their article indicated that absorption appeared to be about 53% at an intake of 190 mg and fell to 17% at an intake of 3000 mg. These studies clearly indicate that the absolute amounts of calcium absorbed increase with increasing intakes.

Age is a factor in calcium absorption. Absoption studies using either labeled calcium or metabolic balance techniques uniformly have revealed that intestinal calcium ab-

Table 2
DIETARY COMPONENTS INTERFERING WITH CALCIUM ABSORPTION

Component	Effect	Action
Vitamin D	Enhancing	Primary regulator of calcium absorption
Protein	Enhancing	Amino acids forming soluble salts of calcium, which are easily absorbed
Lactose	Enhancing	Depends on hydrolyses of lactose (milk sugar)
Phosphate	None	The ratio betwen dietary calcium and phosphorus plays no role in the human nutrition except for infants fed with artificial milk preparations
Phytic acid	Impairing	A phosphate salt of inositol present in bread baked on whole grain flour. Mechanism behind its inhibitory effect unestablished
Dietary fiber	Impairing	Calcium binds to uronic acid present in dietary fiber
Oxalic acid	Impairing	Forms insoluble salts with calcium
Fats	Impairing	Form insoluble soaps with dietary fats principally in patients with chronic intestinal disorders

sorption decreases with aging in both sexes, particularly after 70 years of age.[14] A recent study suggests that inadequate conversion of 25-(OH)D to 1,25-(OH)$_2$D contributes significantly to decreased calcium absorption in elderly normal subjects.[15]

B. Adaption

Most adults can adapt to a low calcium intake by increasing the absorption efficiency, and there is abundant evidence that individuals reject a large fraction of dietary calcium when intake is high.[16] Current data from De Luca and collaborators show that parathyroid hormone through its effect on 1,25 to (OH)$_2$ D$_3$ synthesis is the regulating factor in the adaption mechanism.[9] The adaption takes about a week to be effective. Hegsted et al.[17] using volunteers from the inmates of a Peruvian prison showed that intestinal absorption was adequate to maintain calcium balance even when the diet contained as little as 100 to 150 mg calcium per day.

In the elderly, this adaption mechanism seems to be disturbed mainly because of an impaired synthesis of 1,25 (OH)$_2$ D$_3$.[15] Decreased production of 1,25 (OH)$_2$ D$_3$ could be caused by an age-related decline in the activity of the 1α-hydroxylase enzyme in the kidney.

IV. INFLUENCE OF CALCIUM RETENTION FROM OTHER NUTRIENTS

A. Protein

A relationship between protein intake and urinary calcium excretion has been known for a long time. A high protein intake results in increased urinary losses of calcium. The calciuretic effect of high-protein intake has been attributed by some investigators to the enhancement of calcium absorption.[18,19] This cannot be the sole mechanism, however. Studies performed by Chu et al.[20] suggest that an inhibition of tubular calcium reabsorption, possibly coupled with increased glomerular filtration, is most likely the principal mechanism involved in the calciuretic effect of high protein intake. This group also found that the dermal calcium losses were unaffected by the protein/calcium ratio.

In Table 3 data from several studies measuring the urinary calcium excretion after different calcium and protein levels are shown. The ratio between calcium and protein seems to be very important.

Interesting in this respect are the similarities in the protein/calcium ratio in human milk and cow milk. Although the cow's milk contains four times higher concentrations of both calcium and protein compared to human milk, both types of milk show a similar protein/calcium ratio calculated from weights (mg protein to mg calcium)

Table 3
EFFECTS ON URINARY EXCRETION OF
CALCIUM AT DIFFERENT PROTEIN
INTAKES

Calcium intake (mg/day)	Protein intake (g/day)	Urinary calcium excreion (mg/day)	Ref.
100	6	51	20
100	75	99	20
100	150	161	20
100	<6	51	42
100	78	99	42
800	47	207	41
800	95	303	41
800	147	426	41
900	<6	105	42
900	78	155	42
1300	<6	80	42
1300	78	163	42
1400	48	175	43
1400	141	338	43

amounting to approximately 33. This ratio may vary considerably in different food items, but as averaged a mixed whole-day diet representing a typical Western diet may have a ratio of about 100.[21] Populations consuming less amounts of calcium may have a similar ratio due to a simultaneous lower intake of protein. In a dietary study performed on elderly subjects, the intake of different nutrients was chemically analyzed in 24 hr food collections (duplicate sampling technique).[21,31] A high correlation between calcium and protein intake was found, r = 0.79 and r = 0.59 for men and women respectively, which is highly significant $(p < 0.001)$.

B. Calcium/Phosphorus Ratio

Numerous studies, particularly in vitamin D-deficient animals, have shown the calcium/phosphorus ratio to be another determinant in assessing calcium requirements. The diets of man, especially those consisting largely of vegetable products, almost invariably contain much more phosphorus than calcium. It has, however, been convincingly shown that the phosphorus level in diets has no effect on the calcium retention in man.[22] Therefore, Ca/P ratios in food is no longer considered of practical importance, provided the amount of calcium consumed is sufficiently liberal.[23] However, the absorption of calcium in the first few weeks of life from artificial milk formulas presents special problems.[24] In such preparations, the Ca/P ratio may be important.

C. Minerals and Trace Elements

The dietary calcium level may interfere with the absorption of other divalent metals like magnesium, iron, zinc, and copper.[25] The amounts of dietary calcium are inversely proportional to the absorption of other minerals and trace elements. A high calcium diet therefore, may, in the long run, induce a secondary deficiency of other divalent essential minerals and trace elements. Precipitation of adverse effects in experimental and farm animals have been reported, which can be corrected by the addition of the relevant nutrient to the ratio.[26] No comparable data are available for man.

On the other hand, this "protective" effect of calcium on the absorption of other divalent metals may be valuable for subjects who are exposed to high amounts of toxic metals, for instance, lead and cadmium. Several reports on experimental animals in-

Table 4
AVERAGE DAILY INTAKE OF
CALCIUM PER HEAD BASED ON
FAO FOOD BALANCE SHEETS (mg/
HEAD) IN DIFFERENT COUNTRIES
1957 TO 1959

Country	Intake of calcium
India	347
Japan	368
South Africa	442
United Arabic Republic	449
Chile	520
Turkey	547
Argentina	651
Italy	710
Australia	833
France	930
Scandinavia	1000—1329
Canada	1047
U.S.	1116

dicate impaired retention of lead in the presence of high amounts of calcium in the diets.[27-29] If such data also may be applied to the situation of man, taking in consideration the calcium levels of diet related to the contamination of toxic metals in diet, remains to be proven.

V. CALCIUM INTAKE

Most of the national food composition tables include figures of the calcium content in different food items. This makes it possible to calculate the average daily intake of calcium based on data from food balance sheets or from dietary surveys of populations more or less representative. Values calculated from food balance sheets are very crude and probably over estimations but make comparisons between different countries possible. Table 4 lists data summarized by Food and Agriculture Organization/World Health Organization's Expert Group on Calcium Requirements collected during 1957 and 1959.[39] According to this table, the average calcium intake by Western countries exceed 500 mg daily and may be so high as 1300 mg (Finland).

Dietary surveys show that the daily intake of calcium may vary considerably between individuals, as well as between days. As an illustration, a Swedish study is shown where the duplicate sampling portion technique has been used.[30] Middle-aged subjects each collected seven 24-hr food samples with days between (Figure 1). Among the women consuming around 1500 kcal daily, intakes of calcium below 500 mg were not uncommon.

VI. DIETARY CALCIUM REQUIREMENTS

The daily dietary requirements of calcium have provoked considerable controversy over a number of years. Calcium is a nutrient of which the requirements cannot be decided on their own. Several other nutrients affect its absorption and utilization as discussed above, e.g., lactose dietary fiber, phytates, oxalates, proteins, and vitamin D. In addition, adaptive mechanisms decide the amounts of dietary calcium absorbed.

The identification of the minimal human requirement for any nutrient generally depends on determining the intake level below which a deficiency disorder has been rec-

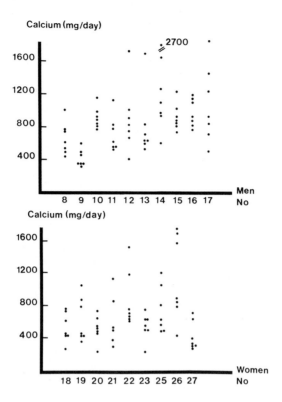

FIGURE 1. Daily calcium intake expressed in milligrams by middle-aged men and women. Each subject collected seven 24-hr duplicate portions of food during intermittent days.[30]

ognized. The problem of defining the calcium requirements is that in humans indisputable symptoms of calcium deficiency do not occur. There is little clear evidence in human subjects of response in either growth or serum calcium levels that can be attributed solely to changes in calcium intake. Finally, the wide range of intakes with which people in various parts of the world maintain themselves without apparent signs of either calcium deficiency or calcium excess makes a statement of human requirements difficult.[32]

Osteoporosis is defined as a calcium deficiency state, but it has not been clearly shown that clinical osteoporosis can be produced in man by calcium deficiency at least within the range of calcium found in adult diets. Furthermore, osteoporosis is now generally recognized to be a multifactorial disorder and even if calcium deficiency is proved to be one of these factors, it almost certainly is not the only one, i.e., not the limiting factor in many cases.

One way to evaluate the daily requirements is to measure daily losses through skin, sweat, urine, and feces. The dermal loss has been estimated to be in the order of 15 to 20 mg/day under sedentary conditions.[33,34] The urinary losses of calcium depend on the amount absorbed and the protein levels in the diet. Normally, 100 to 350 mg calcium is excreted per day.[35] Most of the feces calcium is part of the dietary intake, which for one reason or another never gets absorbed. The remainder, about 130 mg/day, comes from shed epithelial cells and from digestive secretion that pours into the intestinal tract.[36]

Several studies have investigated at which levels of calcium intake an individual is in calcium balance, i.e., the losses through skin, feces, and urine should be equal to

Table 5
RECOMMENDATIONS FOR CALCIUM INTAKE BY ADULTS AND DURING PREGNANCY AND LACTATION IN MG/DAY IN DIFFERENT WESTERN COUNTRIES

	FAO/WHO[39]	U.S.[40]	U.K.[44]	Scandinavia[44]	West Germany[44]	Italy[44]	U.S.S.R.[44]
Adults	400—500	800	500	800	800	500	800
Pregnancy	1000—1200[a]	1200	1200	1200	1200	1200	1500[a]
Lactation	1000—1200	1200	1300	1300	1200	1200	1800—2000

[a] Later half.

the intake. Such studies indicate that calcium intakes above 800 mg are necessary for man to maintain a balance.[37,38] Heany et al., using two independent techniques on a group of perimenopausal women, showed this group to be in a negative calcium balance amounting to 40 to 50 mg/day at an average daily intake of 600 mg calcium.[38] The more calcium these women consumed the lower their negative calcium balance. According to calculations, the requirement of calcium for this group of women should be 1200 mg/day in order to guarantee a positive calcium balance. Marshall et al.[37] found that calcium balance was negative in most subjects at intakes below 600 mg calcium per day. He concluded that intakes as high as 900 mg calcium per day were necessary to guarantee a positive calcium balance.

The calcium requirements increase during pregnancy and lactation. The fetus at term contains about 30 g calcium, most of which is accumulated during the third trimester.[32] During the last few weeks of pregnancy, the mother must provide up to 350 mg of calcium each day for her unborn infant. Human milk contains between 30 to 34 mg/100 mℓ. An average milk production of 850 mℓ/day, thus, increases the demands with 250 to 300 mg calcium daily.[32]

VII. RECOMMENDATIONS

Although no clinical deficiency of calcium ascribed to a low intake has been reported, a nagging suspicion persists that chronic imbalance may lead to osteoporosis in the elderly. Epidemiological and experimental data are confusing, which also are reflected by different calcium levels chosen for recommendations by different expert groups.

The Food and Agriculture Organization/World Health Organization Expert Group suggested that an intake between 400 and 500 mg/day would be adequate.[39] This was based on the fact that both children and adults in populations with intakes of calcium as low as 300 to 400 mg were apparently free from signs of calcium deficiency.

In the U.S. the Food and Nutritional Board (1974) in their Recommended Dietary Allowances suggested a daily intake of 800 mg/day for adults.[40] The higher allowances recommended by this organization were justified on the basis that large percentages of the population had, at that time, intakes of calcium comparable to the higher values. Furthermore, the high protein level in the American diet was taking into account, as well as calcium balance studies indicating daily excretions of that amount.

No special recommendation is given for the elderly. As seen in Table 5, most Western countries follow the higher calcium recommendations adopted by the U.S., but a few, i.e., the U.K. and Italy, apply the Food and Agriculture Organization/World Health Organization recommendations.[39,44]

During pregnancy and lactation, most countries recommend a daily intake of 1200 mg, especially during the second part of these periods.

REFERENCES

1. Cohn, S. H., Vaswani, A., Zani, I., et al., Changes in body chemical composition with age measured by total-body neutron activation, *Metabolism*, 25, 85, 1976.
2. American Academy of Pediatrics, Calcium requirements in infancy and childhood, *Pediatrics*, 62, 826, 1978.
3. Whedon, G. D., The combined use of balance and isotopic studies in the study of calcium metabolism, in *Proc. 6th Int. Congr. Nutr.*, Mills, C. F. and Passmore, R., Eds., Churchill Livingstone, Edinburgh, 1964, 425.
4. Editorial: deaths from domestic falls, *Lancet*, II, 1351, 1972.
5. Food Composition Tables, Energy and certain nutrients in foods, National Food Administration Ordinance, Sweden, 1978.
6. Davidson, S., Passmore, R., Brock, J. F., and Truswell, A. S., *Human Nutrition and Dietics*, 7th ed., Churchill Livingstone, Edinburgh, 1979, 90.
7. McCance, R. A. and Widdowson, E. M., The composition of foods, Medical Research Council Special Report Series 297, Her Majesty's Stationery Office, London, 1973.
8. Varo, P., Mineral element balance and coronary heart disease, *Int. J. Vitam. Nutr. Res.*, 44, 267, 1974.
9. Ribovich, M. L. and de Luca, H. F., Adaptation of calcium absorption: parathyroid hormone and vitamin D metabolism, *Arch. Biochem. Biophys.*, 188, 157, 1978.
10. Vaughon, O. W. and Filer, L. J., The enhancing action of certain carbohydrates on the intestinal absorption of calcium in the rat, *J. Nutr.*, 71, 10, 1960.
11. Charley, P. and Sultman, P., Chelation of calcium by lactose, its role in transport mechanisms, *Science*, 139, 1205, 1963.
12. Malm, O. J., Calcium requirement and adaptation in adult men, *Scand. J. Clin. Lab. Invest.*, 10, (Suppl. 36), 1, 1958.
13. Heaney, R. P., Saville, P. D., and Recker, R. R., Calcium absorption as a function of calcium intake, *J. Lab. Clin. Med.*, 85, 881, 1975.
14. Bullamore, J. R., Wilkinson, R., Gallagher, J. C., Nordin, B. E. C., and Marshall, D. H., Effects of age on calcium absorption, *Lancet*, II, 535, 1970.
15. Gallagher, J. C., Riggs, B. L., Eisman, J., Hamstra, A., Arnand, S. B., and de Luca, H. F., Intestinal calcium absorption and serum vitamin D metabolites in normal subjects and osteoporotic patients, *J. Clin. Invest.*, 64, 729, 1979.
16. Nicolaysen, R., The absorption of calcium, *Acta Physiol. Scand.*, 6, 201, 1943.
17. Hegsted, D. M., Moscoso, J., and Collazox, C., A study of the minimum calcium requirements of adult men, *J. Nutr.*, 46, 181, 1952.
18. McCance, R. A., Widdowson, E. M., and Lehmann, H., The effect of protein intake and the absorption of calcium and magnesium, *Biochem. J.*, 36, 686, 1942.
19. Margen, S., Kaufman, N. A., Costa, F., and Calloway, D. H., Studies in the mechanism of calcium induced by protein feeding, *Fed. Proc. Fed. Am. Soc. Exp. Biol.*, 29, 566, 1970.
20. Chu, J. Y., Margen, S., and Costa, F., Studies in calcium metabolism. II. Effects of low calcium and variable protein intake on human calcium metabolism, *Am. J. Clin. Nutr.*, 28, 1028, 1975.
21. Borgström, B., Nordén, Å., Åkesson, B., Abdulla, M., and Jägerstad, M., Nutrition and old age. Chemical analyses of what old people eat and their states of health during 6 years of follow-up, *Scand. J. Gastroenterol.*, 14, (Suppl. 52), 1979.
22. Malm, O. J., On phosphates and phosphoric acid as dietary factors in the calcium balance of man, *Scand. J. Clin. Lab. Invest.*, 5, 75, 1953.
23. Hegsted, D. M., Calcium and phosphorous, in *Modern Nutrition in Health and Disease*, 5th ed., Goodhart, R. S. and Shils, M. E., Eds., Lea & Febiger, Philadelphia, 1973, 268.
24. Widdowson, E. M., Absorption and excretion of fat, nitrogen and minerals from filled milks by babies one week old, *Lancet*, II, 1099, 1965.
25. Underwood, E. J., *Trace Elements in Human and Animal Nutrition*, 3rd ed., Academic Press, New York, 1971.
26. Anon., Symposium on interaction of mineral elements in nutrition and metabolism, *Fed. Proc. Fed. Am. Soc. Exp. Biol.*, 19, 635, 1960.
27. Morrison, J. N., Quarterman, J., and Humphries, W. R., The effect of dietary calcium and phosphate on lead poisoning in lambs, *J. Comp. Patbol.*, 87, 417, 1977.
28. Six, K. M. and Goyer, R. A., Experimental enhancement of lead toxicity by low dietary calcium, *J. Lab. Clin. Med.*, 76, 933, 1970.
29. Quarterman, J. and Morrison, J. N., The effects of dietary calcium and phosphorous on the retention and excretion of lead in rats, *Br. J. Nutr.*, 34, 351, 1975.

30. Jägerstad, M., Lindstrand, K., Nordén, Å., and Qvist, I., Calcium: A study of food consumption by the duplicate portion technique in a sample of the Dalby population, *Scand. J. Soc. Med.*, Suppl. 10, 71, 1975.
31. Jägerstad, M., Abdulla, M., Svensson, S., and Nordén, Å., Calcium in Nutrition and Old Age. Chemical Analyses of what old people eat and their states of health during six years of follow-up, *Scand. J. Gastroenterol.*, 14, (Suppl. 52), 1979.
32. Paterson, C. R. and Path, M. R. C., Calcium requirements in man: a critical review, *Postgrad. Med. J.*, 54, 244, 1978.
33. Gitelman, H. J. and Lutwak, L., Dermal losses of minerals in elderly women under non-sweating conditions, *Clin. Res.*, 11, 42, 1963.
34. Lentner, C., Lauffenburger, T., Guncaya, J., Dambacker, M. A., and Haas, H. G., The metabolic balance technique. A critical reappraisal, *Metabolism*, 24, 461, 1975.
35. Davis, R. M., Morgan, D. B., and Rivlin, R. S., The excretion of calcium in the urine and its relation to calcium intake, sex and age, *Clin. Sci.*, 39, 1. 1970.
36. Heaney, R. P. and Skillman, T. G., Secretion and excretion of calcium by the human gastrointestinal tract, *J. Lab. Clin. Med.*, 64, 29, 1964.
37. Marshall, D. H., Nordin, B. E. C., and Speed, R., Calcium phosphorous and magnesium requirement, *Proc. Nutr. Soc.*, 35, 163, 1976.
38. Heaney, R. P., Recker, R. R., and Saville, P. D., Calcium balance and calcium requirements in middle-aged women, *Am. J. Clin. Nutr.*, 30, 1603, 1977.
39. Food and Agricultural Organization/World Health Organization, Calcium requirements, WHO Tech. Rep. Ser. 230, 1962.
40. Anon., Food and Nutrition Board: Recommended Dietary Allowances, 8th ed., National Academy of Sciences, Washington, D.C., 1974, 82.
41. Walker, R. M. and Linkswiler, H. M., Calcium retention in the adult human male as affected by protein intake, *J. Nutr.*, 132, 1297, 1972.
42. Margen, S., Chu, J-Y., Kaufmann, N. A., and Callaway, D. H., Studies in calcium metabolism. I. The calciuretic effect of dietary protein, *Am. J. Clin. Nutr.*, 27, 584, 1974.
43. Johnson, N. E., Alcantra, E. N., and Linkswiler, H., Effect of level of protein intake on urinary and fecal calcium and calcium retention of young adult males, *J. Nutr.*, 100, 1425, 1970.
44. Zollner, N., Wolfram, G., and Keller, Ch., Eds., Round table on comparison of dietary recommendations in different European countries, in *Second European Nutrition Conference*, S. Karger, Basel, 1977, 251.

Chapter 5

CALCIUM METABOLISM IN DENTINOGENESIS

Anders Linde

TABLE OF CONTENTS

I. Introduction 56

II. Theories of Mineralization 56

III. Initial Dentinogenesis... 57

IV. Cellular Calcium Localization During Dentinogenesis 58

V. Alkaline Phosphatases and Ca^{2+} Transport 59

VI. Mineral Formation in Dentinogenesis 60

VII. Role of Organic Tissue Components 62
   - A. Lipids.. 62
   - B. Collagen.. 62
   - C. Noncollagenous Proteins.................................. 63
   - D. γ-Carboxyglutamate-Containing Proteins................... 64
   - E. Phosphoproteins... 65
   - F. Proteoglycans 66

VIII. Concluding Remarks 67

Acknowledgments 67

References 68

I. INTRODUCTION

Dentinogenesis means formation of dentin. In the present article, this term will be taken to mean formation of the calcified dentin only. The developmental processes of the dental organ and the formation of the so-called secondary dentin will not be dealt with.

It appears that the mechanisms involved in the formation of the calcified dentin are primarily aimed at influencing the metabolism of calcium in one way or the other and at different levels, so as to direct its way from the blood into its final localization in the mineral crystals.

Due to the similarities of dentin formation with the formation of other hard tissues, dentinogenesis forms an excellent model system for studying the different mechanisms involved in the calcification process. In fact, due to several features, dentinogenesis is a superior experimental system compared to osteogenesis. One example of this is that dentin-forming cells, *odontoblasts,* can be rapidly dissected out to study the different metabolic processes eventually leading to the formation of the calcified dentin.[1] Since the organic components of the extracellular matrix of dentin influence mineral formation, it is of importance that pure samples of dentin can be readily obtained for analysis. The study of dentinogenesis may, thus, be a useful tool for our understanding of how calcium is transferred from blood into calcified tissues, the metabolic mechanisms going on in hard tissue-forming cells, and how this process is regulated.

II. THEORIES OF MINERALIZATION

Earlier, mineral formation was believed to take place through a simple precipitation reaction. The mineral-forming sites were thought to be in equilibrium with the body fluids; when the solubility constant for calcium phosphate is exceeded, e.g., through the local release of phosphate ions by alkaline phosphatase, mineral is formed.[2] It is now realized that, although mass action certainly is of importance, the mechanisms involved are more sophisticated than that, and that both cellular elements and matrix constituents are more directly engaged.

There have been two main questions concerning the role of cells in dentinogenesis. First, is calcium transported through the odontoblasts to the mineralization front, is it transported by an extracellular route only, or does the influx of calcium for mineral formation occur through a combination of these pathways? Secondly, are calcium and phosphate combined within the cells and then released to the mineralization front as calcium phosphate? As will be seen below, none of these questions can be definitely answered yet. It is clear though that the odontoblasts play an active role in dentinogenesis by virtue of their handling of calcium and their synthesis of the dentin organic matrix.

It is important to distinguish between the formation in the tissue of the initial mineral crystals, on one hand, and their subsequent growth on the other, since from a physicochemical point of view these must be regarded as two different processes. It is generally believed that the initiation of each calcium phosphate mineral crystal is facilitated by some kind of *nucleation.* This term refers to the specific stereochemical arrangement within a tissue compartment of specific reactive groups in the proper configuration, having the proper electrical charge distribution or other properties, which in some way lower the energy barrier so as to facilitate and initiate the formation of a solid phase of some calcium phosphate from a solution phase that would be otherwise stable.[3] Several candidates, alone or in combination, have been proposed for the role as *nucleation sites* in dentinogenesis, e.g., collagen, lipids, phosphoprotein, and proteoglycans.

Due to their supposed Ca^{2+} ion or hydroxyapatite affinity, several classes of molecules, e.g., proteoglycans, have been considered as mineralization *inhibitors* or *rate determinants*. A new concept is that of a *facilitated* or *vectorial influx* of Ca^{2+} ions to the mineral forming sites. In this mode the dentin phosphoprotein may influence the rate of mineralization and the crystallographic structure of the mineral being formed.

In view of the findings that calcium is localized to certain subcellular structures within the odontoblasts and that certain proteinaceous components of the organic matrix of dentin are secreted only at the mineralization front, one possibility would also be that nucleation of the mineral phase might occur in some intracellular compartment; the nucleus formed would then be secreted together with its template or nucleator close to the mineralization front, whereafter crystal growth would occur extracellularly.

III. INITIAL DENTINOGENESIS

The first layer, 5 to 10 μm of mineralized dentin, called *mantle dentin*, differs in both its morphology, chemical composition, and mode of formation from the rest of the dentin, the *circumpulpal dentin*.

It seems now to be generally accepted that extracellular matrix vesicles play an important role in the initiation of mineralization in *de novo* calcification. In contrast, no such vesicles appear to be involved in subsequent calcification. Since these vesicles obviously are derived from hard tissue-forming cells, this implies that the initiation of mantle dentin mineralization is under cellular control.

The population of extracellular matrix vesicles in the early predentin is quite heterogeneous, both in size and microscopical appearance. It has been suggested that vesicles with different functions may exist.[4] Thus, it seems appropriate to use the term *mineralization vesicles* to denote the specific matrix vesicles showing a close relationship to the initial mineral crystals.

Extracellular matrix vesicles were identified as loci for the initiation of mineralization of growth plate cartilage[5,6] and, subsequently, in bone.[7] Several investigations have now demonstrated the presence of extracellular vesicles that seem to be involved in mineralization of the first formed predentin, i.e., mantle dentin formation.[8-12] When examined in the electron microscope, the first indications of crystalline material are within 0.1 to 0.2 μm extracellular vesicles. These are regarded as being derived from the newly differentiated odontoblast cell-bodies and their major processes and are surrounded by a triple-layered membrane, about 80 to 100 Å in thickness. The vesicles may provide an enclosed and sequestered microenvironment for the accumulation of calcium and phosphate ions and their interaction with each other and possibly some organic molecules to form mineral.

The mineral has been described as needle-shaped or spicule-shaped. At this initial stage of mineralization, no mineral can be seen outside the vesicles in the collagenous organic matrix. Furthermore, though the odontoblast contains numerous *intra*cellular vescles and other organelles, the crystal-like inclusions seen in the *extra*cellular membrane-bound vesicles are never seen in those or in the cytoplasm.

In the earliest stages of crystal formation, the vesicle membrane is still intact. However, with an increasing number of crystals within the vesicle, these extend beyond the borders of the vesicle and the membrane disappears. Crystals grow randomly from these initial calcification loci to form small globules of mineralized dentin with a spheroidal form. During the growth of the globules, crystals from these extend into, onto, and between collagen fibers. Thus, not until this stage can crystals be seen in association with collagen fibers. The globules eventually coalesce and with mineralized collagen form seams of the first layer of mineralized dentin, the mantle dentin, which is characterized by irregularities and structural unevenness. After this layer of mantle dentin has been formed, the extracellular matrix vesicles are no longer seen.

It is not clear whether an active mechanism exists for the transport of Ca^{2+} into the vesicles. In bone mineralization vesicles, some crystals are parallel to one another.[13] This suggests the presence of a template, presumably macromolecules arranged in a parallel manner.[14] Atomic calcium/phosphorus ratios, obtained using the electron microprobe, indicate that the mineral present in nodules, the small mineral clusters formed with one mineralization vesicle as origin, is predominantly hydroxyapatite. This view is also supported by electron diffraction studies.[15]

Numerous questions remain to be answered before the mineralization vesicle theory is complete. One question is whether the vesicles are in fact already under way with mineralization when being exocytosed. It is not readily apparent how the deposition of a solid phase of calcium phosphate in these structures could influence the initiation of new additional mineral particles in spatially distant and distinct tissue components. Neither is it understood why such vesicles are not needed after the formation of some mineral. One possibility would be that the small mineral patches formed serve some function by binding to their surface macromolecules with an enzymatic or nucleating function, such as e.g., Gla-containing proteins.

IV. CELLULAR CALCIUM LOCALIZATION DURING DENTINOGENESIS

As to the localization of calcium in the odontoblast-predentin region, it is important to correlate the procedure used for localizing calcium and the information that can be expected, e.g., exposure to aqueous solution used in tissue preparation may not only result in the dissolution, redistribution, and loss of the mineral phase from the tissue but may also induce mineral phase transformations. Another problem is that it is often difficult to judge to what stage of dentinogenesis that the respective data pertain.

Autoradiography of dentin-forming tissues at certain time intervals after injection of $^{45}Ca^{2+}$ is supposed to give a dynamic picture of Ca^{2+} metabolism during dentinogenesis. The published data on this subject give conflicting views that may be explained only partially by species differences.

Fromme et al.[16] found that 10 min after an *i.p.* injection, a strong activity of ^{45}Ca was present in the odontoblasts of rat molars. Since, in addition, blood vessels were the only structures in the dental pulp that revealed an activity it was supposed that calcium transport occurred through extracellular pathways to the odontoblasts. At later stages, 30 and 60 min, activity was found in the odontoblast processes, whereas predentin did not reveal any activity. After 2 hr, no activity could be seen in the odontoblasts, their processes, or predentin; the activity was found entirely in the mineralized dentin.

This time sequence contrasts sharply to the one given by Munhoz and Leblond.[17] As soon as 30 sec after an *i.v.* injection of $^{45}Ca^{2+}$ these authors found a band of incorporation into calcium phosphate at the predentin-dentin junction of rat incisors. After 24 hr a radioactive band could be detected at a certain distance into the dentin. An increased activity could also be seen close to the dentin-enamel border.

Nagai and Frank,[18] using similar techniques in cats, were of the opinion that two transfer pathways for Ca^{2+} are of the same importance; one direct pathway from the capillaries, between the odontoblasts and through predentin to reach dentin, and one transcellular transfer through the odontoblasts. Intracellularly, mitochondria and the Golgi complex were found to be labeled. Activity was also found in the cytoplasm of the odontoblast process without being associated with any vesicular structures. Also in this study was a high activity found in dentin adjacent to the dentin-enamel border.

Though it may be risky to infer dynamic processes from static micrographs, precipitation techniques have been preferred by many authors to study odontoblast handling of Ca^{2+}. Indications for the presence of calcium in odontoblasts and odontoblast cell processes have been found by the pyroantimonate method and microincineration.[16,19] When preincubating the tissue so as to allow for the escape of diffusible ions, the pyroantimonate product was found only in the so-called abacus-like bodies and secretory granules of the Golgi region and the distal part of the cell.[20] These data may be interpreted to reflect the binding of Ca^{2+} ions within the Golgi complex of the odontoblasts. Numerous calcium-containing vesicles have been found in the periphery of the odontoblast process immediately beneath the cell membrane frequently associated with coated vesicles.[21] These results suggest that the intracellular calcium in odontoblasts is bound, probably to some organic macromolecules.

Boyde and Reith[22] used rapidly frozen rat incisors that were freeze fractured and lyophilized in combination with energy dispersive X-ray emission analysis, thereby accounting for total calcium content in the odontoblast layer. A strong accumulation of calcium was demonstrated in the distal cell-body of the odontoblast. Phosphorus levels were also increased in this area. In the odontoblast process, calcium and phosphorus levels were high.

Together these studies, with one exception,[17] indicate that some mechanism exists for concentrating calcium within the odontoblasts prior to extracellular mineral formation. Whether Ca^{2+} ions are transported to the mineralization front through the odontoblasts only, or if some extracellular route exists too, remains to be clarified.

Although minor amounts of calcium have been found in odontoblast mitochondria by some authors, there is no evidence to suggest any direct role for these organelles in dentin mineralization. This hypothesis, as suggested by Lehninger,[23] was partially based on the observation that in endochondral bone formation the number of calcium phosphate granules decreased in mitochondria of cells in the area where extracellular mineralization was first seen. It is well known that mitochondria possess a calcium concentrating mechanism capable of creating a strong calcium gradient. A more plausible role for the mitochondria in odontoblasts seems to be as an intracellular Ca^{2+} buffering compartment regulating the cytosolic Ca^{2+}. This would be consistent with a more remote, though possibly important, function as suggested by Shapiro and Greenspan.[24]

V. ALKALINE PHOSPHATASES AND CA^{2+} TRANSPORT

Non-specific alkaline phosphatase (APase, EC 3.1.3.1.) is a group of enzymes with alkaline pH optima acting on a vast number of different phosphate esters including inorganic pyrophosphate and ATP.[25] The same APase isoenzyme seems to be present in different calcifying tissues, odontoblasts having the highest activity.[26] In addition to APase, a more specific Ca^{2+} and Mg^{2+} activated adenosine triphosphatase has been demonstrated in odontoblasts.[27] This *Ca^{2+}-ATPase* may be separated by biochemical means from the APase.[28] Several data indicate that APase and Ca^{2+}-ATPase are intimately associated in the tissue.[28,29]

Earlier studies with electron microscopical histochemistry showed a localization of APase to the plasma membrane in odontoblasts. However, it has been shown that fixation methods used earlier almost totally inhibit APase and Ca^{2+}-ATPase activity in odontoblasts, as measured biochemically.[29] Using techniques to avoid these pitfalls, we found the main APase and Ca^{2+}-ATPase activities in rat incisor odontoblasts to be located in the membranes of intracellular vesicles, obviously derived from the Golgi complex.[29] In contrast to findings by others, no electron dense precipitates, indicating enzyme activity, were found in the odontoblast plasma membrane.

Several theories for the function of APase in calcification have been suggested. Robison's original idea was that this enzyme would provide phosphate ions for the formation of calcium phosphate mineral.[2] Others considered APase to have some special function in the synthesis of different matrix macromolecules. It has also been suggested that APase may split different phosphate esters, thus, keeping these mineralization inhibitory substances away from the area of calcification.[30] It is the view of the author that none of the several theories put forward explains the function of APase during dentinogenesis in a satisfactory way.

Many biological membranes effectuate an active transport of cations and create large concentration gradients using energy derived from APT. Ca^{2+} ions have been shown to be transported across membranes by specific Ca^{2+}-ATPases. The finding of a Ca^{2+}-activated ATPase in dentinogenically active odontoblasts strongly suggests the existence of a similar transmembranous Ca^{2+} pump as part of the calcification process in dentinogenesis. The finding of a close relationship between APase and Ca^{2+}-ATPase in odontoblasts may be indicative of a symbiotic function in the Ca^{2+} ion influx into intracellular vesicles in dentinogenically active odontoblasts. The relationship between these vesicles and those demonstrated to contain calcium[20-22] is not known.[2]

Granström and Linde[31] prepared a microsomal fraction from odontoblasts dissected out from rat incisors. When this nonmitochondrial fraction was incubated with $^{45}Ca^{2+}$, an energy dependent intravesicular accumulation of these ions was found and the properties indicated that a Ca^{2+}-ATPase was involved. Although this study must be regarded as preliminary, the results suggest that odontoblasts have a mechanism for the accumulation of calcium to an intracellular nonmitochondrial vesicular compartment. This mechanism may be of importance for the mineralization process.

VI. MINERAL FORMATION IN DENTINOGENESIS

From numerous studies on other cell types, it is well known that the cytosolic Ca^{2+} concentration is well regulated by a number of membrane-located Ca^{2+} pumps that extrude Ca^{2+} from the cell or concentrate it inside some organelle. Little is known about the odontoblasts in this respect. It is probable that the concentration gradient over the odontoblast cell membrane, like in many other cell types, is at least of the order of 10^3 with a cytosolic free Ca^{2+} concentration of 10^{-6} M or less. A concentration gradient of this magnitude would cause a passive leakage of Ca^{2+} ion into the cell of some importance. As discussed above, there are several reasons to believe that Ca^{2+} is transported through the odontoblast to the mineralization front, and several morphological findings support the idea that Ca^{2+} is in fact concentrated to vesicular structures inside the dentinogenically active odontoblast.

In a number of studies, Höhling and co-workers have used the electron microprobe to investigate the process of hard tissue formation.[32] Using cryostat sections the concentrations of Ca and P in rat incisor predentin, related to dry tissue mass, were on an average 0.28 and 1.1%, respectively. Since these values were considerably higher than those found in tendon, another dense collagenous connective tissue, it was concluded that there is an active accumulation of calcium and phosphate in predentin. Furthermore, when calculated, the ion product of these ions in predentin would be higher than that required for spontaneous hydroxyapatite formation. It was, thus, postulated that this precipitation is prevented by the Ca^{2+} ions being bound to the organic matrix.[33] This would not be the case for phosphate, since almost all of the phosphorus was removed by a brief water or ethanol wash. Evidence was found for well-defined regions in predentin with a "second stage of calcium enrichment", where the Ca mass fraction values were considerably increased.[32,34] These zones often coincided with border regions in which the growth of the irregular mineralization front lagged behind that of neighboring regions. It was suggested that the regions with the secondary calcium enrichment are precursors to the morphologically recognizable appearance of the calcospherites.

It is clear that, in spite of numerous investigations, the seemingly simple question as to which mineral phases are present in bone at different stages is far from its solution. Virtually nothing is known about the situation in dentinogenesis, so that parallels with the closely related osteogenesis have to be drawn. One major difficulty is that many studies of osteogenesis pertain to *de novo* bone formation, a process distinct in several respects from the noninitial dentinogenesis.

The mineral in dentin consists mainly, as in bone, of crystalline *hydroxyapatite* (HAP), $Ca_{10}(PO_4)_6(OH)_2$. Numerous studies of calcium phosphate mineral formation in vitro have focused the interest also to so-called *amorphous calcium phosphate* (ACP). This is a noncrystalline calcium phosphate with only a short range structural order that is formed under certain conditions. In water it converts autocatalytically with time to HAP. A number of different compounds have been demonstrated to affect the formation of ACP and the conversion of ACP to HAP.[35]

The presence of ACP can be seen by infrared spectroscopy as a single broad band at 550 to 600 cm^{-1}, whereas in HAP this is split into two bands.[36] Although this property can be used to study the composition of calcium phosphate mineral, factors such as crystal size and perfection and the presence of other ions must also be taken into account,[37,38] and this restricts the information that can be gained from biological materials such as dentin. Similar factors limit the information that can be obtained from X-ray diffraction analysis regarding the presence of different calcium phosphate minerals.

The finding in vitro of the existence of ACP led to the postulate that, in addition to HAP, bone contains a significant portion of ACP. X-ray diffraction studies of mature bone suggested that the level of ACP was about 35%. Due to the effect of crystal size and lattice defects, e.g., due to carbonate, a value like this must be regarded as highly uncertain. In fact, Posner and Betts,[39] using radial distribution analysis, demonstrated that there was little likelihood that mature bone contained any phase with only a short range order. Instead, it was suggested that ACP should be present only as the initial mineral form and at an early stage be transferred to crystalline HAP. The main reason for this should obviously be the observed processes in vitro.

There is, in fact, some morphological evidence for the existence of amorphous bone mineral. Gay[15] observed in thin sections of bone, prepared by ultracryotomy, large amounts of extracellular noncrystalline mineral appearing as spheres and as an amorphous haze, usually as a background to crystal clusters. This material was found to be morphologically similar to synthetic ACP. The finding of the amorphous mineral chiefly in matrix that seemed to be undergoing mineralization was taken to support the view that ACP precedes and plays a major role in HAP crystal formation in osteogenesis. The ACP was found primarily in interfibrillar areas, and it was speculated that it may be associated with noncollagenous matrix components. In well-mineralized regions, when highly oriented crystals were found, there did not appear to be much ACP.

A number of other calcium phosphate minerals have, at one time or another, been brought into the discussion. Recently, Roufosse et al.[40] found evidence that *brushite*, $CaHPO_4 \cdot 2H_2O$, is the major crystalline phase of calcium phosphate in the least mineralized chick bone, indicating that brushite may be the major component of the earliest mineral deposits in bone. These authors also stressed the fact that there would not necessarily be any direct causal relationship between the formation of brushite and the formation of HAP, partly because there would be no theoretical reason why HAP cannot be formed directly without the prior formation of a distinct intermediate solid phase.

In contrast to bone formation, where e.g., adjacent osteons may reveal strong differences in degree of mineralization, dentin attains virtually its final degree of mineralization shortly after it is deposited. This seems to contradict the existence in dentin of persisting ACP that, only at a later stage, would be converted to HAP. If any ACP at all is being formed in dentinogenesis, it would have to be transformed to HAP *in statu nascendi*. One may speculate that the cells responsible for dentin production are not capable of providing the accurate ion concentrations for the direct formation of HAP so that some additional component would have to be introduced. As will be seen below, the organic tissue components found in dentin might be of some importance in this respect.

VII. ROLE OF ORGANIC TISSUE COMPONENTS

After the mineralization vesicles have fulfilled their presumed role as initiators of mineralization of mantle dentin, mineral formation occurs in the collagenous matrix.

Several of the macromolecules in the dentin matrix may be of importance for mineral formation. The physicochemical properties of some of these components, in fact, strongly support some role in the mineralization process. It must be realized, however that analyses of the different tissue components were performed on components present in the mineralized dentin or bone, i.e., on components that may have been metabolized in one way or the other. Another problem is that it is difficult to directly link these tissue components experimentally to any specific function in dentinogenesis.

A. Lipids

Irving drew attention to the possibility of a role for lipids in calcification, since the content of acidic phospholipids was shown histochemically to be high at the mineralization front in dentin and bone.[41] Furthermore, essential fatty acid deficiency has been shown to have negative effect on dentinogenesis in rats[42] simultaneous with an altered fatty acid composition in dentin.[43] The lipid contents of bovine odontoblast-predentin tissue has been found to be different from that in mineralized dentin.[44]

Wuthier[45] concluded that the sudanophilia detected histochemically at sites of new calcification is caused mainly by acidic lipids. The finding that significant amounts of lipid were refractory to extraction until the tissues were demineralized suggested that the acidic lipids form part of a lipoprotein-mineral complex in the calcified tissue. Boskey and Posner isolated a calcium-phospholipid-phosphate complex from bone that was more prevalent in developing bone compared to mature bone.[46] Additional evidence for an involvement of acidic phospholipids in mineralization was gained when it was found that this complex initiated HAP formation in vitro from a metastable calcium phosphate solution.[47] The complex was regarded not to be composed of HAP seed crystals adsorbed to the acidic phospholipid but, rather, a distinct species which would facilitate initial HAP formation by accumulating calcium and phosphate ions and holding these ions in the proper orientation for initial crystal formation, i.e., through nucleation.

The results from studies on the possible involvement of lipids in the calcification process are clearly intriguing. However, one piece of evidence that is missing are data on the subcellular localization of phospholipid in calcifying tissues. Extracellular mineralization vesicles contain lipids[48] and the possible involvement of lipids in *de novo* calcification, like e.g., mantle dentinogenesis, might be understood. In circumpulpal dentinogenesis, no membranous structures are present where crystals are being formed; hence, a direct involvement of lipids is less easily understood from a morphological viewpoint. A secretion of phospholipid to the sites where crystal formation is initiated remains to be shown.

B. Collagen

Collagen is the major organic constituent of dentin accounting for 80 to 90% of the organic matrix depending on species. During early stages of dentinogenesis, synthesis of Type III collagen takes place; at later stages, Type I collagen is made almost exclusively.[49,50] In contrast, definite amounts of Type III collagen are present in affected dentin from patients with osteogenesis imperfecta.[51] Isolated CNBr peptides from dentin collagen have been demonstrated to have compositions like those from $\alpha_1(I)$ and α_2 chains from skin collagen, showing that dentin contains Type I collagen.[52,53] Evidence has been found for the presence of collagen Type I trimer, i.e., $[\alpha_1(I)]_3$ molecules, in dentin.[54,55]

Dentin collagen is synthesized by the odontoblasts; the pathway of collagen synthesis and secretion by odontoblasts through predentin and into dentin has been elucidated.[56-58]

Thus, it may be concluded that the major collagen in dentin is of Type I and that a certain portion, though small under normal circumstances, might be Type I trimer.

Dentin collagen fibers are arranged as a network. Höhling and co-workers concluded that collagen mineralization starts at the surface and proceeds rapidly into the interior of the fibrils.[59] Chains of dotlike nuclei were initially formed on and within the collagen fibers and oriented along the fiber axis. These chains grow to needles by fusion along this axis and, even later on, to platelike crystals. The distances between the dotlike nuclei in the long direction of the collagen fiber were found to be in good agreement with the distances between the cross-striations and thus, presumably, particular amino acid sequences in the primary structure. It has been hypothesized that mineral formation on the surface might be due to noncollagenous macromolecules. The calcium phosphate needles formed are not in contact but are separated by narrow mineral-free spaces. This lateral separation between the initial chains of the nuclei and the needles has been explained by postulating[59] that mineral nuclei are formed in microchannels between the collagen subfibrils as proposed in the Miller and Parry model.[60] The dimension of these microchannels would influence the size of the initial nuclei; to grow and unite to form crystals they would have to separate the noncovalent attractive forces between neighboring subfibrils. By the event described, growth of apatitic crystals with their c-axis parallel to the collagen fibril axis would occur.

The results of Boyde[61] on ion beam thinned dentin, though puzzling in some respects, may be taken to support the above sequence of events. A large proportion of the mineral phase was found to be organized in the form of hollow cylindrical elements, thus, possibly revealing a microfibrillar collagen structure and a special relationship of collagen to mineral. However, little association was found between the distribution of the mineral phase and the periodicity of collagen. The mineral in dentin, suggested to be in the form of tubules or rods, was possible to trace for distance of several thousands of Å units and may, thus, be extremely long.

C. Noncollagenous Proteins

There are indications for a secretion of noncollagenous proteins (NCPs) into the matrix at the mineralization front.[62] Among the NCPs, several have attracted special attention as being possibly involved in the initiation and the regulation of mineral formation. Unfortunately, many of these theories are founded on a suggested function, based on the structural features of the macromolecule such as e.g., a high negative charge density, rather than on unequivocal experimental evidence. On the other hand, due to their chemical composition it might be expected that some components among the NCPs may exert different specific functions in the calcification process.

In general, the NCPs are solubilized and extracted only after demineralization of dentin.[63] It has earlier been suggested that some NCPs, specifically a large proportion of dentin phosphoprotein, would be covalently linked to collagen and presumably have some specific function in the calcification process.[64,65] It has, however, been shown that very little, if any, glycoprotein and no phosphoprotein is bound in this manner at least in rat incisor dentin.[66]

Linde et al.[63] in reexamining the NCP content of dentin took several precautionary measures to avoid artifactual degradation of the NCPs. These were found to fall into four main categories: acidic glycoproteins, γ-carboxyglutamate-containing proteins, phosphoproteins, and proteoglycans. This pattern is very similar to that of bone;[67] one notable difference is, however, the presence of highly phosphorylated phosphoproteins in dentin not demonstrated in bone.

Little can be found in the literature as to any function of the acidic glycoproteins in biological calcification. Similarly, the presence of plasma proteins in dentin[68,69] is not understood. The other three groups of NCPs have been extensively discussed and will be further dealt with below.

D. γ-Carboxyglutamate-Containing Proteins

The unique calcium binding site, γ-carboxyglutamic acid (Gla), has been extensively investigated in proteins involved in blood coagulation, e.g., prothrombin. This amino acid is formed by a post-translational γ-carboxylation of glutamic acid which is vitamin K-dependent.

Hauschka et al.[70] and Price et al.[71] demonstrated the presence of significant amounts of Gla in bone and isolated Gla-containing proteins from chicken and calf bone. The covalent structures for calf and swordfish Gla-containing proteins reveal a high degree of homology in the sequence containing the three Gla residues and the disulphide bond.[72] These molecules have a relatively low affinity for Ca^{2+} with $K_d = 3.10^{-3} M$.[73] The Ca^{2+} binding is abolished by thermal γ-decarboxylation.[73] The Gla-containing protein of bone has been shown to be synthesized in the bone itself;[74] the Gla protein present in human plasma may thus be derived from bone and presumably liberated during osteoclasia.[72]

Among the total NCPs extracted from rat bone, those containing Gla have been shown to interact with HAP crystals in vitro so as to inhibit their growth.[75] However, Price et al. ruled out any important role for Gla-containing protein in bone mineralization on the grounds that negligible amounts of Gla-containing protein could be demonstrated to be present in rat bone during the early stages of mineralization.[72] Furthermore, the bones of rabbits fed the vitamin K antagonist Warfarin did not show any anomalies in spite of an essentially complete depletion of Gla-containing protein. Gla protein in bone has been proposed to be an informational molecule, comparable to peptide hormones, released into the blood. Plasma levels of bone Gla protein have been reported to be increased in patients with bone diseases characterized by increased bone turnover.[72]

Relatively large amounts of Gla-containing proteins are present in rat incisor dentin.[63] These have a molecular size, about 50 amino acids, and composition similar to the Gla-containing protein in bovine bone. However, recent experiments have demonstrated that a whole family of distinct Gla-containing proteins, at least four, are present in dentin.[76] Preliminary amino acid sequence data show a high level of homology between those that have been studied and the bovine bone protein. At least two Gla proteins in dentin are five residues shorter in the amino terminal region compared to the other. The data suggest that the different Gla-containing proteins are not derived from each other. In fact, our current studies show that a number of similar proteins are present also in bone.[67]

The finding of several structurally related Gla-containing proteins in dentin inevitably leads to the comparison with the blood coagulation mechanism. By means of a cascade of reactions involving Gla-containing proenzymes, an amplification is obtained along with a close control of the process. The role of the Gla-residues is considered to be the binding of the proenzymes to membrane structures through a Ca^{2+} ion linkage. One possibility is, thus, that the numerous Gla-containing, relatively small proteins in dentin and bone are in fact fossilized remnants of large proenzymes with unknown functions. It seems reasonable to assume that some enzymatic amplification reaction, which at the same time exerts a control function, might be functioning during calcification. It is notable that dentin phosphoprotein in some respects has the properties of a negatively charged membrane,[77] and thus might function in the binding of Gla-containing proteins. Another possibility would be a finding to HAP crystals, where the Gla-containing promolecules may have a function that makes the need for mineralization vesicles unnecessary.

E. Phosphoproteins

The major dentin phosphoprotein fraction consists of unusual, strongly polyanionic proteins, constituting about half of the dentin NCPs.[63] When adequately purified,[63,78] about 50% of the amino acid residues are serine, of which 87% are phosphorylated to phosphoserine, and 40% are aspartate; several amino acids and carbohydrate are absent. Rat incisor phosphoprotein contains about 26% phosphate by weight[63] and has a pI = 1.1.[79] Molecular weight data reported in the literature are unreliable, since these have been determined by standard methods developed for globular proteins; the correct value should be 38,000, as obtained by ultracentrifugation of enzymatically dephosphorylated phosphoprotein.[80]

Butler et al.[81] showed that large amounts of dentin phosphoprotein were easily extractable after demineralization by nondegradative methods, while Veis and co-workers[64] presented data indicating that a large proportion was covalently linked to the dentin collagen. It was suggested that the covalently linked phosphoprotein moiety would have some specific role in the nucleation of HAP.[65] However, we have demonstrated that essentially no phosphoprotein is covalently bound to collagen in rat incisor dentin or bone.[66,67,82] The findings indicate that this should be the case also in other species.

Recent studies have shown that there is in fact a number of phosphoproteins present in dentin. The major phosphoprotein fraction in rat incisor dentin, as described above, can after enzymatic dephosphorylation, and only after this, be separated into two distinct components with different amino acid compositions and NH_2-termini.[63] In addition, a number of minor phosphoprotein fractions have been isolated. These are characteristically lower in total serine and phosphate, relatively high in glutamate and contain considerable amounts of most amino acids.[63,83]

In contrast to collagen, which is secreted to form predentin, phosphoprotein is deposited only at the mineralization front.[62] Dimuzio and Veis[84] also found evidence that collagen and phosphoprotein follow different secretory pathways. The absence of phosphoprotein in predentin has been clearly demonstrated by biochemical technique.[85]

As to the biological function of phosphoprotein in dentinogenesis, many investigators have suggested that this polyanion has a role as nucleator for HAP formation, although dentin phosphoprotein has also been suggested to exert an inhibitory function.[86] Highly purified rat incisor dentin phosphoprotein has a strong Ca^{2+} ion affinity.[87] It reveals two classes of binding sites with $K_d = 1.3 \cdot 10^{-7} M$ and $0.85 \cdot 10^{-5} M$. This Ca^{2+} binding is dependent on the phosphate groups, since enzymatic dephosphorylation abolishes the high affinity binding and increases the K_d of the low affinity binding sites. Nawrot et al.[88] showed that the presence of dentin phosphoprotein induced formation of HAP under in vitro conditions when otherwise ACP would have been formed. This was not the case when another phosphate-rich protein, phosvitin, was present. Partial dephosphorylation inhibited this catalytic property of phosphoprotein.

A strong Ca^{2+} affinity of rat incisor dentin phosphoprotein was also demonstrated by Cookson et al. using spectroscopic methods.[78] It was found that the Ca^{2+} ions, though strongly bound, were free to move on the surface of the molecule. Thus, dentin phosphoprotein may bring about a facilitated vectorial influx of Ca^{2+} ions, thereby exerting a strong influence on mineral formation not possible to fit into the classical scheme of nucleation inhibition. This property would be consistent with the promotion of HAP formation at the expense of ACP.[88]

F. Proteoglycans

Proteoglycans (PGs) are large molecules that form the ground substance of connective tissues and cartilage. Long PGs are known to be synthesized at calcification sites and have accordingly played a central role in proposed calcification mechanisms, presumably as inhibitors or nucleators for HAP formation. The PGs consist of a central protein core to which are bound polysaccharide chains, glycosaminoglycans (GAGs), that are strongly polyanionic due to carboxyl and sulfate groups. In cartilage, the PGs are noncovalently bound to chains of hyaluronate with the aid of link proteins forming enormous molecular complexes. It is not known whether this organization exists in hard tissues.

Chrondroitin sulfates are the major GAGs in dentin, but other GAGs such as hyaluronate, keratan sulfate, and dermatan sulfate are also present.[89,90] The content of GAGs in predentin has been reported.[91] Using a density gradient system, Hjerpe and Engfelt[92] studied the PGs of predentin and dentin from rachitic puppies and the differences found were taken as indication for a metabolism of PGs at the mineralization front.

The localization of PG in the odontoblast-predentin region of rat incisor was investigated by Nygren et al.[93] by electron microscopical histochemistry. Sundström[94] demonstrated that sulfated PG, like collagen, is secreted into the predentin and after a couple of hours trapped inside the dentin. In addition, however, a second broader $^{35}SO_4$-containing band was found to be located and retained for some time over the predentin.

In a number of studies, Höhling and co-workers,[32] using the electron microprobe, measured sulfur content in predentin and adjacent dentin. A consistent finding was that the sulfur content, taken to indicate sulfated PG, decreased at the mineralization front or 5 to 10 μm into the dentin. Another evidence for the degradation of PG simultaneous with mineralization is the finding that the content of PGs differ between predentin and dentin.[92] Also, in bone, PG is lost simultaneously with mineral formation.[95] The odontoblast-predentin region possesses the necessary enzymatic armamentarium to degrade PGs.[96,97] Among these enzymes, the protease cathepsin D has been shown to be localized not only in odontoblast lysosomes, but also extracellularly in predentin.[98] A similar finding is the high cathepsin D content in osteoblasts.[99]

The significant features of PGs are their high negative charge density and the large domains occupied in the tissue. The binding of Ca^{2+} ions to the GAG sidechains occurs in a rather nonspecific way; such binding would not be expected to provide specific sites for crystal growth.[100] On the other hand, PGs in cartilage seem to have a mineralization inhibitory function[101] and it is probable that the PGs in predentin exert a similar inhibition of mineral nucleation and growth. Furthermore, it has been shown that PGs and GAGs may influence the extracellular formation of collagen fibrils.[102] The PGs in predentin may well exert such a function. Degradation and removal of the PGs would then occur when the mineralization inhibitory function and the regulatory function for collagen formation are not needed any longer. A less sublime, though nevertheless quite plausible, function would be that the PGs, after having occupied space between collagen fibrils in predentin, are removed to provide room for the mineral crystals.

VIII. CONCLUDING REMARKS

Dentinogenesis comprises the synthesis of dentin organic matrix and, after a certain time lag, its subsequent mineralization during which the composition and properties of this matrix changes. Many factors obviously cooperate in the calcification process, thus, governing the handling of Ca^{2+} ions and their route into the mineral phase. Earlier research in biological calcification often overemphasized the importance of one supposedly crucial factor, depending on the specific interest of that author. Calcium phosphate mineral formation has traditionally rotated around the two notions *nucleation* and *inhibition*. One aim of the present review has been to show that this certainly is an oversimplification.

Mineralization of the dentin matrix is obviously regulated by the cells involved, primarily the odontoblasts, and the chemical characteristics of this matrix. The dental pulp proper does not seem to be directly involved. Not very much is known about the cellular handling of Ca^{2+} ions; more is known about the interactions between these ions and the components of the matrix. One problem is, however, that it is difficult to assess the value of the many investigations of the effects of matrix components on calcium phosphate mineral formation under artificial conditions in vitro. One sometimes gets the impression that anything, under certain conditions, can be shown to influence mineral formation in any direction.

The mechanisms differ between the formation of the first layer of dentin and that deposited subsequently. The first mineral crystals are possibly found in extracellular mineralization (matrix) vesicles. Later, during dentinogenesis these vesicles are not seen, and instead initial mineral nuclei apparently occur in close association with collagen fibrils. Transport of Ca^{2+} ions to the mineralization front occurs to a large extent through the odontoblasts; energy-requiring transport systems may be involved here. In addition, Ca^{2+} ions might also proceed extracellularly, though opinions differ in this respect. The nature of the mineral initially formed in dentinogenesis is not known. One possibility is that it is a calcium phosphate with a short-range order that is more or less rapidly transferred into hydroxyapatite crystals of a fairly uniform size. The factors regulating this size are not known. The noncollagenous matrix components may have several interesting functions. Due to their specific Ca^{2+}-binding properties, the phosphoproteins may exert a catalytic function on the formation of mineral by the facilitated influx of Ca^{2+} ions. The enzymatic degradation of proteoglycans may remove an inhibitory component and provide space for mineral crystals. The possible functions of lipids, acidic glycoproteins, and γ-carboxyglutamate-containing proteins in dentinogenesis await further clarification.

ACKNOWLEDGMENTS

The author has benefited from many discussions on dentinogensis with several colleagues, especially Dr. W. T. Butler and Dr. M. Jontell. Special appreciation goes to Ms. Gunilla Pfannenstill who has struggled with this manuscript. The work of the author, referred to in this review, was supported by the Swedish Medical Research Council.

REFERENCES

1. **Linde, A.**, A method for the biochemical study of enzymes in the rat odontoblast layer during dentinogenesis, *Arch. Oral Biol.,* 17, 1209, 1972.
2. **Robison, R.**, The possible significance of hexosephosphoric esters in ossification, *Biochem. J.,* 17, 286, 1923.
3. **Landis, W. J., Paine, M. C., and Glimcher, M. J.**, Electron microscopic observations of bone tissue prepared anhydrously in organic solvents, *J. Ultrastruct. Res.,* 59, 1, 1977.
4. **Slavkin, H. C., Croissant, R. D., Bringas, P.**, Epithelial-mesenchymal interactions during odontogenesis. III. A simple method for the isolation of matrix vesicles, *J. Cell Biol.,* 53, 841, 1972.
5. **Anderson, H. C.**, Electron microscopic studies of induced cartilage development and calcification, *J. Cell Biol.,* 35, 81, 1967.
6. **Bonucci, E.**, Fine structure of early cartilage calcification, *J. Ultrastruct. Res.,* 20, 33, 1967.
7. **Bernard, G. W. and Pease, D. C.**, An electron microscopic study of initial intramembranous osteogenesis, *Am. J. Anat.,* 125, 271, 1969.
8. **Bernard, G. W.**, Ultrastructural observations of initial calcification in dentine and enamel, *J. Ultrastruct. Res.,* 41, 1, 1972.
9. **Eisenmann, D. R. and Glick, P. L.**, Ultrastructure of initial crystal formation in dentin, *J. Ultrastruct. Res.,* 41, 18, 1972.
10. **Sisca, R. F. and Provenza, D. V.**, Initial dentin formation in human deciduous teeth. An electron microscope study, *Calcif. Tissue Res.,* 9, 1, 1972.
11. **Katchburian, E.**, Membrane-bound bodies as initiators of mineralization of dentine, *J. Anat.,* 116, 285, 1973.
12. **Larsson, Å. and Bloom, G. D.**, Studies on dentinogenesis in the rat. Fine structure of developing odontoblasts and predentin in relation to the mineralization process, *Z. Anat. Entwicklungsgesch.,* 139, 227, 1973.
13. **Gay, C. V., Schraer, H., and Hargest, T. E., Jr.**, Ultrastructure of matrix vesicles and mineral in unfixed embryonic bone, *Metab. Bone Dis. Rel. Res.,* 1, 105, 1978.
14. **Höhling, H. J., Steffens, H., Stamm, G., and Mays, U.**, Transmission microscopy of freeze dried, unstained epiphyseal cartilage of the guinea pig, *Cell Tissue Res.,* 167, 243, 1976.
15. **Gay, C. V.**, The ultrastructure of the extracellular phase of bone as observed in frozen thin sections, *Calcif. Tissue Res.,* 23, 215, 1977.
16. **Fromme, H. G., Höhling, H. J., and Riedel, H.**, Elektronenmikroskopische Studien über die Dentinbildung. II. Mitteilung: Autoradiographische Untersuchungen zur Funktion der Odontoblasten, *Dtsch. Zahnaerztl. Z.,* 27, 6, 1972.
17. **Munhoz, C. O. G. and Leblond, C. P.**, Deposition of calcium phosphate into dentin and enamel as shown by radioautography of sections of incisor teeth following injection of ^{45}Ca into rats, *Calcif. Tisue Res.,* 15, 221, 1974.
18. **Nagai, N. and Frank, R. M.**, Electron microscopic autoradiography of Ca^{45} during dentinogenesis, *Cell Tissue Res.,* 155, 513, 1974.
19. **Fromme, H. G., Höhling, H. J., and Riedel, H.**, Elektronenmikroskopische Studien über die Dentinbildung. I. Mitteilung: Lokalisation von Calcium und alkalischer Phosphatase, *Dtsch. Zahnaerztl. Z.,* 26, 359, 1971.
20. **Reith, E. J.**, The binding of calcium within the Golgi saccules of the rat odontoblast, *Am. J. Anat.,* 147, 267, 1976.
21. **Appleton, J. and Morris, D. C.**, An ultrastructural investigation of the role of the odontoblast in matrix calcification using the potassium pyroantimonate osmium method for calcium localization, *Arch. Oral Biol.,* 24, 467, 1979.
22. **Boyde, A. and Reith, E. J.**, Qualitative electron probe analysis of secretory ameloblasts and odontoblasts in the rat incisor, *Histochemistry,* 50, 347, 1977.
23. **Lehninger, A. L.**, Mitochondria and calcium ion transport, *Biochem. J.,* 119, 129, 1970.
24. **Shapiro, I. M. and Greenspan, J. S.**, Are mitochondria directly involved in biological mineralization?, *Calcif. Tissue Res.,* 3, 100, 1969.
25. **Linde, A. and Granström, G.**, Odontoblast alkaline phosphatases and Ca^{2+} transport, *J. Biol. Bucc.,* 6, 293, 1978.
26. **Granström, G. and Linde, A.**, A comparative study of alkaline phosphatase in calcifying cartilage, odontoblasts and the enamel organ, *Calcif. Tissue Res.,* 22, 231, 1977.
27. **Linde, A. and Magnusson, B. C.**, Inhibition studies of alkaline phosphatases in hard tissue-forming cells, *J. Histochem. Cytochem.,* 23, 342, 1975.
28. **Granström, G., Jontell, M., and Linde, A.**, Separation of odontoblast Ca^{2+}-ATPase and alkaline phosphatase, *Calicif. Tissue Int.,* 27, 211, 1979.
29. **Granström, G., Linde, A., and Nygren, H.**, Ultrastructural localization of alkaline phosphatases in rat incisor odontoblasts, *J. Histochem. Cytochem.,* 26, 359, 1978.

30. Fleisch, H. and Neuman, W. F., Mechanisms of calcification: role of collagen, polyphosphatase and phosphatase, *Am. J. Physiol.*, 200, 1296, 1961.
31. Granström, G. and Linde, A., ATP dependent uptake of Ca^{2+} by a microsomal fraction from rat incisor odontoblasts, *Calicif. Tissue Int.*, 33, 125, 1981.
32. Nicholson, W. A. P., Ashton, B. A., Höhling, H. J., Quint, P., Schreiber, J., Ashton, I. K., and Boyde, A., Electron microprobe investigations into the process of hard tissue formation, *Cell Tissue Res.*, 177, 331, 1977.
33. Ashton, B., Höhling, H. J., Nicholson, W. A. P., Zessack, U., Kriz, W., and Boyde, A., Quantitative analysis of Ca, P and S in mineralizing and non-mineralizing tissues, *Naturwissenchaften*, 60, 392, 1973.
34. Höhling, H. J., Nicholson, W. A. P., Schreiber, J., Zessack, U., and Boyde, A., The distribution of some elements in predentine and dentine of rat incisors, *Naturwissenchaften*, 59, 423, 1972.
35. Blumenthal, N. C., Betts, F., and Posner, A. S., Stabilization of amorphous calcium phosphate by Mg and ATP, *Calcif. Tissue Res.*, 23, 245, 1977.
36. Termine, J. D. and Posner, A. S., Infrared analysis of rat bone: age dependency of amorphous and crystalline mineral fractions, *Science*, 153, 1523, 1966.
37. Blumenthal, N. C., Posner, A. S., and Holmes, J. M., Effect of preparation condition on the properties and transformation of amorphous calcium phosphate, *Mat. Res. Bull.*, 7, 1181, 1972.
38. Blumenthal, N. C., Betts, F., and Posner, A. S., Effect of carbonate and biological macromolecules on formation and properties of hydroxyapatite, *Calcif. Tissue Res.*, 18, 81, 1975.
39. Posner, A. S. and Betts, F., Synthetic amorphous calcium phosphate and its relation to bone mineral structure, *Acc. Chem. Res.*, 8, 273, 1975.
40. Roufosse, A. H., Landis, W. J., Sabine, W. K., and Glimcher, M. J., Identification of brushite in newly deposited bone mineral from embryonic chicks, *J. Ultrastruct. Res.*, 68, 235, 1979.
41. Irving, J. T. The sudanophil material at sites of calcification, *Arch. Oral Biol.*, 8, 735, 1963.
42. Prout, R. E. S. and Tring, F. C., Dentinogenesis in incisors of rats deficient in essential fatty acids, *J. Dent. Res.*, 52, 462, 1973.
43. Prout, R. E. S. and Atkin, E. R., Effect of diet deficient in essential fatty acids on fatty acid composition of enamel and dentine of the rat, *Arch. Oral Biol.*, 18, 583, 1973.
44. Ellingson, J. S., Smith, M., and Larson, L. R., Phospholipid composition and fatty acid profiles of the phospholipids in bovine predentin, *Calcif. Tissue Res.*, 24, 127, 1977.
45. Wuthier, R. E., Lipids of mineralizing epiphyseal tissues in the bovine fetus, *J. Lipid Res.*, 9, 68, 1968.
46. Boskey, A. L. and Posner, A. S., Extraction of a calcium-phospholipid-phosphate complex from bone, *Calcif. Tissue Res.*, 19, 273, 1976.
47. Boskey, A. L. and Posner, A. S., The role of synthetic and bone extracted Ca-phospholipid-PO_4 complexes in hydroxyapatite formatin, *Calcif. Tissue Res.*, 23, 251, 1977.
48. Peress, N. S., Anderson, H. C., and Sajdera, S. W., The lipids of matrix vesicles from bovine fetal epiphyseal cartilage, *Calcif. Tissue Res.*, 14, 275, 1974.
49. Lesot, H., von der Mark, K., and Ruch, J.-V., Localisation par immunofluorescence des types de collagène synthétisés par l'ébauche dentaire chez l'embryon de souris, *C. R. Acad. Sci. Paris*, 286, 765, 1978.
50. Thesleff, I., Stenman, S., Vaheri, A., and Timpl, R., Changes in the matrix proteins, fibronectin and collagen, during differentiation of mouse tooth germ, *Dev. Biol.*, 70, 116, 1979.
51. Gay, S., personal communication, 1980.
52. Butler, W. T., The structure of α1-CB3, a cyanogen bromide fragment from the central portion of the α1 chain of rat collagen. The tryptic peptides from skin and dentin collagens, *Biochem. Biophys. Res.*, 48, 1540, 1972.
53. Volpin, D. and Veis, A., Cyanogen bromide peptides from insoluble skin and dentin bovine collagens, *Biochemistry*, 12, 1452, 1973.
54. Wohllebe, M. and Carmichael, D. J., Type-I trimer and Type-I collagen in neutral-salt-soluble lathyritic-rat dentine, *Eur. J. Biochem.*, 92, 183, 1978.
55. Munksgaard, E. C., Collagen in dentin, *J. Biol. Bucc.*, 7, 131, 1979.
56. Frank, R. M., Etude autoradiographique de la dentinogenese en microscopie électronique a l'aide de la proline tritieé chez le chat, *Arch. Oral Biol.*, 15, 583, 1970.
57. Weinstock, M. and Leblond, C. P., Synthesis, migration and release of precursor collagen by odontoblasts as visualized by radioautography after (^{3}H)proline administration, *J. Cell Biol.*, 60, 92, 1974.
58. Karim, A., Cournil, I., and Leblond, C. P., Immunohistochemical localization of procollagens. II. Electron microscopic distribution of procollagen I antigenicity in the odontoblasts and predentin of rat incisor teeth by a direct method using peroxidase linked antibodies, *J. Histochem. Cytochem.*, 27, 1070, 1979.
59. Höhling, H. J. and Ashton, B. A., Quantitative electron microscopic investigations of mineral nucleation in collagen, *Cell Tissue Res.*, 148, 11, 1974.

60. Miller A. and Parry, D. A. D., The structure and packing of microfibrils in collagen, *J. Mol. Biol.*, 75, 441, 1973.
61. Boyde, A., Transmission electron microscopy of ion beam thinned dentine, *Cell Tissue Res.*, 152, 543, 1974.
62. Weinstock, M. and Leblond, C. P., Radioautographic visualization of the deposition of a phosphoprotein at the mineralization front in the dentin of the rat incisor, *J. Cell Biol.*, 56, 838, 1973.
63. Linde, A., Bhown, M., and Butler, W. T., Noncollagenous proteins of dentin. A reexamination of proteins from rat incisor dentin utilizing techniques to avoid artifacts, *J. Biol. Chem.*, 255, 5931, 1980.
64. Dickson, I. R., Dimuzio, M. T., Volpin, D., Ananthanarayanan, S., and Veis, A., The extraction of phosphoproteins from bovine dentin, *Calcif. Tissue Res.*, 19, 51, 1975.
65. Veis, A., The role of acidic proteins in biological mineralizations, in *Ions in Macromolecular and Biological Systems*, Everett, D. H. and Vincent, B., Eds., Colston Papers No. 29, Scientechnia, Bristol, 1978, 259.
66. Linde, A., Bhown, M. and Butler, W. T., Non-collagenous proteins of dentin. Evidence that phosphoprotein is not covalently bound to collagen, *Biochim. Biophys. Acta*, 667, 341, 1981.
67. Linde, A., submitted, 1982.
68. Thomas, M. and Leaver, A. G., Identification and estimation of plasma proteins in human dentine, *Arch. Oral Biol.*, 20, 217, 1975.
69. Kinoshita, Y., Incorporation of serum albumin into the developing dentine and enamel matrix in the rabbit incisor, *Calcif. Tissue Int.*, 29, 41, 1979.
70. Hauschka, P. V., Lian, J. B., and Gallop, P. M. Direct identification of the calcium-binding amino acid, γ-carboxyglutamate, in mineralized tissue, *Proc. Natl. Acad. Sci. USA*, 72, 3925, 1975.
71. Price, P. A., Otsuka, A. S., Poser, J. W., Kristaponis, J., and Raman, N., Characterization of a γ-carboxyglutamic acid-containing protein from bone, *Proc. Natl. Acad. Sci. USA*, 73, 1447, 1976.
72. Price, P. A., Epstein, D. J., Lothringer, J. W., Nishimoto, S. K., Poser, J. W., and Williamson, M. K., Structure and function of the vitamin K-dependent protein of bone, in *Vitamin K Metabolism and Vitamin K-Dependent Protein*, Suttie, J. W., Ed., University Park Press, Baltimore, 1980, 219.
73. Poser, J. W. and Price, P. A., A method for decarboxylation of γ-carboxyglutamic acid in proteins. Properties of the decarboxylated γ-carboxyglutamic acid protein from calf bone, *J. Biol. Chem.*, 254, 431, 1979.
74. Nishimoto, S. K. and Price, P. A. Proof that the γ-carboxyglutamic acid-containing bone protein is synthesized in calf bone. Comparative synthesis rate and effect of coumadin on synthesis, *J. Biol. Chem.* 254, 437, 1979.
75. Diamond, A. G.and Neuman, W. F., Macromolecular inhibitors of calcium phosphate precipitation in bone, in *Vitamin K Metabolism and Vitamin K-Dependent Protein*, Suttie, J. W., Ed., University Park Press, Baltimore, 1980, 259.
76. Linde, A., Bhown, M., Höglund, A., and Butler, W. T., unpublished data, 1982.
77. Cookson, D. J., Levine, B. A., Williams, R. J. P., Linde, A., Jontell, M., and de Bernard, B., Cation binding by the rat incisor dentine phosphoprotein. A spectroscopic investigation, *Eur. J. Biochem.*, 110, 273, 1980.
78. Munksgaard, E. C., Butler, W. T., and Richardson, W. S., Phosphoprotein from dentin. New approaches to achieve and assess purity, *Prep. Biochem.*, 7, 321, 1977.
79. Jonsson, M., Fredriksson, S., Jontell, M., and Linde, A., Isoelectric focusing of the phosphoprotein of rat incisor dentin in ampholine and acid pH gradients. Evidence for carrier ampholyte-protein complexes, *J. Chromatogr.*, 157, 235, 1978.
80. Jontell, M., Pertoft, H., and Linde, A., *Biochim. Biophys. Acta*, in press, 1982.
81. Butler, W. T., Finch, J. E., Jr., and Desteno, C. V., Chemical character of proteins in rat incisors, *Biochim. Biophys. Acta*, 257, 167, 1972.
82. Jontell, M., Linde, A., and Lundvik, L., Comparative studies of phosphoprotein preparations from rat incisor dentin, *Prep. Biochem.*, 10, 235, 1980.
83. Butler, W. T., Bhown, M., Dimuzio, M. T., and Linde, A., *Coll. Res.*, 1, 187, 1981.
84. Dimuzio, M. T. and Veis, A., The biosynthesis of phosphophoryns and dentin collagen in the continuously erupting rat incisor, *J. Biol. Chem.*, 253, 6845, 1978.
85. Jontell, M. and Linde, A., 1982, submitted.
86. Termine, J. D. and Conn, K. M., Inhibition of apatite formation by phosphorylated metabolites and macromolecules, *Calcif. Tissue Res.*, 22, 149, 1976.
87. Zanetti, M., de Bernard, B., Jontell, M., and Linde, A., Ca^{2+}-binding studies of the phosphoprotein from rat incisor dentine, *Eur. J. Biochem.*, 113, 541, 1981.
88. Nawrot, C. F., Campbell, D. J., Schroeder, J. K., and Van Valkenburg, M. V., Dental phosphoprotein-induced formation of hydroxylapatite during in vitro synthesis of amorphous calcium phosphate, *Biochemistry*, 15, 3445, 1976.

89. Jones, I. L. and Leaver, A. G., Glycosaminoglycans of human dentine, *Calcif. Tissue Res.* 16, 37, 1974.
90. Branford White, C. J. Molecular organization of heparan sulphate proteoglycan from human dentine, *Arch. Oral Biol.,* 23, 1141, 1978.
91. Linde, A., Glycosaminoglycans of the odontoblast-predentine layer in dentinogenically active porcine teeth, *Calcif., Tissue Res.,* 12, 281, 1973.
92. Hjerpe, A. and Engfeldt, B., Proteoglycans of dentine and predentine, *Calcif. Tissue Res.,* 22, 173, 1976.
93. Nygren, H., Hansson, H.-A. and Linde, A., Ultrastructural localisation of proteoglycans in the odontoblast predentin region of rat incisor, *Cell Tissue Res.,* 168, 277, 1976.
94. Sundström, B., New aspects on the utilization of inorganic sulphate during dentin formation, *Histochemistry,* 26, 61, 1971.
95. Baylink, D., Wergedal, J., and Thompson, E., Loss of proteinpolysaccharides at sites where bone mineralization is initiated, *J. Histochem. Cytochem.,* 20, 279, 1972.
96. Engström, C., Linde, A., and Persliden, B. Acid hydrolases in the odontoblast-predentin region of dentinogenically active teeth, *Scand. J. Dent. Res.* 84, 76, 1976.
97. Linde, A. and Persliden, B., Cathepsin D activity in isolated odontoblasts, *Calcif. Tissue Res.,* 23, 33, 1977.
98. Nygren, H., Persliden, B., Hansson, H.-A., and Linde, A., Cathepsin D: ultraimmunohistochemical localization in dentinogenesis, *Calcif. Tiss. Int.,* 29, 251, 1979.
99. Poole, A. R., Hembry, R. M., and Dingle, J. T., Extracellular localization of cathepsin D in ossifying cartilage, *Calcif. Tissue Res.* 12, 313, 1973.
100. Dunstone, J. R., Ion-exchange reactions between acid mucopolysaccharides and various cations, *Biochem. J.,* 85, 336, 1962.
101. Howell, D. S., Calcification mechanisms, *Isr. J. Med. Sci.,* 12, 91, 1976.
102. Toole, B. P. and Lowther, D. A., The effect of chondroitin sulphate-protein on the formation of collagen fibrils in vitro, *Biochem. J.,* 109, 857, 1968.

Chapter 6

MECHANISM OF CALCIUM SECRETION IN THE AVIAN SHELL GLAND

William C. Eastin Jr.

TABLE OF CONTENTS

I. Avian Oviduct 74
    A. Development and Anatomy 74
    B. Oviduct Secretions 74
    C. Shell Gland Anatomy and General Function 74

II. Calcium in Laying Birds 74
    A. Calcium Distribution in the Shell Gland 74
    B. Calcium in the Circulation 75
    C. Effect of Shell Formation on Plasma Calcium 76

III. Control of Calcium Secretion 76
    A. Calcium Secretion and the Laying Cycle 76
    B. Diurnal Timing of Increased Calcium Secretion 76
    C. Effect of Egg Distention on Calcium Secretion 77
    D. Effect of Denervation on Shell Formation 77
    E. Effect of Luminal Calcium on Calcium Secretion 78
    F. Conclusions on Control of Calcium Secretion 78

IV. Effects of Shell Formation on the Shell Gland 78
    A. Shell Gland as a Model for Transport Study 78
    B. Storage of Calcium by the Shell Gland 78
    C. Effects of Shell Formation on Vascular Volume 78
    D. Effects of Shell Formation on Extracellular Water 78
    E. Permeability Changes and Conclusions 78

V. Mechanism of Calcium and Bicarbonate Secretion 79
    A. Electrical Properties of the Shell Gland and Calcium Secretion 79
    B. Predicting Net Electrolyte Fluxes 79
    C. Effects of Electrolyte Concentration on Calcium Secretion 79
    D. Calcium-Bicarbonate Relationship 80

VI. General Conclusions 80

References 81

I. THE AVIAN OVIDUCT

A. Development and Anatomy

In most avian species, only the left Müllerian duct completes development. During the laying season the small inactive oviduct is stimulated to reproductive status. In the domestic chicken, the oviduct develops from about 16 cm long into a convolutional tube some 80 cm in length with distinctive structural differences. These anatomical divisions, beginning anteriorly, are infundibulum, magnum, isthmus, shell gland (uterus), and vagina (Figure 1).

B. Oviduct Secretions

The active oviduct is responsible for secretions to the ovum as it passes through each section (Figure 1). The ovum, after ovulation, enters the funnel-shaped infundibulum (site of fertilization) and during the next 26 hr or so (in the chicken) passes sequentially through the albumen-secreting magnum, the isthmus where it receives shell membranes, and the shell-forming gland before oviposition. The majority (80%) of this transit time is spent in the shell gland.

C. Shell Gland Anatomy and General Function

The shell gland of the domestic hen is made up of a short anterior portion through which the egg passes rapidly, and a dilated posterior section which holds the egg through shell formation.[1] The two parts can be distinguished from one another in gross aspect and in cytology, but attention will be focused on the posterior segment because of its extraordinary physiology. The mucosa of the latter provides the egg shell components:[2] large amounts of calcium carbonate with smaller quantities of magnesium- and calcium-phosphate,[3,4] the organic matrix in which the minerals are embedded which is composed of a protein-acid mucopolysaccharide complex,[5,6] species- or strain-specific egg shell color in the form of porphyrin pigments,[3,7] and a thin outer covering of mucinous protein (cuticle).[4,5] The shell gland is lined internally by mucosal folds of granular, pseudostratified columnar luminal epithelium and coiled tubular glands embedded in the lamina propria and opening to the lumen.

II. CALCIUM IN LAYING BIRDS

A. Calcium Distribution in the Shell Gland

The average eggshell of the domestic hen contains 2 g of Ca (37% of total shell inorganic matter; 98% of the total as $CaCO_3$[4,7]), but several studies,[1,2] beginning with those of Buchner et al.[8] and including the classical work of Richardson,[3] have established that the hen's shell gland neither concentrates Ca during shell formation nor stores it before hand. Thus, histochemical and direct chemical analyses of hen[3,8-10] and pigeon[11] tissues have shown that the oviduct as a whole contains little Ca and the shell gland actually less than the isthmus.[8,12] Whether or not Ca present in shell gland is evenly distributed among luminal and glandular epithelia has not been settled.[1,10] An early study showed little or no difference in the cellular distribution of Ca between laying and nonlaying hens[8] or between those that produce thick shells and those that produce eggs without shells.[10] However, Gay and Schraer,[13] using ^{45}Ca, observed more calcium in the luminal epithelium when an egg was present in the shell gland than when an egg had not yet descended to that point. They concluded that the luminal epithelium is principally responsible for Ca translocation. In any case, Ca occurs in uterine luminal fluid in concentrations determined earlier by Beadle and colleagues[14] to be approximately 20 mg%, but which more exacting techniques of El Jack and Lake[15] indicate is of the order of 100 mg% during shell deposition. Other workers, employing

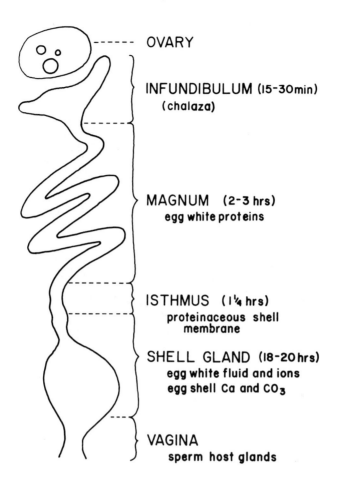

FIGURE 1. Diagram of avian oviduct to show the anatomical sections and secretory contributions each makes to the ovum.

similar methods for Ca determination, have reported concentrations as low as 3 mg%[17] but usually values were greater than 20 mg%.[16,26,35] Although there is a disparity of results, the concentration of free Ca in the shell gland lumen during shell formation would appear to be several times greater than in plasma.

B. Calcium in the Circulation

Ca circulates in diffusible form (ultrafilterable free Ca and ionic salt combinations), and in nondiffusible forms as a colloidal Ca phosphate complex or bound to vitellin (phospho- and lipoproteins). Total Ca increases in plasma several days before first ovulation. The increase is largely in the nondiffusible Ca fraction which rises in conjunction with elevated production of vitellin for yolk formation.[18] At the same time, intestinal absorption of Ca and its storage in labile medullary bone is increased. In actively laying birds, substantial amounts of bone calcium are resorbed for eggshell production as indicated in part by increased osteoclast activity in medullary bone coinciding with shell formation.[19] However, most eggshell Ca (50 to 75%) comes directly from the diet;[20,21] an egg laid 15 min after a hen is fed ^{45}Ca in a single dose contains only traces of radioactivity; the next egg (24 hr) contains fully 30 to 35% of the administered radioactivity, and values thereafter fall sharply to less than 1%.[22] The overall process, including hypercalcemia, vitellogenesis, and Ca absorption and storage in bone is stimulated by estrogen and female androgens. For an indepth discussion of avian Ca metabolism and its hormonal control, the reader is referred to comprehensive treatments of the subject by Gilbert,[23] Simkiss,[24] Urist,[25] and Eastin and Spaziani.[26] For present purposes, attention will be confined to essential features of the Ca cycle related to shell gland physiology.

C. Effect of Shell Formation on Plasma Calcium

Total circulating Ca during sexual inactivity is about 10 mg% in a variety of domestic and wild birds; concentrations increase to 20 to 30 mg% in laying birds, and can be elevated several times higher by estrogen injection.[24,27] Ca levels generally remain high but decrease during shell deposition.[2,24,27] In the domestic hen, an average level of 27 mg% total plasma Ca at ovulation falls significantly to 18 to 22 mg% during shell mineralization.[27,28] The decrease occurs in part in the nondiffusible Ca fraction and is not a result of general systemic hemodilution nor of depressed intestinal absorption of the ion. The decrease results instead from withdrawal of Ca from blood by the shell gland, shown by direct measurement as a drop in arterial-to-venous Ca concentration across the organ.[27,29-31] Using values given above, it can be estimated that the shell gland of a 2 kg hen with total plasma volume of 100 mℓ withdraws 100 to 150 mg of circulating Ca per hour during formation of an egg shell containing 2 g of Ca.[32]

III. CONTROL OF CALCIUM SECRETION

A. Calcium Secretion and the Laying Cycle

Although it is not clear what specifically controls the onset of increased rates of Ca secretion, recent studies have eliminated several hypotheses and, thus, have indirectly provided more focus for our understanding of this phenomenon. Secretion of Ca by the shell gland is low when the shell gland is normally empty, rises when a shell is being formed, and decreases before the egg leaves the shell gland. In the laying chicken, Ca secretion in the shell gland begins to rise about 7 hr after the previous oviposition, reaches a maximum during 14 to 18 hr, declines by 20 hr, and returns to basal levels at 24 hr (2 hr before oviposition) (Figure 2).[7,26,33,34] Thus, the effective higher secretion of Ca only occurs during a period of 12 to 15 hr.[26,34] Significantly, although the egg remains in the shell gland some 20 hr, the organ is forming shell only 60 to 75% of that time at a calculated average Ca delivery rate of a remarkable 133 to 143 mg/hr. Furthermore, that the basic secretory physiology of the shell gland with respect to Ca is dramatically altered when an eggshell is forming has been shown indirectly,[15,35] and directly by in vitro[36] and in vivo measurements.[26] It is also established that functional levels of Ca secretion in the oviduct is specific to the mature shell gland. It is only after first egg laying commences that shell glands acquire the ability to secrete Ca at the higher rates necessary for shell formation.[26]

B. Diurnal Timing of Increased Calcium Secretion

The mechanism(s) that regulates the timing and magnitude of Ca secretion by the shell gland is not yet clear. With respect to timing, data gathered in hens during the egg laying period and the pauses between clutches are particularly revealing. High Ca secretion normally occurs each night between 2000 and 0600 hr. After the last egg in a clutch is laid, the daily period of high Ca secretion is then discontinued (Figure 2 cross-hatched area), remaining at the low basal level typical of shell glands in laying hens prior to arrival of an egg, until after the next ovulation occurs.[26,37] Thus, the mechanism determining the onset and termination of high Ca secretion in the shell gland is not controlled by a rigid diurnal timer that anticipates arrival of an egg and compels Ca secretion even in the absence of ovulation.

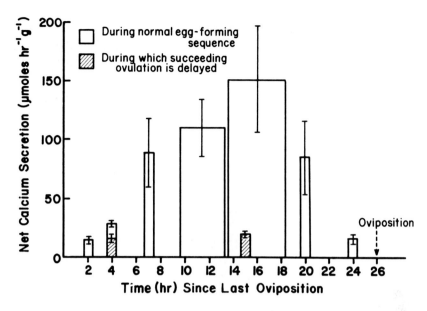

FIGURE 2. Calcium secretion by the shell gland (SG) during a laying cycle (open bars). Different groups of hens were selected for perfusion at indicated intervals after an oviposition. All SGs contained a wax egg and were perfused at 2 ml/min with a solution containing 125 mM NaCl, 75 mM KCl, and 0.33 mM CaCl$_2$. The last three 10-min collections from each SG were analyzed and averaged for net Ca secretion. Heights of bars are mean group secretion rates ± SEM (n = 4-5), dry weight basis. Intervals 10 to 13, 14 to 18, and 20 hr are different from the 2 hr value at the 0.05. 0.001, and 0.02 levels, respectively. Cross-hatched bars are the average of 2 groups of hens (n = 3) that have completed laying a clutch of eggs and in which the next ovulation was normally delayed; measurements were made between last oviposition and subsequent ovulation (see text). (From Eastin, W. C., Jr. and Spaziani, E., *Biol. Reprod.*, 19, 493, 1979. With permission.)

C. Effect of Egg Distention on Calcium Secretion

The possibility that the stimulus for onset of high Ca secretion is distention of the shell gland on arrival of an egg and that high secretion is terminated by reconstriction of the shell gland as the egg departs at oviposition also seems unlikely. In perfusion studies, Eastin and Spaziani[16] compared Ca secretion rates in shell glands with and without surrogate wax eggs and at several times during the laying cycle. Although they were able to stimulate slightly increased rates of Ca secretion by inserting surrogate eggs into chicken shell glands both during the quiescent periods between clutches and also during the time before an ovulated egg would normally arrive in the shell gland, the increases were significantly below the levels seen in shell-forming glands under any condition. Furthermore, these workers found high Ca secretion returned to basal levels before the egg left the shell gland (Figure 2). In regard to causal factors responsible for maximum secretory rates in the shell gland, it is clear that distention by the egg enhances capacity to secrete Ca, but does not explain the high secretory rate of the activated organ.

D. Effect of Denervation on Shell Formation

The nerve supply to the shell gland apparently is not involved, since severing the known autonomic innervation does not permanently affect the egg laying pattern or quality of shell.[26] Also, Sturkie and co-workers[38,39] have shown that transecting the autonomic innervation to the shell gland does not affect oviposition. Similarly, the continuity of muscle of mucosal layers that might convey information longitudinally from upper portions of the oviduct, presumably signaling the imminent arrival of an egg, is not involved.[26,40]

E. Effect of Luminal Calcium on Calcium Secretion

The rate of Ca secretion by the active shell gland does not depend upon the concentration of Ca in the shell gland lumen and in fact, is remarkably resistant to change in luminal Ca concentration.[26] Also, direct sampling of luminal contents during calcification[14,15,35] showed Ca concentrations to be higher than in plasma, suggesting Ca is normally secreted against a Ca gradient.

F. Conclusions on Control of Calcium Secretion

The source of factors that might control Ca secretion by the shell gland appear to have been narrowed through a process of elimination. Taken together, the data seem to indicate that, whatever the source, the controlling factor(s) are most likely carried by the circulatory system. The probability that Ca secretion is directed by circulating levels of one or more hormones has been recently discussed.[26]

IV. EFFECTS OF SHELL FORMATION ON THE SHELL GLAND

A. Shell Gland as a Model for Transport Study

Just as intriguing as the control over increased Ca secretion in the shell gland is the mechanism of transport of shell components. Due to the large amount of Ca and HCO_3 secreted in a relatively short time, the shell gland provides a model of a highly specialized transport system. The precise mechanisms utilized by the shell gland to transfer Ca from the blood and provide HCO_3 to the egg surface in relatively large quantities are not well understood. However, the basic outlines of the process are emerging from both in vitro and in vivo studies.

B. Storage of Calcium by the Shell Gland

The apparent tissue content of Ca in shell gland is lower than other parts of the oviduct and the largest portion of Ca is extracellular.[41] These observations suggest that transfer of the bulk of Ca from blood to shell gland secretory epithelium occurs by extracellular routes.

C. Effects of Shell Formation on Vascular Volume

It was shown that the vascular volume of the shell gland about doubles in the presence of an egg.[41] However a surrogate egg placed in an inactive (nonshell-forming) shell gland did not stimulate noticeable hyperemia nor heavy Ca secretion.[41] Therefore, this rise in blood volume appears to be correlated with the capacity to secrete much larger amounts of Ca.

D. Effects of Shell Formation on Extracellular Water

Shifts in tissue water also have been noted and may be related to tissue blood volume just discussed. Although the shell gland is a constant 86 to 87% water throughout the laying cycle,[41] the distribution is altered in that extracellular water is 51% of fresh tissue weight, but shifts to 63% during the active (shell-forming) phase.[42] In addition, it was found that before shell calcification, total water content of the mucosa was 83% of the wet weight and rose to 87% during shell formation; 40% of the increase was extracellular.[43] These observations indicate an increasingly spacious extracellular compartment and suggest the possibility of paracellular routes for Ca transfer.

E. Permeability Changes and Conclusions

When inulin[14]-C was injected into the heart of chickens during luminal perfusion of the shell gland, it was found to penetrate to the shell gland lumen in surprising quantities and at a rate about fourfold higher in active compared to inactive shell glands.

Similar studies with ^{45}Ca showed that appearance of this isotope in the perfusate with time generally reflected the clearance of inulin^{14}-C and ^{45}Ca from blood.[41] These data were taken to indicate:

1. That the shell gland rapidly shunts Ca from plasma to lumen without accumulating the ion
2. The existence of highly permeable paracellular shunt pathways for passive fluxes in shell gland mucosa
3. These pathways become more penetrable at the time of shell formation

V. MECHANISM OF CALCIUM AND BICARBONATE SECRETION

A. Electrical Properties of the Shell Gland and Calcium Secretion

The electrical properties of the shell gland have been studied both in vivo and in vitro. A transmural electrical potential difference (PD) is common in epithelia which transport ions. In vitro PD measurements have varied with laboratory; Ehrenspeck, Shraer, and Shraer[36] reported very low PDs, (1 mV), and no difference between active and inactive states; others reported higher PDs (5 to 10 mV) and a substantial decrease during the inactive state.[44-46] In vivo the PD was even greater, 10 to 15 mV in active shell gland,[41] but was not measured in inactive shell glands. The PD in vitro could be abolished with metabolic inhibitors or removal of Na from the bathing solution[36,46,47] and ouabain reduced the PD when added to either side but was less effective on the mucosal side.[46,47] In vitro studies abolishing the PD did not affect the net flux of Ca.[47] It appears that maintenance of, or change in, the transmural PD per se is not required for net Ca secretion to the lumen.

B. Predicting Net Electrolyte Fluxes

Information about the physiology of the shell gland has also come from in vivo perfusion studies in which the net movement of the major ions were measured under defined conditions. The results from these studies were used to calculate the electrochemical potential ($\Delta \bar{\mu}$) where $\Delta \bar{\mu} = RT \ln S_1/S_p + zF \Delta \psi$, and where R = the gas constant, T = absolute temperature, S_1 and S_p the ion concentrations in the lumen and plasma, respectively, z the valence, F the Faraday, and $\Delta \psi$ the electrical potential difference.

When concentrations of Na, K, Ca, HCO_3, and Cl, approximating those in plasma, were perfused, the $\Delta \bar{\mu}$ predicted a net transport of Na, K, and Ca to the lumen and Cl and HCO_3 to the plasma.[41] In fact, net HCO_3 movement was to the lumen and net Na to the plasma. The predicted movement of HCO_3, however, is complicated because the source of HCO_3 for secretion appears to be cellular metabolism.[48]

C. Effects of Electrolyte Concentration on Calcium Secretion

When Na in the perfusion fluid was varied, there was a linear relationship between net Na absorption and perfused Na concentration.[41] The observed net Na absorption is not in agreement with the predicted net flux. Net Na flux to the serosal side also was reported for the quail shell gland in vitro.[47] These observations taken together, i.e., net Na movement in vivo and in vitro against a Na $\Delta \bar{\mu}$ gradient, suggest an active transport of Na to the plasma. However, in the isolated mucosa, net transport of calcium was not affected when the tissue was bathed with a Na-free solution.[47] Ouabain, an inhibitor of active Na transport, was found to depress net Ca secretion only if Na was present in the bathing solution. The ouabain effect on Ca is not clear but the presence of Na appears to be necessary.[49]

In in vivo studies, net Ca secretion rates appeared to be highest when the shell gland was perfused with a Ca-free solution and declined somewhat as luminal Ca concentrations were experimentally increased toward plasma concentrations.[41] At all but the highest Ca concentration perfused, the $\Delta\tilde{\mu}$ predicts a favorable gradient for diffusion into the lumen. The presence of a paracellular, low resistance path for diffusion is suggested in view of data that substances as large as inulin in the circulation enter the lumen of the active shell gland, and that this shell gland is more penetrable than inactive glands to inulin. Whether, in fact, Ca crosses the epithelium via these routes has not been established. However, the Ca secretion rate continued high and constant when perfused Ca was raised as much as 4.5-fold over plasma concentrations.[41] A PD of 15 mV (lumen negative) could provide enough driving force to produce net Ca secretion against a lumen-to-plasma concentration ratio of 3.16. If the assumption is made that increasing amounts of Ca do not change the resistance of membranes or in any way destroy the tissue, then the observed absence of a linear relationship between calcium secretion and perfused calcium concentration suggests that calcium is actively transported. This conclusion is compatible with results of in vitro studies.[36,37]

Ehrenspeck, Schraer, and Schraer,[36] and Pearson and Goldner[47] reported a net flux of Ca toward the lumen in the absence of a Ca $\Delta\tilde{\mu}$ gradient; both sides of the tissue were bathed in identical solutions and the PD was maintained at zero. In addition, that Ca transport is linked with a source of metabolic energy is directly indicated by findings in vitro that net Ca flux was abolished in chicken shell glands by the metabolic inhibitor, KCN,[36] and in shell glands of the Japanese quail by anoxia.[49]

D. Calcium-Bicarbonate Relationship

Evidence from published in vitro and in vivo work suggests that active Ca secretion in the shell gland is linked to the secretion of HCO_3 and to the activity of carbonic anhydrase.[29,50] Thus, artificially raising HCO_3 concentrations in in vivo luminal perfusions caused an increase in Ca secretion but a decrease in HCO_3 secretion.[41] When the isolated quail shell gland was bathed on both sides with HCO_3-free solutions, net Ca secretion was reduced 60% by a decreased serosal to mucosal flux.[47] Acetazolamide (2 mM), an inhibitor of carbonic anhydrase, perfused through the shell gland lumen of chickens completely abolished both Ca and HCO_3 secretions, while not significantly affecting net fluxes of Na, K, and Cl ions.[41] Similarly, Pearson and Goldner[49] observed a decrease in Ca flux of 55% in isolated quail shell gland with 1 mM acetazolamide, whether or not HCO_3 was present in the solutions. Ehrenspeck, Schraer, and Schraer[36] reported acetazolamide caused increased Ca fluxes in both directions across the isolated chicken shell gland, but these investigators employed 50 mM concentrations of the inhibitor. At very high concentrations, acetazolamide may cause nonspecific effects on the shell gland tissue. Finally, the association between Ca and HCO_3 secretions also was evident in in vivo experiments in which shell gland lumens perfused with a solution containing ouabain showed depression in both secretions.[41] However, the mechanism of the ouabain effect on net Ca and HCO_3 secretion is not clear.

VI. GENERAL CONCLUSIONS

Although the results of studies designed to explain the control and mechanism of Ca and HCO_3 secretion in the avian shell gland are not conclusive, they provide a better understanding of the transport physiology of this organ. Ca secretion increases to coincide with the period the egg is normally in the shell gland forming a shell and decreases when the shell is complete, but well before (2 hr) ovulation. In addition, this increase does not occur on days of no ovulation; but so far, the evidence relative to the controlling factor(s) of elevated Ca secretion has all been negative. Thus, although Ca secretion rates in shell glands are always greater when distended in egglike fashion than when not, rates in active shell glands whether distended or not are substantially higher than in inactive controls. Timing of increased Ca secretion is apparently not cued by the previous oviposition but is associated with occurrence of ovulation. In addition, severing the shell gland from the upper oviduct does not affect secretory cycles[26] and severing known autonomic nerves to the shell gland does not permanently affect egg laying or egg shell quality.[26,38] Although not establishing the controlling factors, these results have been helpful in that they provide more focus on other hypotheses on the control of Ca secretion.[26]

Whether the bulk of the 5 g of Ca and HCO_3 required for an eggshell is normally provided by active transport or by passive diffusion is still not settled. The existence of paracellular routes for passive diffusion has been demonstrated, but whether Ca actually crosses the shell gland epithelium via these pathways is not known. However, evidence from transport studies indicates active transport systems for both Ca[36,41,47] and HCO_3[41] in this tissue under somewhat artificial conditions. The strongest support for active transport comes from direct samplings of luminal fluid during shell formation which have shown Ca concentrations to be several-fold higher than in blood.[14,15,17,26] For the luminal Ca concentration to increase under normal conditions, in spite of the driving force for passive absorption during the time the egg shell is being formed, suggests the bulk of Ca is actively secreted.

REFERENCES

1. Aitken, R. N. C., The oviduct, in *Physiology and Biochemistry of the Domestic Fowl*, Vol. 3, Beu, D. J. and Freeman, B. M. Eds., Academic Press, New York, 1971, 1237.
2. Simkiss, K. and Taylor, T. G., Shell formation, in *Physiology and Biochemistry of the Domestic Fowl*, Vol. 3, Bell, D. J. and Freeman, B. M. Eds., Academic Press, New York, 1971, 1331.
3. Richardson, K., The secretory phenomena in the oviduct of the fowl including the process of shell formation examined by the microincineration technique, *Phil. Trans. R. Soc. London*, B225, 149, 1935.
4. Romanoff, A. L. and Romanoff, A. J., *The Avian Egg*, John Wiley & Sons, New York, 1949, 174.
5. Baker. J. R. and Balch, D. A., A study of the organic material of the hen's eggshell, *Biochem. J.*, 82, 352, 1962.
6. Simkiss, K. and Tyler, C., A histochemical study of the organic matrix of hen eggshells, *Q. J. Microbiol. Sci.*, 98, 19, 1957.
7. Tyler, C., Avian eggshells: their structure and characteristics, *Int. Rev. Gen. Exp. Zool.*, 4, 82, 1969.
8. Buckner, G. D., Martin, J. H., and Peter, A. M., Chemical studies of the oviduct of the hen, *Am. J. Physiol.*, 71, 349, 1925.
9. Common, R. H., Observations on the mineral metabolism of pullets, *J. Agric. Sci.*, 28, 347, 1938.
10. McCallion, D. J., A cytological and cytochemical study of the shell gland of the domestic hen, *Can. J. Zool.*, 31, 577, 1953.
11. Clavert, J., Contribution a l'etude de la formation des oeufs telo lec ith iques des oiscaux. Mechanismes de l'edification de la coquille, *Bull. Biol. Fr. Belg.*, 82, 19, 1948.
12. Taylor, T. G. and Hertelendy, F., Changes in the blood calcium associated with egg shell calcification in the domestic fowl. II. Changes in diffusible calcium, *Poult. Sci.*, 40, 115, 1961.
13. Gay, C. V. and Schraer, H., Autoradiographic localization of calcium in the mucosal cells of the avian oviduct, *Calcif. Tissue Res.*, 7, 201, 1971.
14. Beadle, B. W., Conrad, R. M., and Scott, H. H., The composition of the uterine secretion of the domestic fowl, *Poul. Sci.*, 17, 498, 1938.
15. El Jack, M. H. and Lake, P. E. The content of the principal inorganic ions and carbon dioxide in uterine fluids of the domestic hen *(Gallus domesticus)*, *J. Reprod. Fertil.*, 13, 127, 1967.
16. Rieser, J. W., Smith, J. D., and Burke, W. H., Composition of turkey uterine fluid, *Poul. Sci.*, 51, 203, 1972.

17. Solomon, S. E., Fluctuations in total calcium and magnesium in the plasma and uterine fluid of the domestic fowl, *Br. Poul. Sci.*, 12, 165, 1971.
18. McDonald, M. R. and Riddle, O., The effect of reproduction and estrogen administration on the partition of calcium, phosphorus, and nitrogen in pigeon plasma, *J. Biol. Chem.*, 159, 445, 1945.
19. Bloom, M. A., Domm, L. V., Nalbandov, A. U., and Bloom, W., Medullary bone of laying chickens, *Am. J. Anat.*, 102, 411, 1958.
20. Driggers, J. C. and Comar, C. L., The secretion of radio-active calcium in the hen's egg, *Poul. Sci.*, 28, 420, 1949.
21. Jowsey, J. R., Berlie, M. R., Spinks, J. W. T., and O'Neil, J. B., Uptake of calcium by the laying hen and subsequent transfer from egg to chick, *Poul. Sci.*, 35, 1234, 1956.
22. Comar, C. L. and Driggers, J. C., Secretion of radio-active calcium in the hen's egg, *Sciences*, 109, 282, 1949.
23. Gilbert, A. B., Formation of the egg in the domestic chicken, in *Advances in Reproductive Physiology*, Vol. 2, McLaren, E., Ed., Academic Press, New York, 1967, 111.
24. Simkiss, K., Calcium metabolism and avian reproduction, *Biol. Rev.*, 36, 321, 1961.
25. Urist, M. R., The effects of calcium deprivation upon the blood, adrenal cortex, ovary and skeleton in domestic fowl, *Recent Progr. Horm. Res.*, 15, 455, 1959.
26. Eastin, W. C., Jr. and Spaziani, E., On the control of calcium secretion in the avian shell gland (uterus), *Biol. Reprod.*, 19, 493, 1978.
27. Hodges, R. D., pH and mineral ion levels in the blood of the laying hen *(Gallus domesticus)* in relation to eggshell formation, *Comp. Biochem. Physiol.*, 28, 1243, 1969.
28. Hertelendy, F. and Taylor, T. G., Changes in blood calcium associated with eggshell calcification in the domestic fowl. I. Changes in total calcium, *Poult. Sci.*, 40, 108, 1961.
29. Hodges, R. D. and Lorcher, K., Possible sources of the carbonate fraction of eggshell calcium carbonate, *Nature (London)*, 216, 609, 1967.
30. Lorcher, K. and Hodges, R. D., Some possible mechanisms of formation of the carbonate fraction of eggshell calcium carbonate, *Comp. Biochem. Physiol.*, 28, 119, 1969.
31. Winget, C. M., Smith, A. H., and Hoover, G. N., Arterio-venous differences in plasma calcium concentration in the shell gland of the laying hen during shell formation, *Poul. Sci.*, 37, 1325, 1958.
32. Taylor, T. G. and Stringer, D. A., Eggshell formation and skeletal metabolism, in *Avian Physiology*, 2nd ed., Sturkie, P. D., Ed., Cornell University Press, Ithaca, N.Y., 1965, 481.
33. Burmeister, B. R., A study of the physical and chemical changes of the egg during its passage through the isthmus and uterus of the hen's oviduct, *J. Exp. Zool.*, 84, 445, 1940.
34. Bradfield, J. R. G., Radiographic studies on the formation of the hen's eggshell, *J. Exp. Biol.*, 28, 125, 1951.
35. Mongin, P. and Sauveur, B., Composition du fluide uterin et de l'albumen durant le sejour de l'ouef dans l'uterus chez la poule domestique, *C. R. Acad. Sci. Paris*, 270D, 1715, 1970.
36. Ehrenspeck, G., Schraer, H., and Schraer, R., Calcium transfer across isolated avian shell gland, *Am. J. Physiol.*, 220, 967, 1971.
37. Pearson, T. W., Pryor, T. J., and Goldner, A. M., Calcium transport across avian uterus. III. Comparison of laying and nonlaying birds, *Am. J. Physiol.*, 232, E437, 1977.
38. Sturkie, P. D. and Freedman, S. L., Effects of trans-section of pelvic and lumbrosacral nerves on ovulation and oviposition in fowl, *J. Reprod. Fertil.*, 4, 81, 1962.
39. Sturkie, P. D. and Mueller, W. J., Reproduction in the female and egg formation, in *Avian Physiology*, Sturkie, P. D., Ed., Springer-Verlag, New York, 1976, 302.
40. Pearl, R. and Surface, F., The nature of the stimulus which causes a shell to be formed on a bird's egg, *Sciences*, 29, 741, 1909.
41. Eastin, W. C., Jr. and Spaziani, E., On the mechanism of calcium secretion in the avian shell gland (uterus), *Biol. Reprod.*, 19, 505, 1978.
42. Cohen, I. and Hurwitz, S., Intracellular pH and electrolyte concentration in the uterine wall of the fowl in relation to shell formation and dietary minerals, *Comp. Biochem. Physiol.*, 49A, 689, 1974.
43. Mongin, P. and Carter, N. W., Studies on the avian shell gland during egg formation: agueous and electrolytic composition of the mucosa, *Br. Poul. Sci.*, 18, 339, 1977.
44. Kemeny, A., Laklia, J. B., and Lencses, G., The isolated shell gland as an experimental object, *Acta Physiol. Acad. Sci. Hung.*, 40, 313, 1971.
45. Cohen, I. and S. Hurwitz, The electrical potential difference and the short circuit current of the uterine mucosa of hens in relation to egg shell formation, *Poul. Sci.*, 52, 340, 1973.
46. Hurwitz, S., Cohen, I., and Bar, A., The transmembrane electrical potential difference in the uterus (shell gland) of birds, *Comp. Biochem. Physiol.*, 35, 873, 1970.
47. Pearson, T. W. and Goldner, A. M., Calcium transport across avian uterus. I. Effects of electrolyte substitution, *Am. J. Physiol.*, 225, 1508, 1973.

48. **Bernstein, R. S., Nevalainen, T., Schraer, R., and Schraer, H.**, Intracellular distribution and role of carbonic anhydrase in the avian *(Gallus domesticus)* shell gland mucosa, *Biochem. Biophys. Acta*, 159, 367, 1968.
49. **Pearson, T. W. and Goldner, A. M.**, Calcium transport across avian uterus. II. Effects of inhibitors and nitrogen, *Am. J. Physiol.*, 227, 465, 1974.
50. **Diamanstein, T.**, Uber die lokale Rolle der carboanhydratase in Hinblick anf die Eischalenverkalung, *Arch. F. Gefluegelkd.*, 30, 309, 1966.

Calcium in Pathology

Chapter 7

DISORDERS OF CALCIUM METABOLISM IN CHILDREN

Sushma Palmer

TABLE OF CONTENTS

I.	Introduction		88
II.	Calcium Homeostasis		88
III.	Dietary Intake, Absorption, and Excretion		89
IV.	Hypocalcemic States		90
	A.	Neonatal Hypocalcemia	90
		1. Early Neonatal Hypocalcemia	90
		2. Delayed Neonatal Hypocalcemia	91
	B.	Hypocalcemia in Later Life	91
V.	Hypercalcemia		92
	A.	Acute Disuse Hypercalcemia	92
	B.	Hypervitaminosis D	92
	C.	Idiopathic Hypercalcemia of Infancy	93
	D.	Sarcoidosis	93
VI.	Diagnosis and Nutritional Management		93
	A.	Hypocalcemia	93
		1. Diagnosis	93
		2. Treatment	95
	B.	Hypercalcemia	96
		1. Diagnosis	96
		2. Treatment	99
VII.	References		101

I. INTRODUCTION

The bulk of evidence pertaining to the key role of calcium in human metabolism is derived from studying disorders which are characterized by an imbalance in calcium homeostasis. Calcium is the fifth most abundant inorganic element in the body and a major component of the human skeleton. Calcium homeostasis is dependent on a complex interplay of biochemical and hormonal factors involving the physiology of calcium, magnesium, phosphorus, acid-base balance, vitamin D, the parathyroid hormone (PTH), and calcitonin (CT). Each of these affects calcium metabolism and is affected by it. A number of endogenous mechanisms and dietary factors influence calcium absorption and utilization. An imbalance in any one of these may disturb calcium homeostasis. The objective of this chapter is to discuss two major metabolic abnormalities — hypocalcemia and hypercalcemia — which are symptomatic of disturbances in calcium homeostasis. This chapter also provides guidelines for diagnosis and nutritional management of these conditions in children.

II. CALCIUM HOMEOSTASIS

The organs of greatest importance in maintaining calcium homeostasis are the bone, the intestine, and the kidneys.[1] Almost 99% of the body's calcium is contained in the skeleton, the other 1% is in the intravascular, interstitial, and intracellular fluids.[2] Most of the calcium in the blood is present in the plasma where it exists in three forms: the ionized physiologically active form, the nondiffusible form which makes up 30 to 50% of plasma calcium and is mainly bound to albumin, and the nonionized form which makes up 5 to 15% of the total calcium in the plasma and exists as chelates and as complexes mainly with citrate.[1,3] For most purposes, the total serum calcium is a good indicator of calcium status. However, in recent years, the importance of ionized calcium levels as indicators of calcium status has become more recognized.[4]

Ionized calcium is the only physiologically active form of calcium in the plasma. It influences the release of neurochemical transmitters at synaptic junctions; the synthesis, secretion, and metabolic effects of protein hormones; and the release and activation of intracellular and extracellular enzymes.[5] Calcium is also necessary for blood coagulation, myocardial function, muscle contractibility, and the integrity of the intracellular cement substance. Calcium may also act as a membrane stabilizer in the transport function of cell membranes. In hypercalcemic states, membrane permeability and cell excitability are reduced, and the reverse occurs in the case of hypocalcemia.[1]

The level of ionized calcium in the plasma is maintained within a narrow range by the combined actions of PTH, 1,25-dihydroxycholecalciferol (1,25-DHCC) — an active metabolite of vitamin D — and possibly by calcitonin, which is a group of peptides. The parathyroid hormone increases renal tubular reabsorption of calcium and both PTH and vitamin D increase intestinal calcium absorption and bone resorption of calcium.[6,7] Together these hormones lead to a rise in serum calcium. Calcitonin, however, inhibits calcium resorption from the bone, thereby lowering the serum calcium level.[7] The rates of secretion of PTH and 1,25-DHCC are affected by the plasma level of ionized calcium. Hypocalcemia stimulates PTH and 1,25-DHCC production. Hypercalcemia diminishes these, whereas it enhances the secretion of CT.[7] Intestinal and renal cells affect the transcellular flow of calcium, whereas bone and muscle cells use it for mineralization and contraction.[8]

Other hormones associated with calcium homeostasis are the androgens and estrogens; the glucocorticoids, which counteract the effects that vitamin D has upon intestinal calcium absorption; the thyroid hormone which influences bone maturation and the transfer of calcium from serum to bone; and the pituitary growth hormone which produces hypercalciuria without changing serum calcium levels.[5]

Calcium and phosphorus are intimately related because of the presence of hydroxyapatite, a calcium-phosphate compound in the bone.[9] Magnesium appears to be necessary for the calcium-mobilizing effect of PTH on the bone. In addition, the intestinal absorption of calcium and magnesium may be interrelated.[10] Changes in the acid-base balance also affect calcium homeostasis by affecting the amount of protein-bound and ionized calcium in extracellular fluid.[11]

Disturbances in neonatal calcium homeostasis should be considered in the context of disorders of calcium, phosphorus, magnesium, and acid-base balance, as well as disorders of PTH, CT, and vitamin D function.[1]

III. DIETARY INTAKE, ABSORPTION, AND EXCRETION

Calcium can enter the body only through the diet. However, there are several mechanisms which result in its continuous loss from the body. Variations in individual adaptation to a low calcium intake and the presence of a number of substances in the diet which interfere with calcium absorption make it difficult to resolve the issue of optimal calcium intake. Approximately 100 to 200 mg of calcium are excreted daily in the urine. Another 140 to 175 mg daily are lost in the bile and pancreatic juices, and dermal losses average another 20 mg daily. Approximately 25 to 30 mg of calcium are deposited in the fetal skeleton in the third trimester of pregnancy, and 500 to 700 mg are excreted in the milk daily during lactation.[12] In the case of a positive calcium balance, the excess calcium is deposited in the skeleton. In negative balance, calcium must be mobilized from the skeleton to maintain the homeostatic concentration in extracellular fluid.[13]

The rate of absorption of calcium should be considered in establishing the dietary requirements for calcium. Calcium absorption is affected by age, endocrine function, nutritional status, and by the level of dietary protein, phosphorus, fat, and vitamin D. Ingested calcium is incompletely absorbed from the gut. Depending on the intake, 70 to 80% of dietary calcium is excreted in the feces.[14] Oxalates and phytates which are present in some foods, for example, in tea and cereals, may combine with calcium and form insoluble salts and hinder calcium absorption, but only if the calcium intake is minimal.[15] Vitamin D appears to be required for efficient calcium absorption.[16] In animals, the ratio of calcium to phosphorus in the diet appears to be important in the absorption of calcium, as well as in the loss of calcium from bone, and a ratio of 2:1 is considered to be optimal.[15] The evidence for requirement of a rigid ratio in the human diet is less definitive. Humans appear to tolerate much wider ratios ranging from 2:1 to 1:2 in the diet. Some scientists believe that ideally the ratio should be near or above 1[17] or that the calcium to phosphorus ratio is not important if dietary calcium levels are sufficient. Changes in protein intake however, appear to affect calcium metabolism in that a high protein diet using purified proteins appears to increase the urinary excretion of calcium,[18] possibly by decreasing the fractional reabsorption of calcium by the kidney.[19] Allen et al.[19] suggest that it is unlikely that an increase in calcium intake can prevent a negative calcium balance and the probable bone loss which is induced by increased protein intake. A high protein intake where the protein is derived from meat, however, may not affect calcium excretion.[20] A sudden reduction in calcium intake in individuals who consume large amounts of calcium may temporarily lead to a negative calcium balance due to a relatively low rate of absorption in such individuals. However, adaptation to a low calcium intake eventually seems to occur.[21]

The Recommended Dietary Allowances[20] for calcium have been the subject of considerable controversy in recent years. For breast-fed and other infants, the level is currently set at 60 mg/kg/day; for children between the ages of 1 and 10 years, at 1200 mg/day; and for 10- to 18-year olds, at 800 mg/day. The requirement is increased to 1200 mg daily during pregnancy and lactation.[20]

IV. HYPOCALCEMIC STATES

Ideally, hypocalcemia should be defined in terms of the concentration of the physiologically active moiety of serum calcium, i.e., ionized calcium; but, techniques for measuring serum ionized calcium are not yet standardized for clinical application. Therefore, this value is usually estimated from the total serum calcium. Bergman and Isaksson[22] reported that the normal serum concentrations of total calcium are approximately 10 mg/dℓ from 1 to 16 years of age and that the lowest values (9.9 mg/dℓ and 9.6 mg/dℓ) occur at 6 to 9 years of age and in adulthood. Brown et al.[23] reported 8.6 mg/dℓ in infants and 10 mg/dℓ as normal serum concentrations of calcium at 0.5 to 6 years of age. Harrison and Harrison[4] suggest that hypocalcemia should be suspected when the total serum calcium levels fall below 7.5 mg/dℓ and the ionized calcium level is <2.5 mg/dℓ.

Reduction in serum ionized calcium can result from a failure or deficiency of compensation by increased PTH output in response to appropriate stimuli, i.e., a form of hypoparathyroidism[4] or due to a lack of responsiveness of the target tissue to PTH. This includes the condition, pseudohypoparathyroidism. Symptoms of hypocalcemia can theoretically occur when the renal tubular reabsorption of calcium is inhibited, calcium uptake by the bone is greater than mobilization from the bone, and when mobilization of calcium out of the bone is inhibited whereas uptake into the bone continues.[7] In practice, however, more than one mechanism is usually involved.

The major classes of hypocalcemia are summarized in Table 1.

A. Neonatal Hypocalcemia

Hypocalcemia occurs most commonly in the neonatal period. It probably results from multiple causes which overwhelm the infant's capability to maintain the extrauterine calcium homeostasis. A decrease in calcium supply due to poor dietary intake, hypomagnesemia, increase in phosphorus from endogenous or exogenous sources, and the use of alkali for the treatment of acidosis are important causative factors. Further, increased physiological stress such as in suppressed parathyroid gland function and reduced response to a decrease in serum calcium, low vitamin D stores, or a defect in vitamin D metabolism and increase in CT levels, elevated circulating adrenal glucocorticoids are important hormonal considerations. Additional predisposing factors are toxemia of pregnancy, placenta previa, hyperparathyroidism and neonatal jaundice, respiratory distress, acidosis, anoxia, asphyxia, or cerebral injury[24,25] A severe deficiency of vitamin D and calcium in the maternal diet may precipitate congenital rickets.[26] Transient idiopathic or permanent hypoparathyroidism may also be responsible for hypocalcemia in the neonatal period.[25]

There are two types of neonatal hypocalcemia: *early neonatal hypocalcemia,* which occurs within the first 48 hr before feedings have begun, and *delayed neonatal hypocalcemia,* which occurs near the end of the first week of life or later in infants receiving a high-phosphorus formula.[4,27]

1. Early Neonatal Hypocalcemia

The greatest incidence of early neonatal hypocalcemia is in premature infants ($\sim$ 30% have serum calcium <7 mg/dℓ), in infants of insulin-dependent diabetic mothers ($\sim$ 50% incidence), and in infants with birth asphyxia ($\sim$ 30% incidence). In infants who weigh less than 1500 g, the incidence of hypocalcemia is $\sim$ 50%.[4,28] A deficiency of adequate parathyroid hormone function appears to be the major factor in early neonatal hypocalcemia. The symptoms may be precipitated by excessive secretion of calcitonin.[4]

Table 1
MAJOR CLASSES OF HYPOCALCEMIA

Causes/Predisposing Factors	Age of Onset/Type	Major Physiological Disturbance
Prenatal complications Complications of Delivery Postnatal complications	0—48 hr after birth Early neonatal	Calcitonin—PTH imbalance
High phosphate feedings (formula) Hypomagnesemia	Several days to several weeks postnatally Delayed neonatal	Transient physiological hypoparathyroidism
Aplasia or hypoplasia of parathyroids Thymic aplasia (DiGeorge Syndrome)	Neonatal-postnatal-childhood Congenital hypoparathyroidism	Failure of adequate secretion of PTH
Genetic autoimmune injury Metastatic neoplasm Chromosomal anomaly Surgery	Infancy-childhood-later Acquired hypoparathyroidism Idiopathic hypoparathyroidism	Failure of adequate secretion of PTH
Inborn errors of metabolism Acquired metabolic disturbances (Vitamin D deficiency, phosphate poisoning, MG deficiency)	Infancy-childhood Pseudohypoparathyroidism I and II	End-organ resistance to PTH
Dietary vitamin D deficiency/ malabsorption	Early infancy (2—3 months) Pseudohypoparathyroidism	End-organ resistance to PTH
Block in vitamin D metabolism and deficiency of 1,25-dihydroxy vitamin D Hypomagnesemia, hyponatremia, hypokalemia, infections	Infancy-childhood Pseudohypoparathyroidism	End-organ resistance to PTH

2. Delayed Neonatal Hypocalcemia

This is usually a term reserved for full-term infants who develop symptoms after several days to several weeks and are receiving high-phosphate feedings from formulas rather than from human milk. Harrison and Harrison[4] also suggest that infants receiving a cow's milk formula in the first ten days often consume a higher total volume of milk than breast-fed infants and, therefore, have a higher phosphate intake. Cow's milk contains 1200 mg calcium or 1000 mg of phosphate per liter compared to human milk which contains 340 mg calcium or 150 mg phosphate per liter. Many infants receiving high-phosphate feedings may have relative hypoparathyroidism. They may be unable to handle the phosphate load due to low PTH levels and also a physiologically low glomerular filtration rate. This may lead to hyperphosphatemia and hypocalcemia.[4,25]

B. Hypocalcemia in Later Life

The causes of hypocalcemia in later life are listed in Table 1. It may occur due to failure of adequate secretion of PTH or due to end-organ resistance to PTH. Hypoparathyroidism (congenital or acquired) is a recognized cause of postnatal hypocalcemia. It may result from damage to the parathyroid gland because of thyroidectomy or due to excessive iodine administration. It may be associated with chromosomal abnormalities or may be familial in nature. Hypocalcemia can also occur due to pseudohypoparathyroidism. In such cases, the parathyroid glands apparently secrete normal or increased amounts of PTH, but hypocalcemia and hyperparathyroidism occur due to failure of response of the target cells or due to an inborn error of metabolism, as in pseudohypoparathyroidism I and II. Hypomagnesemia may induce hypocalcemia either due to a primary defect in intestinal absorption of magnesium or in patients with malabsorption and malnutrition. In states of vitamin D deficiency leading to intestinal malabsorption of calcium and hypocalcemia, there is skeletal and partial renal tubular unresponsiveness to PTH.[25] Hypocalcemia secondary to vitamin D deficiency is seen in infants who are either breast-fed or formula-fed but are not receiving vitamin D supplements. These infants are usually 2 to 3 months old, may have a vitamin D-deficient diet, may suffer from vitamin D malabsorption, may have a defect in the conversion of vitamin D to its active metabolite (as in vitamin D-dependent rickets), or in unusual circumstances, there may be a deficiency of vitamin D in the maternal diet.[4]

Hypocalcemia is observed in children recovering from acidosis and hypernatremia, in cases of sepsis, chronic renal insufficiency, renal tubular acidosis, in hypoproteinemia, and in phosphate poisoning due to excessive laxatives containing phosphate.[25] It seldom occurs due to dietary deficiency of calcium unless vitamin D is also deficient and there is resistance to PTH. Excessive administration of anticonvulsant drugs, particularly prolonged multiple drug therapy, is now being recognized as a cause of hypocalcemia and rickets.[29]

V. HYPERCALCEMIA

Total serum calcium levels in excess of 11 mg/dℓ and ionized calcium levels >5.6 mg/dℓ characterize hypercalcemia biochemically. It can occur due to excessive parathyroid secretion, excessive intake or absorption of vitamin D or calcium for the gut, hypersensitivity to small doses of vitamin D, neoplastic disease, and other causes which are outlined in Table 2.[25,30,31] Other causes of hypercalcemia are a decreased rate of calcium removal from the serum, increased serum albumin level or other calcium-binding proteins, and increased amounts of calcium complexed to anions.[25]

Hypercalcemia may occur due to primary hyperparathyroidism in childhood and may manifest itself clinically in fatigue, weakness, constipation, rickets, abdominal pain, and colic. In cases of acute hypercalcemia, the total serum calcium level may exceed 15 mg/dℓ and lead to nausea, vomiting, dehydration, lethargy, and coma.[32,33] Excessive topical or systemic administration of vitamin A may also cause hypercalcemia due to increased skeletal resorption.[34] Subcutaneous fat necrosis has an unknown etiology and may occur in term infants who were in abnormal uterine positions. In some cases, hypercalcemia may occur due to large deposits of calcium.[25]

A. Acute Disuse Hypercalcemia

Acute demobilization in adolescents, e.g., in the case of bone injury, may cause hypercalcemia due to an abrupt decrease in bone deposition, but continued bone resorption and calcium mobilization.[34] In such cases, there is a lack of muscle pull and a cessation of weight-bearing on the bone. This results in bone demineralization, and a net flow of calcium and phosphorus from the skeleton into the extracellular fluid.[4]

B. Hypervitaminosis D

Hypervitaminosis D can result from a disturbance of calcium homeostasis when concentrated forms of vitamin D, e.g., irradiated ergosterol or irradiated cholecalciferol are ingested. It is likely that the unmetabolized vitamin D, i.e., 25-hydroxy vitamin D (25-OHD) can accumulate and produce manifestations of hypercalcemia. It appears

Table 2
MAJOR CLASSES OF HYPERCALCEMIA

Predisposing Factors	Cause(s) of hypercalcemia
Congenital hyperparathyroidism	
Immobilization (acute disuse osteoporosis)	
Neoplastic disease (Metastatic tumors with osteolysis, tumors secreting prostaglandins, PTH-like hormone-secreting tumors, myelomas)	Demineralization of bone, calcium mobilization with or without increased calcium absorption
Hyperthyroidism	
Hypersensitivity to vitamin D (Idiopathic hypercalcemia, sarcoidosis)	
Excessive intake of vitamins A, D, and/or calcium	Excessive PTH secretion, and increased intestinal absorption of calcium
Hyperparathyroidism	
Phosphate depletion	

that for older children, the consumption of 2000 units of vitamin D per kilogram for many months may lead to hypercalcemia and hypercalciuria. The toxic action of vitamin D is thought to be due to an increase in the intestinal absorption of calcium.[4]

C. Idiopathic Hypercalcemia of Infancy

Idiopathic hypercalcemia of infancy is thought to be a subgroup of hypervitaminosis D-induced hypercalcemia. Hypercalcemia may result from increased sensitivity to overdosage with vitamin D, although the etiology of this disorder is not clear. It may occur in infants receiving milk fortified with vitamin D or dietary vitamin D supplements, although large doses do not appear to be necessary. Excessive prenatal exposure to vitamin D or hypersecretion of PTH may also be etiological factors.[35] There is some suspicion that infants with idiopathic hypercalcemia may have intrauterine maldevelopment. Elfin-like facies, such as a pug nose, epicanthal folds, thickened lips, a flat nasal bridge, and immature facial appearance are characteristic of these infants.[36] Other symptoms include failure to thrive, emaciaton, hypotonia, loss of appetite, vomiting, constipation, polyuria, and polydipsia, which occur between 6 weeks to 9 months of age.[25] The condition may disappear spontaneously, especially the hypercalcemia, but often mental retardation occurs and is permanent. Occasionally, the condition is fatal.[31]

D. Sarcoidosis

Another condition in which hypercalcemia is present (but rarely in the younger age groups) is sarcoidosis. This has its onset in childhood and is thought to be due to unusual sensitivity to vitamin D.[4]

VI. DIAGNOSIS AND NUTRITIONAL MANAGEMENT

A. Hypocalcemia
1. Diagnosis

The clinical symptoms in neonatal hypocalcemia include irritability, tremor, laryngospasm, twitching, and seizures, collectively known as tetany. Some infants, however, may exhibit lethargy, anorexia, and vomiting.[25] The symptoms of tetany are almost invariably present when the ionized calcium level is ≤2.5 mg/dℓ and less often they appear at 2.5 to 3 mg/dℓ.[4,37] Since the normal range of ionized calcium is 4.0 to 4.5 mg/dℓ or a mean of 4.4 mg/dℓ, it appears that a reduction of one third in the ionized calcium fraction is needed before symptoms will appear.[4] According to Woodhouse,[7] symptoms of tetany usually develop when the ionized calcium level falls below 4.3 mg/dℓ or the total serum calcium is 7.0 to 7.5 mg/dℓ.

In neonatal hypocalcemia, maternal history of complications during pregnancy, e.g., toxemia, diabetes mellitus, diet, drugs, and nutritional status during pregnancy need to be assessed. In older infants, dietary analysis to assess phosphate intake is necessary. In infants on high phosphate feedings, the only manifestations of hypocalcemia in the first few days may be poor feeding, vomiting, lethargy, and cyanotic episodes. A few may have convulsive seizures.[4] In children, a history of malnutrition, malabsorption, rickets, chronic renal disease, and thyroid surgery should alert the clinician. The family history should be examined for occurrence of tetany, seizures, hypoparathyroidism, and a pediatric examination should be performed to observe clinical signs of hypocalcemia, chromosomal anomalies, and underlying diseases, such as pseudohypoparathyroidism.[25]

Biochemically, neonatal hypocalcemia is diagnosed by a low serum calcium level, and elevated phosphate concentration. Since ionized calcium measurements are not routinely available, the clinician may have to rely on total serum calcium levels.[4] If tetany occurs in the neonate, the serum level of magnesium should also be determined. While calcium measurements are in progress, hypocalcemia may be tested via intravenous administration of 10% calcium gluconate, 2 mℓ/kg.[4] If hyperexcitability and nervous system symptoms appear, and if they are due to hypocalcemia, there should be an immediate improvement with calcium gluconate. Once hypocalcemia has been diagnosed, the differential diagnosis rests between transient physiological hypoparathyroidism and the less common syndromes, congenital hypoparathyroidism and congenital vitamin D deficiency. In both transient and congenital hypoparathyroidism, the serum phosphate concentrations are elevated, usually >8 mg/dℓ, and are occasionally as high as 12 mg/dℓ. In congenital vitamin D deficiency, the serum phosphate concentrations are usually normal or reduced. Serum levels in the mother also provide diagnostic help. The usual age of onset of physiological hypoparathyroidism is between 5 and 10 days of age, but may be as late as 6 weeks.[4]

In mothers of infants with neonatal hypocalcemia, the serum levels of calcium, phosphate, and immunoreactive PTH (iPTH) should be measured to uncover underlying hyperparathyroidism. In older children, in addition to the above tests, serum proteins and alkaline phosphatase need to be measured. Concentrations of iPTH are raised in hypocalcemia which is associated with disorders other than primary hypoparathyroidism or hypomagnesemia, such as pseudoparathyroidism, malabsorption, deficit of vitamin D, renal tubular acidosis, and renal insufficiency. The diagnosis in some cases may depend on a combination of the history and biochemical findings. In hypocalcemia secondary to vitamin D deficiency, serum phosphorus is in the low range, serum alkaline phosphatase is usually increased, and serum iPTH is increased. In vitamin D deficiency due to a decreased intake or malabsorption, serum 25-OHD is decreased. In vitamin D-dependent rickets, serum 25-OHD is normal or increased.[4,31]

In acquired idiopathic hypoparathyroidism, the onset of recognizable symptoms of hypocalcemia is in later childhood. Sometimes enamel development is impaired and, occasionally, cataracts may be formed due to prolonged undiagnosed hypocalcemia. Somatic growth may be normal. Convulsive seizures are usually present. Autoimmune hypoparathyroidism can be distinguished from idiopathic hypoparathyroidism by the presence of multiple sites of tissue injury in addition to the loss of parathyroid gland

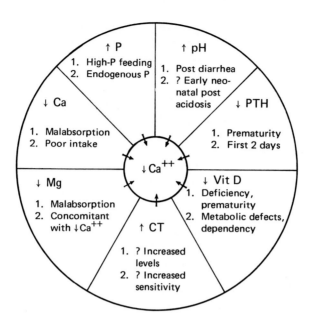

FIGURE 1. Possible pathophysiologic variables relating to neonatal and infantile hypocalcemia. (From Tsang, R. C., *Developmental Nutrition,* Coliver, K., Cox, B., Johnson, T., and Moore, M., Eds., Ross Laboratories, Columbus, Ohio, 1979, 91.)

function. A familial autosomal recessive pattern of inheritance is common in this syndrome.[4]

Figure 1 may be helpful in the differential diagnosis of some forms of hypocalcemia.

2. Treatment

The objectives of treatment in hypocalcemic states are to lower serum phosphate concentration and to increase the intestinal absorption of calcium. This can usually be accomplished by adding sufficient calcium to the diet to precipitate calcium phosphate in the intestinal lumen. The calcium to phosphate ratio in the diet should be increased to 4:1. A soluble source of calcium such as calcium gluconate (9% calcium by weight) or calcium lactate (13% calcium by weight) is preferable, usually in a powder form, because tablets do not dissolve as readily in milk. Calcium chloride is not recommended because of the possibility of gastric irritation.[4,31]

Hypocalcemic seizures in neonates can be treated with 10% intravenous calcium gluconate, 2 mg/kg, while monitoring the pulse to prevent bradycardia and cardiac arrest. When the intravenous feedings have been discontinued, tube feedings which are supplemented with calcium lactate or gluconate can be used. Gradually oral calcium feedings can be substituted.[4,25] In early neonatal hypocalcemia, 10% calcium gluconate is usually given intravenously, 1 or 2 mg/kg every 6 to 8 hr, a total of 4 to 6 mℓ/kg in a 24-hr infusion. (1 mℓ of 10% calcium gluconate supplies 9 mg Ca.) It should not be given intravenously.[1,4]

Table 3[38] lists the calcium and phosphorus content of foods which can be used to achieve the appropriate dietary ratio in addition to supplementation with calcium salts.[31]

If hypocalcemia occurs due to high phosphorus feedings or phosphorus poisoning, a reduction of serum phosphorus can be accomplished by reducing dietary phosphate, adding calcium salts, and by reducing intestinal absorption of phosphate which will occur if the dietary calcium to phosphate ratio is increased as described above.[1,4]

After the addition of calcium salts, the serum calcium level has to be monitored carefully to avoid hypercalcemia. When serum calcium levels are normalized, the calcium supplement is gradually reduced and finally withdrawn. This may take a few days to several weeks. If neonatal hypocalcemia persists beyond 21 days, congenital idiopathic transitory hypoparathyroidism should be suspected. In such cases, oral calcium and vitamin D in doses of 10,000 to 25,000 IU/day may be required. In cases of vitamin D deficiency hypocalcemia, oral administration of calcium salts is ineffective and vitamin D in relatively high doses is required. Doses of 8,000 to 16,000 IU of vitamin D (0.2 to 0.4 mg/day) for 10 days may be needed and then gradually reduced.[4]

In hypocalcemia due to fat malabsorption or atresia of the bile ducts 4000 to 8000 IU of vitamin D may continue to be required. In infants with liver disease, 1,25-DHCC in physiological doses of 0.5 μg/day can be used, or high doses of vitamin D are also effective. In vitamin D-dependent rickets, children may require very large doses of vitamin D because of a block in the hydroxylation of the vitamin in the intestine, but the exact dose has to be determined on an individual basis.[4]

In uncontrolled cases, permanent hypoparathyroidism may be present. Vitamin D therapy in hypoparathyroidism is directed primarily toward restoration of serum levels of 1,25-DHCC and optimal calcium absorption.[39] Regardless of which metabolite of vitamin D is used, the treatment should start with a low dose of the vitamin with a stepwise increase and careful monitoring of serum calcium levels. Oral calcium in conjunction with vitamin D is useful in the clinical management of hypoparathyroidism. Doses of 50 mg elemental calcium per kilogram or 530 mg calcium gluconate can be used in three to four divided doses. Overdosing with vitamin D should be avoided. Supplements of elemental magnesium in the amounts of 7 to 15 mg/kg may be needed if serum magnesium is <1.5 mg/dℓ.[39] Hypocalcemia accompanying thyroidectomy will usually respond to intravenous infusions of calcium gluconate.[25]

B. Hypercalcemia
1. Diagnosis

Hypercalcemia results from a variety of causes, including hyperparathyroidism. It may appear clinically with nonspecific symptoms of varying severity. These may include generalized weakness, headaches, tiredness, anorexia, nausea, irritability, and sometimes, vomiting, constipation, thirst, dry mouth, occasional nocturia, and drowsiness.[30] In addition, there may be depressed deep tendon reflexes, dehydration, hypertension bradycardia, hypokalemia, aminoacidemia, nephrocalcinosis, renal calculi, and metastatic calcification.[25]

Watson[30] stresses the difficulty in identifying marginal cases of hypercalcemia because of variations in the normal range of serum calcium in various laboratories. The normal range for fasting plasma calcium was reported to be 8.9 to 10.2 mg/dℓ by one laboratory[40] and 8.9 to 10.1 mg/dℓ by Purnell et al.[41] Harrison and Harrison[4] recommend that for accurate measurements of serum calcium, the blood should be drawn at least 6 hr after the last calcium feeding and with a minimum of venous stasis. Serum iPTH and calcium levels are helpful in the differential diagnosis of hypercalcemia.

In diagnosing infantile hypercalcemia, note should be made of traumatic birth, maternal thyroid surgery, and affected siblings. The typical facies of infants with idiopathic hypercalcemia provide diagnostic clues.[25] The early clinical symptoms in hypercalcemia due to vitamin D overdosage and hyperabsorption of calcium are the same as in other forms of hypercalcemia. These include anorexia, constipation, vomiting, polydipsia, polyuria, weight loss, growth failure, dehydration, drowsiness, coma, central nervous system manifestations, hypertension, and renal insufficiency.[4] The earliest biochemical evidence in vitamin D overdosage is an increase in the urinary excretion of calcium. Note should be made of patients who are receiving pharmacological doses of vitamin D. The daily calcium output in the urine is fairly steady in children. On a

Table 3
CALCIUM AND PHOSPHORUS CONTENT OF FOODS[38]

MEATS AND FISH

(100-g servings, cooked)

Food (Ca content <30 mg)	Ca (mg)	P (mg)	Food (Ca content < 30 mg)	Ca (mg)	P (mg)
Bacon (raw, 2 slices)	3	54	Halibut	15	210
Beef	10	213	Perch	27	190
Chicken	10	280	Swordfish	21	210
Egg (one)	27	103	Whitefish	—[a]	210
Frankfurter (one)	4	76	Tuna (canned)	10	290
Ham	7	175	Clams (raw or steamed)	70	150
Lamb	15	291	Crab (steamed)	44	180
Liver, chicken	10	144	Haddock (baked)	44	272
Liver, calf	15	500	Lobster (meat)	62	184
Luncheon meat (1 slice)	3	47	Salmon (canned)	280	380
Turkey	7	240	Scallops	110	350
Veal	11	140	Shrimp (canned)	100	225
Flounder	23	244	Oysters (raw or broiled)	81	240

VEGETABLES

(100-g serving or 1/2 cup or 3 oz)

Food (20—30 mg Ca)	Ca (mg)	P (mg)	Food (30—50 mg Ca)	Ca (mg)	P (mg)
Asparagus (canned)	22	53	Beans (green or wax)	31	23
Brussels sprouts (cooked)	22	64	Beets (fresh)	32	31
Cauliflower (cooked)	26	53	Cabbage (raw)	33	18
Corn (canned)	8	55	Cabbage (cooked)	43	20
Cucumber (fresh)	19	20	Carrots	34	32
Eggplant	14	31	Celery (raw)	40	28
Green Pepper (one)	7	16	Potato (sweet)	30	44
Lettuce	19	7	Kidney beans	46	165
Mushrooms (raw, ½ cup)	9	68	Squash (winter, summer)	35	35
Peas	28	110	**Food (high calcium content)**	**Ca (mg)**	**P (mg)**
Peas (dried, cooked)	11	90			
Potatoes, white	6	40			
Tomato, raw	12	25	Artichokes (cooked)	51	69
Chicory, as garnish	0	0	Beet greens (cooked)	72	18
Onions, as garnish	0	0	Broccoli	82	64
Parsley, as garnish	0	0	Chard (cooked)	80	31
			Collard greens (cooked)	240	44
			Dandelion greens	147	44
			Kale, cooked	140	42
			Mustard greens	128	30
			Spinach (canned)	160	35
			Turnip greens	180	34

FRUITS

(120 g or ½ cup)

Food (5—20 mg Ca)	Ca (mg)	P (mg)	Food (5—20 mg Ca)	Ca (mg)	P (mg)
Apple	7	11	Honeydew melon	11	18
Apple juice (canned)	10	18	Lemon juice, Lime juice	8	16
Applesauce	5	7	Lemon (one, peeled)	19	12
Apricots (fresh, canned)	18	25	Orange juice	12	23
Avocado (1/2)	11	45	Peaches (fresh)	10	19

Table 3 (continued)
CALCIUM AND PHOSPHORUS CONTENT OF FOODS[38]

Food	Ca (mg)	P (mg)	Food	Ca (mg)	P (mg)
Banana (small)	8	25	Peach nectar	5	14
Blueberries	13	13	Pears (fresh)	12	18
Cantaloupe	14	17	Pears (canned)	6	9
Cherries (fresh, canned)	18	14	Pear nectar	4	6
Cranberry juice	6	4	Pineapple (canned)	20	8
Fruit cocktail	11	16	Pineapple juice (canned)	19	12
Grapes (green, seedless)	12	20	Plums (fresh)	12	18
Grape juice	14	15	Prune juice (1/3 cup)	12	34
Grapefruit (1/2)	16	16	Tomato juice	8	22
Grapefruit juice (canned)	10	18	Watermelon	8	11

Food (Ca >20 mg)	Ca (mg)	P (mg)	Food (Ca >20 mg)	Ca (mg)	P (mg)
Blackberries	31	18	Raisins	66	110
Dates	50	63	Raspberries	27	27
Lemons with peel	61	15	Rhubarb	100	20
Orange (fresh, small)	36	20	Strawberries	21	21

CEREALS AND BREADS

Food	Ca (mg)	P (mg)	Food	Ca (mg)	P (mg)
Bran flakes (3/4 cup)	21	108	Rolls (plain, one)	21	24
Cornflakes (2/3 cup)	1—5	6	Pancake (one)	58	70
Cream of wheat (instant, 3/4 cup)	14	140	Muffin (one)	42	60
			Biscuit (one)	34	49
Oatmeal (3/4 cup)	17	102	Saltines (2)	1	5
Puffed oats (1/2 cup)	22	51	Graham crackers (2)	3	10
Puffed rice (1/2 cup)	1	7	Rice (cooked, 1/2 cup)	12	71
Puffed wheat (1/2 cup)	2	24	Macaroni (cooked, 1/2 cup)	6	35
Shredded wheat (1/2 cup)	11	97	Noodles (cooked, 1/2 cup)	8	47
Bread (Vienna, French, Italian, 1 slice)	5	23	Spaghetti (cooked, 1/2 cup)	7	42
Bread (white, rye, whole wheat, 1 slice)	24	24—52	Pizza (1 piece)	144	127

DAIRY PRODUCTS AND DESSERTS

Food	Ca (mg)	P (mg)	Food	Ca (mg)	P (mg)
Milk (240 g, whole, skim, chocolate)	290	233	Yogurt (240 g)	290	233
			Butter, margarine (1 tsp.)	1	1
Cream (table, 28 g)	32	24	Angel cake (1 slice)	5	13
Cheese (cheddar, 28 g)	180	115	Sugar cookies (one)	6	8
Cheese (cottage, 1/2 cup)	65	173	Jello (1/2 cup)	0	0
Cheese (Swiss, 28 g)	324	197	Apple pie (1 section)	3	7
Ice cream (1/2 cup)	97	27	Cupcake (one)	55	88

MISCELLANEOUS

Food	Ca (mg)	P (mg)	Food	Ca (mg)	P (mg)
Tea, coffee (1 cup)	4	9	Mustard (1 tsp.)	4	4
Vegetable oils	0	0	Mayonnaise (1 T)	3	4
Jam, jelly (1 T)	4	2	Salad dressing (1 T)	12	11
Maple syrup (1 T)	20	2	Tomato catsup (1 T)	3	8
Honey (1 T)	1	1	Peanut butter (1 T)	9	61
Molasses (1 T)	33	9	Potato chips (30 g)	11	39
Almonds (15 g)	38	80	Pickles, sweet (1 T)	5	2
Peanuts (ten)	13	74	Soy sauce (1 T)	15	19
Pecans, shelled (15 g)	10	40			

Table 3 (continued)
CALCIUM AND PHOSPHORUS CONTENT OF FOODS[38]

Note: For calcium restricted diets the food lists above can be used to select meats, vegetables, fruit, breads, and desserts in order to plan a diet with a specific calcium and phosphorus content. A low calcium diet contains 100—250 mg calcium and 700—800 mg phosphorus. The following foods are high in calcium and should be avoided:

- Milk, cream cheese, ice cream, cream sauces, cream soups
- Whole wheat bread, biscuits, pancakes, waffles, enriched cereals
- Shellfish, fresh salmon or other fish with a high calcium content
- Dried fruits, canned fruits and beverages sweetened with calcium saccharin
- Desserts, cake or pies made with milk
- Mashed or creamed potatoes, fresh sweet potatoes
- Molasses, maple syrup, nuts, chocolate, peanut butter
- Baking powder, baking soda

^a Data not available.

moderately high intake of calcium (1000 to 1400 mg/day), a child under the age of 4 years usually excretes 25 to 50 mg calcium per day, and at 5 to 14 years, 75 to 100 mg of calcium per day. This amounts to 2 to 4 mg/kg body weight per day or 0.25 mg calcium per milligram of creatinine. In addition to hypercalciuria, serum calcium is increased, serum phosphorus is variable and is not reduced as in primary hyperparathyroidism. Alkaline phosphatase is usually decreased and serum iPTH is low. Levels of 25-OHD in serum are many times greater in subjects receiving high doses of vitamin D. This constitutes definitive diagnosis of excessive vitamin D intake as the cause.[4]

The diagnosis of sarcoid hypercalcemia requires an examination of lymph nodes, lungs, liver, and the spleen. The presence of hyperglobulinemia which is characteristic of this syndrome is also helpful. A complete dietary analysis and history of drugs is essential to distinguish between hypercalcemia due to chronic excessive doses of vitamins A or D. Serum iPTH levels are low or undetectable in vitamin D intoxication and sarcoidosis.[4]

The diagnosis of immobilization hypercalcemia can be made on the basis of gradual onset of typical symptoms which originate with a fraction of the femur and sudden transition from an active physical life to being bedridden. Serum calcium in such cases usually reaches 14 to 16 mg/dℓ. Gradually, urinary calcium excretion increases to as much as 400 mg/day and iPTH may also be increased.[4]

In hypercalcemic states resulting from primary or persistent hyperparathyroidism, parathyroid hyperplasia after renal transplants, ectopic PTH secretion from a neoplasm, therapy with thiazide diuretics, and in familial benign hypercalcemia, iPTH levels in the serum are helpful diagnostic clues. In cases of ectopic PTH secretion due to a neoplasm, serum iPTH levels are relatively lower for a given serum calcium concentration than in cases of primary hyperparathyroidism.[42] Familial benign hypercalcemia can be diagnosed when the total and ionized serum calcium levels are elevated and the serum iPTH is normal and does not change despite changes in serum calcium levels. Such patients are clinically normal. Similar tests in other asymptomatic family members confirm the diagnosis.[25] The differential diagnosis of tumor-associated hypercalcemia depends on the nature of the primary tumor. In primary hyperparathyroidism in addition, serum ionized calcium is high (>5.6 mg/dℓ) on repeated measurements. Other biochemical tests show hypophosphatemia, although not consistently, increased alkaline phosphatase, anemia, elevated sedimentation rate, hyperuricemia, hypomagnesemia, and hyperchloremic acidosis.[4,25]

2. Treatment

The basic objectives in the treatment of hypercalcemia are a speedy reduction in the serum calcium concentration while attempting to eliminate the underlying cause. Except in some cases, e.g., primary hyperparathyroidism where parathyroidectomy is indicated, dietary management is usually effective. In idiopathic hypercalcemia of infancy, dietary fluids should be increased, vitamin D eliminated, and calcium intake restricted to below 100 mg daily (see Table 3 for calcium content of foods).[38] In the treatment of mild hypercalcemia (serum calcium, 12 to 13 mg/dℓ) it is necessary to restore the electrolyte and water balance, to use a low calcium diet[31] (<200 mg/day calcium, see Table 3), and restrict the intake of vitamin D.[43] Acute hypercalcemia (serum calcium >15 mg/dℓ) requires diuresis with isotonic saline, intravenous furosemide, or intravenous chelating agents. In unresponsive cases, hemodialysis or peritoneal dialysis against low calcium solutions may be necessary.[25]

In patients with toxicity due to vitamins A or D, vitamin supplements and the major sources of the vitamin need to be eliminated from the diet. The exact treatment of hypervitaminosis D hypercalcemia depends on the degree of toxicity. To begin with, concentrated supplements of vitamin D and vitamin D-fortified foods should be eliminated from the diet. Unfortified dietary sources of vitamin D do not make a major contribution. Excessive exposure to sunshine, however, may have a significant impact and should be avoided. Calcium intake should be decreased by eliminating milk and milk products[31] (see Table 3). If the symptoms are severe, it may become necessary to rapidly decrease the serum calcium level. Occasionally, calciuretic diuretics such as furosemide can be used for this purpose.[4] These inhibit tubular reabsorption of calcium, sodium, and water and increase calcium output. The intake of sodium, potassium, and water also need to be increased. A low calcium, low vitamin D diet should be prescribed for several weeks or months. In severe cases, simultaneously, the intake of phosphorus may need to be increased via buffered solutions providing 1 to 2 g phosphorus per day given orally. Attempts have also been made to use calcitonin to increase the uptake of calcium by the bone. In patients receiving large doses of vitamin D for prolonged periods for conditions such as hypophosphatemic rickets, it is important to monitor the progress of the condition. Urinary excretion of vitamin D is usually slow. Serum and urinary calcium must be monitored closely for months. When the urinary excretion of calcium reaches 2 mg/kg/day, any vitamin D supplements should be reduced. When it reaches 4 mg/kg/day, the vitamin D should be discontinued completely and started at a lower dose when the urinary calcium excretion reaches below 2 mg/kg/day. Hypercalcemia in case of sarcoidosis can be alleviated by strict elimination of vitamin D from the diet and protection of the body from the sun.[4]

In acute disuse osteoporosis in adolescents confined to bed due to injury, a daily program of physical activity, restriction of vitamin D intake, and elimination of dairy products from the diet (to lower the dietary calcium levels) is necessary.[31] The intake of sodium and water (intravenously, if needed) should be increased to prevent dehydration and increase the urinary output of calcium. This can be accomplished by using 200 to 250 mg/kg/day of isotonic saline[44] which should be monitored every 6 hr to prevent overtreatment. Other foods high in calcium should also be eliminated[31] (see Table 3). In each case, serial measurements of serum calcium levels should be made.

Since both neonatal hypocalcemia and idiopathic hypercalcemia may be associated with mental retardation, early diagnosis of these conditions is important. The prognosis for normal mental development in infants with neonatal hypocalcemia as reported by Cockburn et al.[27] is good, although 11% of their patients at 1 year of age were gauged as being developmentally "slow", with single instances of mild ataxia, spasticity, and dystonia. In idiopathic hypercalcemia, however, despite frequent spontaneous correction of the hypercalcemic state, mental retardation occurs often and may be permanent. Metastatic calcification may be noted in the heart, lungs, and kidneys.[25] This condition in its more severe form may be fatal.

REFERENCES

1. **Tsang, R. C.**, Calcium and phosphorus, in *Developmental Nutrition,* Coliver, K., Cox, B., Johnson, T., and Moore, M., Eds., Ross Laboratories, Columbus, Ohio, 1979, 91.
2. **Potts, J. T., Jr. and Deftos, L. J.**, Parathyroid hormone, calcitonin, vitamin D, bone and bone mineral metabolism, in, *Duncan's Diseases of Metabolism,* 7th ed., Bondy, P. K. and Rosenberg, L. E., Eds., W. B. Saunders, Philadelphia, 1974.
3. **Fanconi, A. and Rose, G.**, The ionized, complexed and protein-bound fractions of calcium in plasma, *Q. J. Med.,* 27, 463, 1958.
4. **Harrison, H. H. and Harrison, H. C.**, *Disorders of Calcium and Phosphate Metabolism in Children and Adolescents,* W. B. Saunders, Philadelphia, 1979, 47.
5. **Root, A. W. and Harrison, H. E.**, Recent advances in calcium metabolism. I. Mechanisms of calcium homeostasis, *J. Pediatr.,* 88, 1, 1976.
6. **Kodicek, E.**, Recent advances in vitamin D metabolism, *Clin. Endocrinol. Metab.,* 1, 305, 1972.
7. **Woodhouse, N.**, Hypocalcemia and hypoparathyroidism, *Clin. Endocrinol. Metab.,* 1, 323, 1974.
8. **Kodicek, E.**, The story of vitamin D. From vitamin to hormone, *Lancet,* 1, 326, 1974.
9. **Raisz, L. G.** Physiologic and pharmacologic regulation of bone resorption, *N. Engl. J. Med.,* 282, 909, 1970.
10. **Bergman, L., Kjellmer, I., and Selstam, U.**, Calcitonin and parathyroid hormone: relation to early neonatal hypocalcemia in infants of diabetic mothers, *Biol. Neonat.,* 24, 151, 1974.
11. **Kaplan, E. L., Hill, B. J., Lock, S., et al.**, Acid-base balance and parathyroid function:metabolic alkalosis and hyperparathyroidism, *Surgery,* 70, 198, 1971.
12. **Anon,** Calcium in bone health, *Dairy Council Digest,* 47, 31, 1976.
13. **Lutwak, L.**, Dietary calcium and the reversal of bone demineralization, *Nutr. News,* 37, 1, 1974.
14. **Wilkinson, R.**, Absorption of calcium, phosphorus, and magnesium, in, *Calcium Phosphate and Magnesium Metabolism,* Nordon, B. E. C., Ed., Churchill Livingstone, Edinburgh, 1976, 36.
15. **Hegsted, D. M.**, Calcium and phosphorus, in, *Modern Nutrition in Health and Disease,* 5th ed., Goodhart, R. and Shils, M., Eds., Lea & Febiger, Philadelphia, 1973, 268.
16. **DeLuca, H. F.**, Vitamin D: the vitamin and the hormone, *Fed. Proc.* 33, 2211, 1974.
17. **Life Sciences Research Office,** *Evaluation of Health Aspects of Phosphates as Food Ingredients,* SGOGS-32, Federation of the American Society for Experimental Biology, Bethesda, Md., 1975, 1.
18. **Margen, S., Kaufman, N. A., Costa, F., and Calloway, D. H.**, Studies in the mechanism of calciuria induced by protein feeding, *Fed. Proc. Fed. Am. Soc. Exp. Biol.,* 29, 566, 1970.
19. **Allen, L. A., Oddoye, E. A., and Margen, S.**, Protein induced hypercalciuria: a longer term study, *Am. J. Clin. Nutr.,* 32, 741, 1979.
20. **Food and Nutrition Board,** National Research Council, Recommended Dietary Allowances, 9th ed., Natl. Acad. Sci., Washington, D.C., 1980, 125.
21. **Davidson, L., Passmore, R., and Brock, J. F.**, *Human Nutrition and Dietetics,* 5th ed., Williams & Wilkins, Baltimore, 1972, 99.
22. **Gergman, L. and Isaksson, B.**, Plasma calcium fractions in normal subjects from birth to adult ages, *Acta Paediatr. Scand.,* 60, 630, 1971.
23. **Brown, D. M., Boen, J., and Bernstein, A.**, Serum ionized calcium in newborn infants, *Pediatrics,* 49, 841, 1972.
24. **Tsang, R. C., Wen-Chen, I., and Friedman, M. A.**, Parathyroid function in infants of diabetic mothers, *J. Pediatr.,* 86, 399, 1975.
25. **Root, A. W. and Harrison, H.**, Recent advances in calcium metabolism. III. Disorders of calcium homeostasis, *J. Pediatr.,* 88, 177, 1976.
26. **Moncrieff, M. and Fadahunsi, T. O.**, Congenital rickets due to maternal vitamin D deficiency, *Arch Dis. Child.,* 48, 810, 1974.
27. **Cockburn, F., Brown, J., Belton, N., and Forfar, J.**, Neonatal convulsions associated with primary disturbance of calcium phosphorus and magnesium metabolism, *Arch. Dis. Child.,* 48, 99, 1973.
28. **Tsang, R. C., Steichen, J. J., and Brown, D. R.**, Perinatal calcium homeostasis: neonatal hypocalcemia and bone demineralization, *Clin. Perinatol.,* 4(2), 385, 1977.
29. **Hahn, T., Hendin, B., Scharp, C., Boisseau, V., and Haddad, J.**, Serum 25-hydroxycalciferol levels and bone mass in children on chronic anticonvulsant therapy, *N. Engl. J. Med.,* 292, 550, 1975.
30. **Watson, L.**, Primary hyperparathyroidism, *Clin. Endocrinol. Metab.,* 3, 215, 1974.
31. **Palmer, S.**, Calcium, in *Pediatric Nutrition in Developmental Disorders,* Palmer, S. and Ekvall, S., Eds., Charles C Thomas, Springfield, Ill., 1978, 402.
32. **Daum, F., Rosen, J. F., and Boley, S. J.**, Parathyroid adenoma, parathyroid crisis, and acute pancreatitis in an adolescent, *J. Pediatr.,* 83, 275, 1973.
33. **Frier, B. M. and Marrian, V. J.**, Uncomplicated hyperparathyroidism, *Arch. Dis. Child.,* 49, 808, 1974.

34. Root, A., Bongiovanni, A., Eberlein, W. R., and Mitchie, A., Measurements of the kinetics of calcium metabolism in children and adolescents utilizing nonradioactive strontium, *J. Clin. Endocrinol. Metab.*, 26, 537, 1966.
35. Rasmussen, H., Differentiation of parathyroid hyperplasia from adenoma, *N. Engl. J. Med.*, 280, 1416, 1969.
36. Bongiovanni, A., Eberlein, W., and Jones, L., Idiopathic hypercalcemia of infancy with failure to thrive. Report of three cases with a consideration of the possible etiology, *N. Engl. J. Med.*, 257, 951, 1957.
37. Rosen, J. F., Roginsky, M., Nathenson, G., et al., 25-hydroxy vitamin D—plasma levels in mothers and their premature infants with neonatal hypocalcemia, *Am. J. Dis. Child.*, 127, 220, 1974.
38. U.S. Department of Agriculture, Nutritive Value of American Foods in Common Units, Agriculture Handbook No. 456, Agricultural Research Service, U.S. Department of Agriculture, Washington, D.C., 1975.
39. Tsang, R. C., Noguchi, A., and Steichen, J. J., Pediatric parathyroid disorders, in Symposium on Pediatric Endocrinology, *Pediatr. Clin. North Am.*, 26(1), 223, 1979.
40. Davies, D., Dent, C., and Watson, L., Idiopathic hypercalciuria and hyperparathyroidism, *Br. Med. J.*, 1, 108, 1971.
41. Purnell, D., Smith, L., Scholz, D., Elveback, L., and Arnaud, C., Primary hyperparathyroidism: a prospective clinical study, *Am. J. Med.*, 50, 670, 1971.
42. Benson, R., Riggs, B., Pickard, B., and Arnoud, C., Radioimmunoassay of parathyroid hormone in hypercalcemic patients with malignant disease, *Am. J. Med.*, 56, 821, 1974.
43. Goldsmith, R., Treatment of hypercalcemia, *Med. Clin. North Am.*, 56, 951, 1972.
44. Tsang, R. C., Donovan, E. F., and Steichen, J. J., Calcium physiology and pathology in the neonate, *Pediatr. Clin. North Am.*, 23(4), 611, 1976.

Chapter 8

ROLE OF CALCIUM METABOLISM IN RENAL STONE FORMATION*

Paul Otto Schwille and Dieter Scholtz

TABLE OF CONTENTS

I.   Introduction 104

II. Theories of Stone Formation 104

III. Genetic Influences 105
     A.   General Remarks 105
     B.   The Distribution According to Sex 106

IV. The Role of Environmental Factors 107
     A.   Hygiene 107
     B.   Climate 107
     C.   Nutrition 108
          1.   The Mineral Content of Various Foodstuffs 108
          2.   Oxalate 109
          3.   Purine and Proteins 111
          4.   Carbohydrates 111
          5.   Fatty Acids 111

V.   Established Disorders of Calcium Metabolism with Associated Lithiasis 112

VI. Clinical Diagnosis of the Metabolic Subgroups of Calcium Urolithiasis 113
     A.   Hypercalciuria 113
          1.   Resorptive Hypercalciuria (= Primary Hyperpara
               thyroidism) 113
          2.   Renal Hypercalciuria 113
          3.   Absorptive Hypercalciuria 113
     B.   Normocalciuria 114
     C.   Uric Acid Lithiasis and the Intermediate Group 114
     D.   Comments Concerning the Function of the Parathyroid Glands in
          Idiopathic Urolithiasis 114

VII. Techniques for Measurement of Intestinal Absorption of Calcium (CaA) ... 115

VIII. Factors Influencing Intestinal Calcium Absorption 116

IX. Concluding Remarks 117

References 118

---

*   Research on this and related subjects are supported by "Deutsche Forschungsgemeinschaft" (Grant Schw 210/1-3).

I. INTRODUCTION

In the past few years, many investigators from the fields of clinical and experimental medicine, as well as from basic research disciplines such as biochemistry, physical chemistry, and mineralogy have devoted their efforts to urolithiasis research, particularly to the unsolved problems of the origin of stones containing calcium. This trend seems to be justified for two reasons: (1) urolithiasis represents a very common disorder, probably due to circumstances based on geographical differences; the frequency of occurrence in man varies between 1 and 15% of the total population;[1-5] and (2) approximately 70% of all stones contain calcium and, so far, it has not been clarified, whether the precipitation of calcium ions forms the expression of a pathological metabolism as in relatively rare defined disease states with accompanying lithiasis, or constitutes an associated symptom of malregulations which, per se, are without clinical significance. This latter might well be the case in the majority of individuals suffering from so-called "idiopathic lithiasis".

Table 1 explains about the four clinically important species of stones and their mineralogical names. Additionally, other concrements, such as xanthine, 2,8-dihydroxyadenine, and carbonates exist, which occur clinically very seldom. The interesting pathophysiology of these stones cannot be covered here, however, (for details see handbooks on urology and metabolism).

The following sections contain attempts to evaluate factors which are currently discussed concerning research on the origin of the stones. The effects and feed-back of exogenous factors and of food components on the calcium metabolism, the intestinal and renal handling of calcium, particularly in the subgroups of idiopathic calcium lithiasis, the present state of knowledge on the role of parathyroid hormone, as well as the techniques and the evaluation of methods to measure the intestinal absorption of calcium are emphasized.

II. THEORIES OF STONE FORMATION

Mechanisms presently accepted which result in nidus formation (= crystalline microconcrement) have been presented frequently, recently by Schwille et al.[6] and Pak.[7]

The *precipitation-crystallization* theory of an excessive saturation of the urine in stone-forming phases (calcium oxalate, calcium phosphate, etc.) favors spontaneous precipitation and transformation of the latter into different forms of crystals (oxalates, phosphates, etc.) out of which larger aggregations of crystals and concrements develop. According to the definition, the process will then proceed selfsustaining (= homogenous nucleation) if the respective formation product of a given salt in the urine is exceeded. For this reason, a highly oversaturated urine is considered as the critical dimension. In another situation, the formation of crystals, e.g., calcium oxalate, can be triggered by the simultaneous presence of large quantities of a different salt, e.g., sodium urate, which the degree of saturation of calcium oxalate is still comparatively low (= heterogenous nucleation, so-called "seed effect"). Both processes can be located intraluminally, if the time for the formation of a sufficiently large nidus (stoppage at narrow passages is required) is shorter than the transit time of the intratubular liquid. This has, however, so far not been formulated in sufficient detail experimentally.

Inhibitors of the crystal formation and aggregation in the urine such as citrate, pyrophosphate, magnesium, or glycosaminoglycanes, e.g., heparine, are low or missing completely in comparison with promotors of supersaturation (calcium, oxalate, phosphate, alkaline pH). At the present time, it seems to be justified to include this *"inhibitor theory"* into the "supersaturation theory".

Table 1
URINARY CONCRETIONS OF HUMANS, THE CLINICAL NAME (STONE SPECIES) AND THE MINERALOGICAL DENOMINATION

Stone species		Mineralogical denomination
I. Calcium lithiasis	Whewellite	= Calcium oxalate monohydrate
	Weddellite	= Calcium oxalate dihydrate
	Hydroxyapatite	= Penta-calcium-hydroxyphosphate
	Brushite	= Di-calcium-phosphate
	Whitlockite	= Tri-calcium-phosphate
II. Uric acid lithiasis		Uric acid (nonionized)
		Uric acid dihydrate
		Ammonium hydrogen urate
		Sodium hydrogen urate monohydrate
III. Cystine lithiasis		Cystine
IV. Infected lithiasis	Struvite	= Magnesium ammonium phosphate-hexahydrate

Organic matrix substances are found in all stones and they might form the basis for the assumption that from so-called "subepithelial plaques" stones are formed if the plaques are released into the lumen. The calcium-binding property of the matrix may be due to γ-carboxyglutamic acid, which is found in renal calcium but no other stones.[8] If *this matrix* theory should be supported by the results of future investigations, more extensive parallel phenomena concerning the formation of stones and of the bone structure may be revealed. Factors controlling the formation and the metabolism of the bone matrix are presently better known than those influencing the formation of the matrix of urinary stones. Bone formation has a close correlation with the hormonal control of the calcium metabolism (for details see handbooks for metabolism and endocrinology).

III. GENETIC INFLUENCES

A. General Remarks

Renal stones are occurring worldwide in domestic animals and in man. According to our knowledge, little is mentioned in the literature of veterinary medicine concerning genetic factors, which could be useful in the search for causes of stone formation in man.

In Vienna, Austria, and its vicinity, the anatomically small races of dogs (Pekinese, Maltese, miniature poodles) predominantly suffer from urolithiasis, but the total number of carriers of urinary stones is estimated to be much larger than the number of cases of the disease becoming actually known.[9] A well-known exception since 1940 is the tendency of the Dalmation dog towards uric acid lithiasis. This dog race exhibits sustained hypouricemia in the presence of a urinary urate clearance greater than glomerular filtration rate, most likely not owing to a generalized defect in transport of urate across renal tubular and other tissue membranes, but owing to a defective urate metabolism of the liver.[10] At present, it is unknown whether in Dalmation dogs urinary calcium is also elevated as it is in hypouricemic humans (see below).

For a long time, a genetic factor for urolithiasis has been suspected in man, but definite proof is still missing. Latest investigations of families with increased frequency of hypercalciuria and of calcium lithiasis prove a family-dependent form of hypercalciuria, which is transmitted autosomally dominant. In first degree relatives of either sex nephrolithiasis is frequent.[11] The trias of hypouricemia, hypercalciuria, and a reduced bone density is regarded as a genetically transmitted syndrome also. According

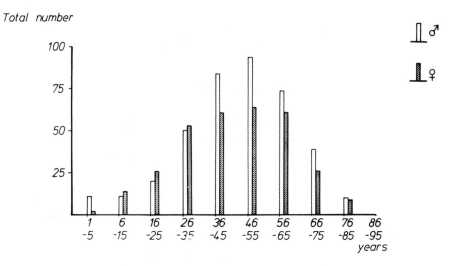

FIGURE 1. Frequency of renal stones (no bladder stones) in relation to age and sex in the Stone Clinic of the University of Erlangen (FRG, January 1974 to July 1977 (n = 709).

to our knowledge, this is the first description of a combined disorder of purine and calcium metabolism.[12] Stone formation is frequent in this context and in another variety of hereditary hyperuricosuric urolithiasis.[13]

B. The Distribution According to Sex

Disregarding the composition of the stones and the age of the patients, all observers agree on the increased frequency of urolithiasis in men as compared with women.

In our stone clinic, predominance of men develops in the fourth decade of life, with the women dominating at lower ages (Figure 1). No reasons are known so far for this phenomenon, but two facts might be of interest for further studies of the underlying causes of urolithiasis: (1) stones with a predominant content of calcium phosphate (>75% of the total stone substances) are formed more frequently by women,[14] the maximum of phosphate lithiasis occurring in the third and fourth decade of life; and (2) primary hyperparathyroidism, the most frequent disorder of the calcium metabolism with urolithiasis, manifesting all over the world three times as frequently in women than in men, has its peak in the sixth decade, and a high percentage of the almost regularly formed mixed concrements (oxalate, phosphate) consists of neutral calcium phosphate (apatite). These findings might support a hypothesis, that in women during the childbearing age, an additional factor favoring the formation of stones might be present, which is not fully developed in men or missing completely, with the conditions reversing after the onset of menopause.

Studies across collectives of healthy subjects and patients suffering from urolithiasis are intended to show whether the traditional calciotropic hormones (parathyroid hormone, calcitonin) play a key role in these phenomena. It has been postulated that serum parathyroid hormone increases with age[15] and that calcitonin is lower in postmenopausal osteoporosis.[16] Combined disorders of the metabolism of the bones and urolithiasis are known to clinicians ("bones and stones"; see Section V) and the possibility for a common base for mineralization in the bones and in the kidneys has been mentioned above. Since highly sensitive radioimmunoassays are available for calcitonin[17] and homologous materials allow the application of similar methods for human parathyroid hormone,[15] reliable data on interrelationships can be expected in the future.

IV. THE ROLE OF ENVIRONMENTAL FACTORS

A. Hygiene

Endemic occurrence of urolithiasis, particularly in children, has for a long time in the Near East (Turkey) and Southeast Asia (Thailand) been considered as an expression of underdeveloped conditions for life. Lately, it has undeniably been proved that this formation of ammonium acid urate is based on a chronic deficiency in phosphate (see Nutrition below) due to a diet poor in phosphate (rice, flour) and stone disease is stopped by oral phosphate supplementation.[18] Poor physical hygiene, therefore, seems only as an exceptional cause, if at all, of urolithiasis, e.g., in cases due to infections (see Table 1) in the urinary system when the presence of germs of the *Proteus-, Pseudomonas-, Coli* group is a precondition.

B. Climate

Until a few years back, a direct influence of heavy perspiration due to heat was known only to affect the composition of the urine (low volume, deeply acidic pH) and to result in a tendency for formation of concrements containing uric acid and calcium. This is valid for various hot geographic regions, e.g., Israel and Bulgaria, with a higher rate of uric acid stones (up to 35% of the population), as well as for professional exposition at the work station.[19] The main reason is that many people maintain too low of an intake of fluids.[20] In the past years, it was clearly recognized that the duration of exposition to UV-radiation from sunlight for the synthesis of vitamin D_3 (here abbreviated as D) in the skin from 7-dehydrocholesterol constitutes, besides oral administration of D, one of the two factors determining an optimal supply of D and is responsible for a normal calcium metabolism in man. This branch of research was greatly facilitated by the fact that more polar metabolites of D, which are either formed in the liver (25-hydroxy-D) or in the kidneys (1a,25-dihydroxy-D; 24R,25-dihydroxy-D; 25,26-dihydroxy-D; 1a,24R,25-trihydroxy-D) became measureable. The 25-hydroxy-D metabolite, meanwhile, is accepted as a yardstick for the degree of exogenous supply of D, and its concentration in the serum is recorded during various clinical disorders, including urolithiasis. Within solid buildings and in locations with low atmospheric ultraviolet radiation, its concentration is lowered (normal 10 to 35 ng/mℓ in this laboratory), in the open air or in sunny regions it is higher,[21] without, however, a direct influence on the calcium metabolism. The reason for this is the fact that the namely metabolite is, e.g., only 1/500 times as effective on the intestinal calcium transport as, e.g., 1,25-dihydroxy-D. In spite of the fact that higher 25-hydroxy-D in the serum can be found together with elevated calciuria, the hepatic metabolite is considered to be normal in calcium lithiasis by most of the laboratories, notwithstanding considerable hypercalciuria in many cases.

Interrelations of the varying intensity of the UV radiation according to the different seasons of the year,[22] the resulting levels of 25-hydroxy-D, the degree of hypercalciuria, crystalluria, and frequency of occurrence of stones in the various seasons are, e.g., in Great Britain, considered controversially.[23-25] We have found no peak for urine calcium and oxalate excretion during summer and no indications for a higher urolithiasis morbidity during a specific period of the year.[26] The contradictions will be solved, when D-metabolites, which are metabolically differently effective, are measured in parallel series, together with the intestinal calcium and oxalate absorption and the renal excretion (see also Section VI.). Improvements of the method of analysis and recording of the serum concentration in the metabolic subgroups of nephrolithias (see Table 1, except infected lithiasis; Table 2) will permit better founded conclusions.[112]

Deficiency in 25-hydroxy-D indicates certain forms of rickets (children) or osteomalacia (adults) which may also occur together with calcium lithiasis (see Section V).

Table 2
CLINICAL DISORDERS FREQUENTLY ASSOCIATED WITH CALCIUM UROLITHIASIS

Diagnosis	Suggested portion of urolithiasis morbidity[a] (%)
Primary hyperparathyroidism in unselected stone-formers[b]	4—5
Hypercalcemia, not originating from parathyroid glands:	1—2
Malignancies with generalized bone metastases	
Sarcoidosis	
Drug overconsumption (vitamin D, dihydrotachysterol)	
Pseudohyperparathyroidism	
Medullary sponge kidney	3—4
Renal tubular acidosis type I	1
Enteric hyperoxaluria	1
Hypercalciuric rickets	0.5—1
Oxalosis	0.1
Wilsons's Disease	0.05—0.1

[a] Synonymous with documented stone.
[b] Requiring medical care.

C. Nutrition

It is generally assumed that in times of war and distress urinary stones would occur less frequently. This is considered as an expression of a lower food intake or of lower intake of calories with foodstuffs of lesser nutritional value. Analogous to this simple model in the past seventies the same idea was pursued based, however, on overnutrition, i.e., elevated or exclusive intake of proteins originating from animals,[27] of other proteins,[28] of carbohydrates,[29] and of fats, considered as a risk factor of urolithiasis in man and animals. Particularly, hypercalciuria (proteins) and hypomagnesiuria (carbohydrates) were considered in this context. With respect to stone-forming, humans demographic prospective population studies in all countries are, therefore, urgently required to uncover the true risk for stone formation in the presence and the past.

1. The Mineral Content of Various Foodstuffs

The intake of tap water probably introduces geographically quite differing quantities of calcium and magnesium into the body,[30] although differences regarding the urinary excretion of these ions in areas with hard and soft water were refuted.[31] Contradicting the earlier opinion that the formation of stones was proportional to the calcium content of the water consumed, recent reports claim the inverse to be true, i.e., higher stone incidence in regions with soft water.[32] While this observation is not yet confirmed, it may indicate a relationship between the minerals in the drinking water, biological control functions of the mineral metabolism, which until now have not been closely investigated, and processes triggering the formation of stones.

A diet supplying excessive amounts of calcium (milk and milk products) can be ruled out as the sole cause of hypercalciuria, frequent in patients suffering from stones. At least three different principles constitute the base for that disorder (see Section VI) and, among our patients, hypercalciuria due to diet represents at the most 10% of all cases of calcium lithiasis.[6]

All investigators have so far ruled out an exogenous deficiency of calcium, recognizable clinically in a negative calcium balance or alternately in a lowered calcium content of the bones, while the function of the intestines and the kidneys is otherwise normal. Considering all aspects, the causes for the so-called "idiopathic calcium lithiasis" (= no basic disorder recognizable) cannot simply be based on variations of exogenous calcium and probably also not of magnesium.

Since from in vitro studies magnesium is considered as a low molecular weight inhibitor,[33] and since at least in a large urolithiasis subgroup a deficiency of magnesium can be demonstrated in the urine,[34] a high supply of magnesium with the drinking water seems to be desirable by many investigators. However, in the first place, mineral water almost never contains sufficient amounts of magnesium, as related to the content in calcium, and secondly, there is no proof that low magnesium in the urine during lithiasis indicates a deficiency of the total body magnesium or forms only an additional local phenomenon of other renal-tubular transport processes without feedback on the processes of nucleation. Finally, magnesium therapy of calcium lithiasis was found ineffective,[35] leaving open the true role of a low urinary magnesium in stone disease.

Recently, sodium in the diet, the distal-tubular transport of which in the kidney is closely associated with calcium,[36] has been considered as a determinant of the plasma phosphate.[37] The narrow linkage of the sodium and calcium transport mechanisms can be proven in cases of calcium lithiasis also.[38,39] In all groups of urolithiasis, plasma phosphate is uniformly lower in comparison to healthy subjects[34] (see also Table 3), but contrary to the above authors,[37] we found in the subgroup "renal hypercalciuria" additionally increased urinary sodium,[34] similar to the results obtained by others.[39] These contradictions render the interpretation of the role of exogenous sodium during urolithiasis difficult, but an elevated function of the parathyroid glands can definitely be ruled out as the cause of the hypophosphatemia.[34,39] In experiments with animals, some symptoms characteristic for calcium lithiasis of man are observed during exogenous phosphate deficiency: oxalate concrements in the urinary system, hypophosphatemia, hypercalciuria;[41] however, under similar conditions, others found low renal gluconeogenesis and ATP content,[42] indicators of a profound disturbance of metabolism. In addition, skeletal symptoms of rickets develop,[112,113] but the hyperphosphaturia characteristic for stone patients is missing. In particular, the last finding which is independent of the parathyroid glands and nephrogenous cAMP[34] lends little credibility to exogenous deficiency of phosphate as a causal factor of stone formation in man.

A phosphorus content of at least 0.8% or more in the diet has in rats a high, predictable potential for the development of calcium phosphate lithiasis, while for lower proportions of phosphorus, the comparatively high ratio of calcium vs. phosphorus and also magnesium is supposed to act inhibitory on lithiasis formation.[43] Even in such elegant animal experiments,[43] as in man,[14] no correlation was observed between the formation of stones and minerals in serum and the status of the parathyroid glands of the former animals is unknown.

2. Oxalate

The calcium precipitating effect of oxalate has no biological significance in the serum, since its oxalate concentration is very low.[44] Of the total quantity of oxalate of 100 to 800 mg ingested per day via the food intake,[45] approximately 5 to 6% are absorbed intestinally by healthy subjects[46] (7 to 10% of a synthetic meal).[112] The remaining oxalate precipitates with free calcium ions within the intestinal lumen to insoluble calcium oxalate and is excreted together with the feces. It is presently suspected that intestinal hyperabsorption of calcium is followed by a hyperabsorption of oxalate, since under these conditions, increasingly free oxalate ions are available intraluminally and probably diffuse passively along the concentration gradient of calcium through the intestinal wall.[48] Elevated urine oxalate in cases of unclassified calcium lithiasis has, however, not been found by numerous laboratories,[26,48,49] which seems to be incompatible with a key role for oxalate as the trigger of homogenous nucleation (see Section II). However, conditions in patients with documented calcium malabsorption and simultaneous hyperabsorption of oxalate (resections of the small intestine, intestinal inflammations, etc.) confirm the important role of oxalate. Due to the higher

Table 3
DATA OF CONTROLS AND PATIENTS WITH UROLITHIASIS, CHARACTERISTIC FOR THE VARIOUS SUBGROUPS

Age (years)	CONTROLS n = 34 ♂/♀ = 19/15 37/19—63	A - HC n = 15 ♂/♀ = 4/11 44/29—66	R - HC n = 23 ♂/♀ = 10/13 46/29—66	N - C n = 60 ♂/♀ = 34/26 43/20—71	UA - I n = 17 ♂/♀ = 15/2 55/40—69[c]	UA - II n = 12 ♂/♀ = 10/2[b] 52/39—63[b]	pHPT n = 20 ♂/♀ = 6/14 55/34—71[c]
Calcium/creatinine ratio[d] (mg/mg)	0.08/0.02 — 0.16	0.12/0.07 — 0.15[a]	0.20/0.16 — 0.25[c]	0.08/0.01 — 0.15	0.07/0.02 — 0.19	0.08/0.04 — 0.14	0.25/0.11 — 0.58[c]
Calcium/creatinine ratio[e] (mg/mg)	0.21/0.05 — 0.37	0.48/0.39 — 0.55[c]	0.38/0.19 — 0.93[c]	0.25/0.04 — 0.38[b]	0.16/0.06 — 0.35	0.12/0.05 — 0.26	0.43/0.20 — 0.77[c]
Sodium/creatinine ratio[d] (μequiv./mg)	101/29 — 216	103/38 — 297	152/68 — 303[b]	118/24 — 564	104/42 — 223	92/35 — 180	142/11 — 377[c]
Magnesium/creatinine ratio[d] (mg/mg)	45/16 — 101	37/25 — 61	49/27 — 91	37/10 — 77[a]	50/25 — 79	37/17 — 64	51/26 — 141[c]
Phosphate excretion[d] (μg/min)	279/44 — 765	308/100 — 542	351/81 — 752[a]	333/114 — 764	298/134 — 888	328/195 — 525	484/129 — 770[c]
pH[d]	6.26/5.29 — 7.91	6.33/5.40 — 7.41	6.25/5.28 — 7.20	6.43/5.58 — 7.38	5.35/5.00 — 6.67[c]	5.41/5.15 — 5.65[c]	6.71/5.30 — 7.86
Citrate excretion[f] (mg)	521/20 — 1041	446/198 — 835	454/73 — 1018	427/20 — 1103[a]	208/33 — 860[c]	329/86 — 539[b]	469/148 — 1100
Serum Calcium (mg/dℓ)	9.86/8.72 — 10.58	9.86/9.09 — 10.84	9.90/9.14 — 10.36	9.82/8.95 — 10.46	9.82/8.94 — 10.35	9.93/9.45 — 10.39	12.75/10.16 — 16.65[c]
Serum phosphate (mg/dℓ)	3.76/2.54 — 5.03	3.0/2.47 — 4.26[c]	3.23/2.50 — 4.91[c]	3.18/1.89 — 4.54[c]	2.84/2.05 — 4.22[c]	3.29/2.46 — 3.99[a]	2.40/1.49 — 4.67[c]
Phosphate threshold (mg/dℓ)	4.07/2.40 — 5.30	3.23/2.40 — 4.40[c]	3.13/2.05 — 4.90[c]	3.18/2.00 — 4.55[c]	2.90/1.80 — 4.20[c]	3.17/2.10 — 4.10[c]	1.89/0.80 — 2.80[c]
Serum PTH[g] (pg-equiv/mℓ)	201/<150 — 630	281/<150 — 555	256/<150 — 866	<150/<150 — 820	<150/<150 — 547	<150/<150 — 345	1624/555 — 7238[c]
Nephrogenous cAMP[d] (μM/g) creatinine	2.2/ — 0.6 — 6.3	1.6/ — 0.3 — 3.3	1.8/0.4 — 4.8	1.2/ — 0.7 — 4.3[b]	1.5/ — 0.01 — 2.8	2.1/ — 1.2 — 3.5	4.3/0.6 — 9.3[c]

Note: Values represent medium/range.

[a] $p < 0.05$.
[b] $p < 0.01$.
[c] $p < 0.001$ vs. controls using the test of Mann and Whitney.
[d] Values in the fasting urine.
[e] Values in the urine 3 hr after the oral calcium load.
[f] Values in the 24-hr urine.
[g] Parathyroid hormone determined by using bovine PTH (1-84) as standard and tracer, and the guinea pig antibovine PTH antibody RD 24 (Wellcome and Beckenham), final dilution 1:150,000 (significance not tested in this group).

molecular weight as compared with calcium, small increases of the urinary concentration of oxalate are already capable to shift the urine into the metastable range of saturation with respect to calcium oxalate. Measuring of the intestinal absorption of oxalate in all subgroups and simultaneous consideration of the clinical degree of seriousness of the disorder will be the precondition for a final elucidation of the interrelationships of oxalate and calcium in calcium lithiasis (see Section VI). Since oxalate is synthesized endogenously (probably about 50% of the total urine oxalate per day) from precursors (e.g., glyoxylic acid), the urine will never become completely free of oxalate due to dietary measures. This was demonstrated by us when feeding oxalate-free formula diets to healthy subjects.[50] In spite of reduced excretion of oxalate, the activity product of calcium oxalate increased in such subjects, possibly not solely due to the calcium content of the diets, but also as a consequence of consecutively increased absorption of calcium.[51]

3. Purine and Proteins

Recent investigations[52] indicate a relationship between exaggerated consumption of nucleoproteins, the resulting hyperuricosuria and high risk of stone formation. Hyperuricosuria may initiate heterogenous nucleation (see Section II) of calcium oxalate by lowering the formation product of the latter,[53] and it also causes precipitation of uric acid crystals, on the surface of which calcium oxalate crystals are formed by the principles of epitaxy. During a daily protein consumption of 48 g, healthy volunteers excrete 168 to 182 mg calcium in the urine, but during a protein consumption of 142 g 322 to 345 mg calcium.[28] The causes for this striking increase in urinary calcium and in the resulting risk of stone formation[27] are multifactorial: acid ash, metabolic acidosis, great amounts of D,L-methionine, sodium content, increase of glomerular filtration rate, inhibition of tubular reabsorption. A direct relationship is assumed between total acid production of the kidneys, sulfate excretion, and urinary calcium.[28] According to animal experiments, the production and excretion of sulfate seems to play the determinant role.[54]

4. Carbohydrates

Without other changes in the diet composition, the increased intake of refined carbohydrates results in a larger number of urines with elevated calcium concentration (> 9 mmol/ℓ, or 360 mg/ℓ).[29] Similar observations were reported earlier,[55] and probably, despite raised calcium excretion, magnesium excretion is not changed, resulting in a higher risk of stone formation[55,56] (see Section II).

5. Fatty Acids

Obviously, little is known about the relationship between the composition and amounts of exogenously administered fatty substances, the mineral metabolism, and stone formation. During a diet containing soybean oil and small amounts of cholesterol, a decrease of serum calcium was found in rats.[57] In one of our own experiments with rats using a similar diet, nephrocalcinosis developed in the juxta-medullar cortical region of the kidney, accompanied by low inorganic phosphate content in the renal tissues (cortex, medulla) and by hypomagnesemia.[58]

From these incompletely cited various possibilities to modify the mineral metabolism by exogenous factors, unfortunately, one cannot deduce if one or a combination of several of these factors necessarily result in stone formation. We also do not know if a long time of exposure to such factors is the prerequisite for the development of the latter.

V. ESTABLISHED DISORDERS OF CALCIUM METABOLISM WITH ASSOCIATED LITHIASIS

Except oxalosis, all have in common a manifest alteration of calcium homeostasis, but only in a few cases combined disturbances (calcium, phosphate, oxalate) are present. Not always is urolithiasis an accompanying symptom, especially not of hyperparathyroidism (only in about 70% of cases), of hypercalciuria due to malignancy, pseudohyperparathyroidism (= ectopic production of parathyroid hormone), or hypercalciuric rickets. To our knowledge, reliable indications of stone frequency in the latter three are not available. Stone formation is an expression either of a considerable oversaturation of the urine due to hypercalciuria, hyperphosphaturia, hyperoxaluria, or a lack of inhibitors (citrate), together with a less acidic urine (defective generation of hydrogen ions, loss of bicarbonate). Whereas etiology, pathophysiology, diagnosis, therapy, and prognosis are described in numerous handbooks of endocrinology and metabolism, in monographies and reviews, some particularities should be pointed out.

Medullary sponge kidney — This X-ray morphologically impressive disease, described a few decades ago,[59] probably is due to a multifactorial etiology (faulty oncogenic development, uric acid plugging in fetal life etc.), and therefore, possibly represents not a clinical entity.[60] According to some authors, it is more than incidentally frequently associated with primary hyperparathyroidism.[62,63] We found normal plasma levels and urinary excretion of cAMP in two cases from our stone clinic and in four cases from London,[114] although all individuals had renal type hypercalciuria (see Chapter 6) and urolithiasis resp. nephrocalcinosis. It is also assumed that long-standing hypercalciuria and nephrocalcinosis as in primary hyperparathyroidism may be the cause for formation of microcysts. Since this disease is diagnosed more frequently now, an intense investigation of the mineral metabolism, development of an appropriate animal model included, seems to be necessary.

Renal tubular acidosis type — This type of acidosis in its classical form is more frequent in children than in adults. In contrast, in adult stone-formers, signs of an incomplete manifestation of this disorder are found: defective urinary acidification despite normal acid-base values in the blood.[63] Characteristic clinical findings in these stone-formers are early onset, frequent stone recurrencies, and frequent necessity for surgical intervention. Such patients apparently suffer from a metabolically very active form of stone formation[64] in relation to those without signs of urinary hypoacidity.

Morbus Wilson — Many individuals with this disease, which frequently is associated more than with calcium urolithiasis in 16% of the cases also have a defective urinary acidification during an oral acid load, low Tm, and threshold concentration for bicarbonate, visualized by high urinary losses of the latter.[65] Hypercalciuria of the absorptive type (see Chapter 6) occurs rarely; hyperuricosuria occurs frequently.[65]

Hypercalciuric rickets — Hypercalciuria and profound disturbances of mineral metabolism with skeletal signs of osteomalacia were observed earlier.[66] Recently, in cases showing similar clinical signs, calcium lithiasis was found too.[67] Systematic, quantitative, histomorphometric data from the bone state of idiopathic stone-formers, classified on the basis of their calciuria (see Chapter 6) were not available until now; but there are some indications that the combination of bone disease in the presence of calcium urolithiasis, normal urinary cAMP, hyperphosphaturia, hypercalciuria, a combination described by Bordier[40] as hypercalciuria type 3, is identical with renal hypercalciuria. The latter apparently is more frequent than assumed until now, and we found no signs of secondary hyperparathyroidism[34] in contrast to other authors.[68]

VI. CLINICAL DIAGNOSIS OF THE METABOLIC SUBGROUPS OF CALCIUM UROLITHIASIS

According to the present state of knowledge in pathophysiology, calcium lithiasis has to be classified into various subgroups, based on the calcium excretion in the urine. The diagnostic procedure follows the checking sequence suggested by Pak et al.,[68] which we apply in a modified form.[69] It essentially includes evaluation of the excretion of calcium in the urine collected over a period of 24 hr (carried out without dietary restrictions), in the 2-hr fasting urine, and coarse assessment of the intestinal calcium absorption from determining the urinary calcium/creatinine ratio 3 hr after the oral administration of 1000 mg of elementary calcium in a test meal. Classification into one of the groups listed below is feasible. Two main groups of calcium excretion are recognized.

A. Hypercalciuria

This disorder is present if in the 24-hr urine, more than 300 mg of calcium (185; 66—309)* are excreted by men or more than 250 mg of calcium by women (189; 104—255), or more than 4 mg/kg body weight by either sex. At the present time, hypercalciuria is further subdivided into the following subgroups.

1. Resorptive Hypercalciuria (= Primary Hyperparathyroidism)

In most of the cases, an adenoma of one or sometimes several parathyroid glands forms the cause, in very rare instances a carcinoma. In the serum, calcium and parathyroid hormone are elevated and phosphate is lowered; in the fasting urine, the excretion of calcium and phosphate and nephrogenous cyclic AMP (cAMP) are elevated (see Table 3). The tubular transport maximum for phosphate (mg per 100 mℓ glomerular filtratation rate) is low and also the phosphate threshold concentration in serum (mg/dℓ). The same constellation is found in the 24-hr urine.

2. Renal Hypercalciuria

This disorder is characterized by increased calcium excretion in the fasting urine (see Table 3) resulting from diminished net tubular reabsorption, the cause of which is unknown. Contrary to Pak,[68] who observed an increased activity of the parathyroid glands caused by this urinary loss of calcium, we (see Table 3) and others[39] find no indications for secondary hyperparathyroidism in those individuals. In serum calcium fractions, parathyroid hormone and in urine cAMP are in the normal range. However, an increased excretion of phosphate is found in all subgroups (see Table 3) and reduced phosphate threshold as well.

3. Absorptive Hypercalciuria

Type I — This type of hypercalciuria is based on an increased calcium absorption in the small intestine, independent of the quantity of calcium in the food intake. Calcium is unchanged in the serum and in the fasting urine, but markedly elevated in the urine 3 hr after oral administration of calcium. Elevated levels of 1,25-dihydroxy-D in the blood are found in part of the patients only.[70]

Type II — In this type of hypercalciuria, an intestinal hyperabsorption of calcium and a hypercalciuria resulting from this fact exists during excessive intake of calcium, e.g., 1000 mg calcium per day, only. If the daily calcium quantity is reduced to 400 mg/day, excretion of calcium is normal.[68] So far, no definite proof exists whether the present dietary habits of the population justify a separate recognition of this type of hypercalciuria or not.

* Median; interdecile range — 80 of healthy subjects (n = 42).

B. Normocalciuria

In approximately 60% of the patients with calcium lithiasis, which are seen by us, no disturbance of the calcium excretion is found. The cause for the formation of stones in these cases seems to be a relative hypercalciuria due to the absence of inhibitors of stone formation, as which citrate[71] and magnesium (see Table 3), have been proved and what is speculated for pyrophosphate.[72] Additional possible causes for the formation of stones in these patients are an elevated urinary excretion of uric acid, sodium urate,[53] and oxalate. The latter is found frequently in patients with repeated stone recurrencies and is designated as so-called "mild hyperoxaluria".[73]

C. Uric Acid Lithiasis and the Intermediate Group

Except for the low pH (< 5.5) of the urine, known to all clinicians, metabolic criteria of this so-called "idiopathic uric acid lithiasis" (uric acid lithiasis during primary gout being the opposite) have been investigated and the opinions on the renal acidogenesis in these patients are divergent.[74,75] We showed that particularly urinary uric acid is normal and serum uric acid only slightly elevated,[69] that urinary citrate is the lowest observed in urolithiasis,[24] and that beside that phenomenon a hyperphosphaturia is present as in the subgroups of idiopathic calcium lithiasis mentioned above (see Table 3; UA-I).

We succeeded in proving the existence of an intermediate group between the two extreme populations forming calcium-, resp. uric acid lithiasis. The main distinguishing features of this group are episodes of formation of either pure uric acid stones or pure calcium oxalate stones while the urinary pH is permanently deeply (<5.5) acidic (see Table 3; UA-II). Urinary citrate assumes an intermediate position also.[34] The existence of this group was recently confirmed by other investigators[115] or it can be deduced from the composition of the collective patients described by others.[76]

D. Comments Concerning the Function of the Parathyroid Glands in Idiopathic Urolithiasis

Normal or even reduced function of the parathyroid glands (see subgroup normocalciuria; Table 3), expressed in terms of nephrogenous cAMP, as well as by serum parathyroid hormone, measured in a radioimmunoassay with antigenic recognition sites for all circulating fragments of the human hormone molecule (so-called "intact hormone assay") has not been elaborated clearly in the past. From this complex, important questions affecting future research arise:

1. Which factors are responsible for a reduced tubular phosphate absorption during low tubular generation of cAMP which appears due to a deficiency in parathyroid hormone?
2. Which factors suppress the secretion of parathyroid hormone or of its bioactive fragments, and why are there until now no indications for defective activity of parathyroid hormone in other target organs (bones, gastrointestinal tract) of stone-formers?
3. If the same metabolic combination of the mineral metabolism is common to all forms of idiopathic urolithiasis, including uric acid lithiasis, which factors determine them once the development of a deeply acidic urinary pH, predisposing for the formation of concrements from free uric acid, and which ones determine on the other hand the basis for the formation of calcium concrements during a urinary pH higher than 5.5 in most of the cases?

We hope, that the intermediate group included in Table 3 (UA-II; see above) will prove to be a useful aid during coming work in the field of metabolic interrelations.

These above facts and the findings mentioned in the current literature concerning the direct influences on the activity of the parathyroid glands, independent of free calcium ions in the serum,[77-79] indicate, that the small intestine may be of prime importance in understanding stone-formers' calcium regulation. The various more or less elaborate techniques for measuring intestinal absorption of calcium are, therefore, briefly explained below.

VII. TECHNIQUES FOR MEASUREMENT OF INTESTINAL ABSORPTION OF CALCIUM (CaA)

Measurement of the calcium balance represents the oldest method to determine the CaA.[80] For this investigation the patient needs to be hospitalized for 14 days. After achieving an equilibrium of the metabolism by furnishing a diet constant especially in calcium and phosphorus, the calcium received via the food intake and the calcium excreted with the feces and the urine are measured. The balance is intake minus fecal output and urinary output, the net absorption of calcium is calculated from the intake minus fecal excretion. This procedure leads to an underestimation of calcium absorption, since the fecal calcium from endogenous sources (= secretion of absorbed calcium into the intestine) is not quantitated separately.

For the calculation of the true CaA endogenous calcium needs, therefore, to be measured also by means of additional administration of radioactive calcium. In general, this method is very sophisticated and offers no advantages in comparison with methods described below, particularly if additionally tracer calcium needs to be utilized.

Determination of the calcium/creatinine ratio in the urine after oral administration of calcium (see above) represents a coarse, indirect yardstick for the CaA. This method, however, allows only to discriminate between an increased and a normal CaA in most of the cases. Since the effort required is low and there is no need for tracer administration, the method is well suited for routine measurements where no exact determination of the degree of CaA is required.

Determination of the CaA with radioactive calcium is presently used extensively, since these methods are simple as compared with the first described method and can be performed in an ambulatory manner in most of the cases.

Procedures utilized are

1. Preoral administration of a single tracer (^{47}Ca or ^{45}Ca) and measuring of the radioactivity.
   A. In the feces:[81] for a complete collection, feces need to be collected for at least one full week, until the excretion of calcium has dropped below 1% of the administered dose, or a nonabsorbable marker, e.g., ^{51}Cr needs to be supplied additionally for correction purposes. The method described disregards the endogenous fecal calcium and the CaA is, therefore, estimated too low.
   B. By means of a whole body counter:[81] this method determines the percentage of the administered tracer retained in the body after 8 days. The observed retention is affected by the kinetics of calcium uptake by bone and its subsequent excretion.
   C. In the blood almost immediately or up to 48 hr after application of the tracer:[82] the same comments as in (B.) apply concerning the precision of this method. Due to the simplicity it is, however, well suited for routine determination of the CaA.
2. Use of two tracers, i.e., the oral and intravenous administration of radioactive calcium. This technique is based on the assumption that fractional loss of the

absorbed portion of the oral dose is identical with that of the i.v. dose and also permits quantification of the fractional dose absorbed. The following methods of application are used.

A. The same isotope is administered in an interval of a few days. This methodology is based on the assumption that the calcium metabolism of the individual does not change during this period of time.

B. Simultaneous application of two different tracers (^{47}Ca and ^{45}Ca). Either the ratio of the two isotopes measured (= fraction of oral dose) in the urine and serum or the total counts, measured by means of a whole body counter after 6 hr, i.e., after complete absorption, is used as a measure to determine the kinetics of the absorption, the radioactivity applied is measured repeatedly in the blood. Using the so-called "deconvolution method", the absorption during the different time intervals and the fractional CaA are determined.[83]

According to recent investigations, the latter method is presently regarded as the best of all tracer methods for determining the CaA.[84,85]

Additionally, this method is comparatively simple and, particularly in the 3-hr short form, suitable for out-patient investigation. This method should, therefore, be preferred in further investigation of the causes of a changed CaA. As in the other methods using tracers, one must assume an undisturbed passage from the mouth through the stomach to the small intestine for the validity of this method. Such a restriction is, however, necessary only if one intends to perform measurements exclusively at the site of the calcium absorption in the small intestine, and influences of the intestine, e.g., the motility, have to be excluded. Under these circumstances, the following method might be indicated.

With the perfusion technique, a triple-lumen tube is placed under fluoroscopy control into the small intestine and solutions of known ionic concentration, with and without ^{47}Ca, together with a nonabsorbable marker (polyethylene glycol) are infused via the upper lumen and, subsequently, aspired through lumina lower down for analysis of their mineral and marker content.[86] After a correction of the water flux by means of the marker, the differences in the calcium concentrations of the infusate and the aspirate can be used to calculate the net absorption and the unidirectional calcium fluxes in the different segments of the small intestine. It needs to be considered, however, that the particular flux direction (intraluminal or extraluminal) depends on the concentration of the solution used for the infusion. This method is the most exact for the direct measurement of the CaA at the site of absorption. On the other hand, the method is very elaborate and uncomfortable for the patient.

VIII. FACTORS INFLUENCING INTESTINAL CALCIUM ABSORPTION

Hormones, food components, and intestinal factors are known to affect the CaA. Since different techniques of determination are used under varying circumstances and, also, since many of these factors mentioned in man are partly open for discussion or not known. 1,25-dihydroxy-D, which is formed in the kidneys, represents the most important hormonal regulator for the CaA. The hormone increases the CaA either via triggering the synthesis of a specific, calcium-binding protein in the brushborder membranes of the intestinal mucosal cell, or via still unknown mechanisms. The existence of this protein has so far been proved in rats and chicken only. The formation of 1,25-dihydroxy-D is correlated positively with the serum parathyroid hormone level and negatively with the serum inorganic phosphate and calcitonin.[87]

The hypothesis, that hypophosphatemia, caused by a reduced phosphate threshold or a deficiency in phosphate, results in increased synthesis of 1,25-dihydroxy-D, has been proved so far only in animals[88] and in women.[89] This relationship is denied by other investigators,[90] particularly in patients with calcium lithiasis. The CaA is additionally affected by 5,6-trans-25-hydroxy-D and by the synthetic 1-a-hydroxy-D.

Concerning the influence of parathyroid hormone on the CaA, an increased absorption during primary hyperparathyroidism[91] and reduced absorption after surgical removal of the parathyroid glands was observed. In animals, which showed a reduced CaA after parathyroidectomy, exogenous administration of parathyroid hormone restored CaA to normal and, in intact animals, the CaA became elevated by parathyroid hormone.[92] A permissive influence of parathyroid hormone on the increased rate of synthesis of 1,25-dihydroxy-D is presently considered the underlying mechanism of action for this phenomenon.[87]

Reduced CaA in streptozotocin-treated diabetic rats[93] and in diabetic humans[94] is normalized by insulin, most likely via a normalization of the lowered synthesis of 1,25-dihydroxy-D. An influence on CaA of healthy men has not been investigated up to now.

The effect of glucagon on the CaA is regarded divergently, since it is supposed to increase[95] or to lower the CaA.[96]

In animals after acute administration of gastrin, an inhibition,[97] stimulation,[98] or no effect[99] on the CaA is observed. In healthy men, a reduction[100] of the CaA is noticed. Long-standing hypergastrinemia, as it occurs, e.g., after highly selective vagotomy, results in no change in animals,[99] but results in a raised CaA in patients with duodenal ulcers.[101] It is uncertain, whether this change is due to a direct influence of gastrin on the CaA or to an indirect influence via its trophic actions on intestinal mucosa.[102] Somatostatin, which controls the secretion of several peptide hormones, is supposed to have an inhibiting influence on the CaA in man,[103] but no effect in animals.[104]

The following is known about the various food components (see also Chapter 4): consumption of animal protein was found to raise CaA,[105] but recent investigations were unable to confirm such an effect.[106] From perfusion studies it is known, that sugars, e.g., lactose, xylose, or sorbitol increase the CaA in men.[107] Fats in the form of triglycerides with medium length of the molecular chain increase the CaA in rats,[108] while causing no reduction of the CaA in man, contrary to long chain type triglycerides.[109]

Of the different intestinal factors, intraluminal pH-changes might have an influence on the CaA, but are rated low, however.[110] The effects of changes in the intestinal motility, e.g., emptying of the stomach, transport in the small intestine, etc., are still largely unknown. If CaA in stone disease varies, dependent on one or more of the above factors is now being studied by this laboratory.

IX. CONCLUDING REMARKS

This overview to questions of current interest in investigations of the calcium metabolism of stone-formers is necessarily incomplete. It cannot, e.g., comment on results obtained with the various therapeutic regimens which elaborate highly important insights into calcium metabolism obtained either from retrospective or prospective studies, especially when comparing data of individuals undergoing repeat stone recurrency or becoming free of stones by the drug. At this stage, the reader should be cautious regarding the impact of calcium metabolism in general as an aggravating factor to idiopathic urolithiasis, and our own impression is that, in fact, it is overestimated: from a double-blind trial on the effect of thiazide swallowed over 1 year, we learned that there was a marked reduction of urinary calcium in participants taking the drug and not as much in those taking placebo, but yet the stone recurrency rate was the same in both populations.[111] This probably is a simple instance leading to assume that further stone research is faced with many more factors, besides calcium, if we want to control the disease at some future day.

REFERENCES

1. Sierakowsky, R., Finlayson, B., Landes, R. B., Finlayson, C. D., and Sierakowsky, N., The frequency of urolithiasis in hospital discharge diagnoses in the United States, *Invest. Urol.*, 15, 438, 1978.
2. Ljunghall, S. and Hedstrand, H., Epidemiology of renal stones in a middle-aged male population, *Acta Med. Scand.*, 197, 439, 1975.
3. Schneider, H. J. and Hesse, A., Zur Epidemiologie des Harnsteinleidens, *Therapiewoche*, 26, 5881, 1976.
4. Boshammer, K., Büscher, H. K., Cottet, J., Gaca, A., Henning, O., and van der Vuurst de Vries, J. H. J., Die Steinerkrankungen, in *Handbuch der Urologie*, Vol. 10, Alken, C. E., Dix, V. W., Weyrauch, H. M., and Wildbolz, E., Eds., Springer-Verlag, Basel, 1961, 34.
5. Johnson, C. M., Wilson, D. M., O'Fallon, W. M., Malek, R. S., and Kurland, R. T., Renal stone epidemiology: a 25-year study in Rochester, Minnesota, *Kidney Int.*, 16, 624, 1979.
6. Schwille, P. O., Scholz, D., and Sigel, A., Urolithiasis — ein Überblick zur Pathophysiologie, Diagnostik und Therapie, *Wien. Klin. Wochenschr.*, 92, 411, 1980.
7. Pak, C. Y. C., Calcium urolithiasis: is it analogous to bone formation?, *Calcif. Tiss. Res.*, 26, 195, 1978.
8. Lian, J. B., Prien, F. L., Glimcher, M. J., and Gallop, P. M., The presence of protein-bound carboxyglutamic acid in calcium-containing renal calculi, *J. Clin. Invest.*, 59, 59, 1151, 1977.
9. Pobisch, R., Urolithiasis bei Hund und Katze, *Wien. Tieraerztl. Monatsschr.*, 56, 93, 1969.
10. Wyngaarden, J. B. and Kelley, W. N., Miscellaneous forms of hypouricemia, in *Gout and Hyperuricemia*, Wyngaarden, J. B. and Kelley, W. N., Eds., Grune & Stratton, New York, 1976, 415.
11. Coe, F. L., Parks, J. H., and Moore, E. S., Familial idiopathic hypercalciuria, *N. Engl. J. Med.*, 300, 337, 1979.
12. Sperling, O., Weinberger, A., and Oliver, J., Hypouricemia, hypercalciuria, and decreased bone density: a hereditary syndrome, *Ann. Int. Med.*, 80, 482, 1974.
13. Sperling, O., Hereditary hyperuricosuric urolithiasis, Proc. 4th Int. Symp. Urolith. Res., Williamsburg, Va., 1980.
14. Scholz, D., Schwille, P. O., Ulbrich, D., Bausch, W. M., and Sigel, A., Composition of renal stones and their frequency in a stone clinic, relationship to parameters of mineral metabolism in serum and urine, *Urol. Res.*, 7, 161, 1979.
15. Gallagher, J. C., Riggs, B. L., Jerpbak, C. M., and Arnaud, C. D., The effect of age on serum immunoreactive parathyroid hormone in normal and osteoporotic women, *J. Lab. Clin. Med.*, 95, 373, 1980.
16. Milhaud, G., Benezech-Leferre, M., and Moukhtar, M. S., Deficiency of calcitonin in age-related osteoporosis, *Biomedicine*, 26, 272, 1978.
17. Chesnut, C. H., Baylink, D. J., Sisom, K., Nelp, W. B., and Roos, B. A., Basal plasma immunoreactive calcitonin in postmenopausal osteoporosis, *Metabolism*, 29, 559, 1980.
18. Valyasevi, A. and Dhanamitta, S., Field preventive program for bladder stone disease in Thailand, Proc. 4th Int. Symp. Urolith. Res., Williamsburg, Va., 1980.
19. Hering, F., Calciumoxalatnephrolithiasis als Berufskrandheit, *Urol. Ausg. A*, 16, 227, 1977.
20. Bateson, E. M., Renal tract calculi and climate, *Med. J. Aust.*, 2, 111, 1973.
21. Poskitt, E. M. E., Cole, T. J., and Lawson, D. E. M., Diet, sunlight, and 25-hydroxyvitamin D in healthy children and adults, *Br. Med. J.*, 1, 221, 1979.
22. Holmberg, J. and Larsson, A., Seasonal variation of vitamin D_3 and 25-Hydroxyvitamin D_3 in human serum, *Clin. Chim. Acta*, 100, 173, 1980.
23. Hallson, P. C., Kassidas, G. P., and Rose, G. A., Seasonal variation in urinary calcium and oxalate in normal subjects and patients with idiopathic hypercalciuria, in *Urolithiasis Research*, Fleisch, H., Robertson, W. G. Smith, L. H., and Vahlensieck, W., Eds., Plenum Press, New York, 1976, 459.
24. Robertson, W. G., Gallagher, J. C., Marshall, D. H., Peacock, M., and Nordin, B. E. C., Seasonal variations in urinary excretion of calcium, *Br. Med. J.*, 4, 436, 1974.

25. Robertson, W. G., Peacock, M., Marshall, R. W., Speed, R., and Nordin, B. E. C., Seasonal variation in the composition of urine in relation to calcium stone formation, *Clin. Sci. Mol. Med.,* 49, 597, 1975.
26. Schwille, P. O., Paulus, M., Scholz, D., Sigel, A., and Wilhelm, E., Urin-Oxalat bei rezidivierender Calcui-Urolithiasis ohne und mit Überfunktion der Nebenschilddrusen und bei Gesunden, *Urol. Ausg. A,* 18, 215, 1979.
27. Robertson, W. G., Heyburn, P. J., Peacock, M., Hanes, F. A., and Swaminathan, R. S., The effect of high animal protein intake on the risk of calcium stone formation in the urinary tract, *Clin. Sci.,* 57, 285, 1979.
28. Schuette, S. A., Zemel, M. B., and Linkswiler, H. M., Studies on the mechanism of protein-induced hypercalciuria in older men and women, *J. Nutr.,* 110, 305, 1980.
29. Thom, J. A., Morris, J. E., Bishop, A., and Blacklock, N. J., The influence of refined carbohydrate on urinary calcium excretion, *Br. J. Urol.,* 50, 459, 1978.
30. Donaldson, D., Pryce, J. D., Rose, G. A., and Tovey, J. E., Tap water calcium and its relationship to renal calculi and 24 hr urinary calcium output in Great Britain, *Urol. Res.,* 7, 273, 1979.
31. Dauncy, M. J. and Widdowson, E. M., Urinary excretion of calcium, magnesium, sodium, and potassium in hard and soft water areas, *Lancet,* 1, 711, 1972.
32. Landes, R. R., Melnick, J., Sierakowsky, R., and Finlayson, B., An inquiry into relation between water hardness and the frequency of urolithiasis, in *Nutritional Imbalances in Infant and Adult Disease: Minerals, Vitamin D and Cholesterol,* Seeling, M. S., Ed., Spectrum Publ., Jamaica, N.Y., 1977, 9.
33. Smith, L. H., Meyer, J. L., and McCall, J. T., Chemical nature of crystal inhibitors isolated from human urine, in *Urinary Calculi,* Cifuentes Delatte, L. Rapado, A., and Hodgkinson, A., Eds., S. Karger, Basel, 1973, 318.
34. Schwille, P. O., Scholz, D., Schwille, K., Engelhardt, W., Goldberg, I., and Sigel, A., Metabolic features of different subgroups of urolithiasis with special regard to citrate in urine, serum, ultrafiltrate, and the function of parathyroid glands, *Nephron,* in press, 1982.
35. Fetner, C. D., Barilla, D. E., Townsend, J., and Pak, C. Y. C., Effects of magnesium oxide on the crystallization of calcium salts in urine in patients with recurrent nephrolithiasis, *J. Urol.,* 120, 399, 1980.
36. Ullrich, K. J., Frömter, E., and Murer, H., Prinzipien des epithelialen Transports in Nieren und Darm, *Klin. Wochenschr.,* 57, 977, 1979.
37. Schellenberg, B., Tschöpe, W., Ritz, E., Wesch, H., and Schlierf, G., Urinary sodium excretion in renal stone formers, *Klin. Wochenschr.,* 58, 575, 1980.
38. Schwille, P. O., Schlenk, I., Samberger, N. M., and Bornhof, C., Fasting urinary excretion of magnesium, calcium, and sodium in patients with renal calcium stones, *Urol. Res.,* 4, 33, 1976.
39. Sutton, R. A. L. and Walker, R. V., Response to hydrochlorothiazide and acetazolamide in patients with calcium stones, *N. Engl. J. Med.,* 302, 709, 1980.
40. Bordier, P., Ryckewart, A., Gueris, J., and Rasmussen, H., On the pathogenesis of so-called idiopathic hypercalciuria, *Am. J. Med.,* 63, 398, 1977.
41. Werness, P. G., Knox, F. G., and Smith, L. H., Low phosphate diet in rats: a model for calcium oxalate lithiasis, in Proc. 4th Int. Symp. Urolith. Res., Williamsburg, Va., 1980.
42. Kreusser, W. J., Descoeudres, C., Oda, Y., Massry, S. G., and Kurokawa, K., Effect of phosphate depletion on renal gluconeogenesis, *Min. Electr. Metab.,* 3, 312, 1980.
43. Chow, F. H. C., Taton, G. F., Boulay, J. P., Lewis, L. D., Remmenga, E. E., and Hamar, D. W., Effect on dietary calcium, magnesium and phosphorus on phosphate urolithiasis in rats, *Invest. Urol.,* 17, 273, 1980.
44. Akcay, T. and Rose, G. A., The renal and apparent oxalate in plasma, in *Oxalate in Human Biochemistry and Clinical Pathology,* Rose, G. A., Robertson, W. G., and Watts, R. W. E., Eds., Wellcome, London, 1979, 82.
45. Zarembski, P. M. and Hodgkinson, A., The determination of oxalic acid in the food, *Analyst, (London),* 87, 698, 1962.
46. Archer, H. E., Dormer, A. E., Scowen, E. F., and Watts, R. W., Studies on the urinary excretion of oxalate by normal subjects, *Clin. Sci.,* 16, 405, 1957.
47. Hodgkinson, A., Evidence of increased oxalate absorption in patients with calcium-containing renal stones, *Clin. Sci. Mol. Med.,* 54, 291, 1978.
48. Revusova, V., Zvara, V., and Gratzlova, J., Urinary oxalate excretion in patients with urolithiasis, *Urol. Int.,* 26, 277, 1971.
49. Butz, M., Teupe, B., and Kohlbecker, G., Oxalsäure im Serum und Urin von Rezidiv-Oxalatsteinbildungen, *Ver. Btsch. Ges. Urol.,* 29, 308, 1978.
50. Scholz., Paulus, M., Schwille, P. O., and Sigel, A., Über den Einfluss verschiedener Formula-Diäten auf die Aktivitatsprodukte steinbildender Substanzen im Urin von Gesunden, in *Pathogenese und Klinik der Harnsteine V,* Gasser, G. and Vahlensieck, W., Eds., Springer — Verlag, Basel, 1977, 342.

51. Scholz, D., Schwille, P. O., and Sigel, A., Responses of gastrointestinal hormones and intestinal calcium absorption during an oral carbohydrate meal, Proc. 4th Int. Symp. Urolith. Res., Williamsburg, Va., 1980.
52. Coe, F. L., Moran, E., and Kavalich, A. G., The contribution of dietary purine overconsumption to hyperuricosuria in calcium oxalate stone formation, J. Chron. Dis., 29, 793, 1976.
53. Pak, C. Y. C., Barilla, D. E., Holt, K., Brinkley, L., Tolentino, R., and Zerwekh, J., Effect of oral purine load and allopurinol on the crystallization of calcium salts in urine of patients with hyperuricosuric calcium urolithiasis, Am. J. Med., 65, 593, 1978.
54. Whiting, S. J. and Draper, H. H., The role of sulphate in the calcuria of high protein diets in adult rats, J. Nutr., 110, 212, 1980.
55. Lemann, J. Jr., Litzow, J. R., and Lennon, E. J., Possible role of carbohydrate-induced calciuria in calcium oxalate kidney stone formation, N. Engl. J. Med., 280, 232, 1969.
56. Lemann, J. Jr., Lennon, E. J., Piering, W. R., Prien, E. L., and Ricanati, E. S., Evidence that glucose ingestion inhibits net renal tubular reabsorption of calcium and magnesium in man, J. Lab. Clin. Med., 75, 578, 1970.
57. Diggs-Pace, R., Hsu-Chen, L., and Prothro, J., Effect of dietary calcium and fat on calcium and lipid content of rat tissue, Nutr. Rep. Int., 7, 121, 1973.
58. Schwille, P. O., Brandt, G., Brunner, P., Ulbrich, D., and Kömpf, W., Pankreasinseln, Plasma-Glucagon und renale Verkalkungen unter verschiedener Grunddiät bei der Ratte, Urol. Ausg. A, 14, 306, 1975.
59. Lenarduzzi, G., Reporto pielografico poca commune: Dilatazione della vie urinarie intrarenali, Radiol. Med., 26, 346, 1939.
60. Ekström, T., Engfeldt, B., Lagergren, C., and Lindwall, N., Medullary Sponge Kidney: A Röntgenologic, Clinical and Histopathologic and Biophysical Study, Almquist and Wiksell, Stockholm, 1959.
61. Gremillion, K. H., Kee, J. W., and McIntosh, D. A., Hyperparathyroidism and medullary sponge kidney. A chance relationship?, JAMA, 237, 799, 1977.
62. Rao, S. D., Frame, B., Block, M. A., and Parfitt, A. M., Primary hyperparathyroidism: a cause of hypercalciuria and renal stones in patients with medullary sponge kidney, JAMA, 237, 1353, 1977.
63. Backman, U., Danielson, B. G., Johansson, G., Ljunghall, S., and Wikeström, B., Incidence and clinical importance of renal tubular defects in recurrent renal stone formers, Nephron, 25, 96, 1980.
64. Dahlberg, P. J., van den Bergh, C. J., Kurtz, S. B., Wilson, D. M., and Smith, L. H., Clinical features and management of cystinuria, Mayo Clin. Proc. 52, 533, 1977.
65. Wiebers, D. O., Wilson, D. M., McLeod, R. A., and Goldstein, N. P., Renal stones in Wilson's disease, Am. J. Med., 67, 249, 1979.
66. Dent, C. E., Rickets and osteomalacia, nutritional and metabolic, Proc. R. Soc. Med., 63, 401, 1970.
67. Salti, J. S. and Hemady, K., Hypercalciuric rickets: a rare cause of nephrolithiasis, Nephron, 25, 222 1980.
68. Pak, C. Y. C., Ohata, M., Lawrence, E. C., and Snyder, W., The hypercalciurias: causes, parathyroid function, and diagnostic criteria, J. Clin. Invest., 54, 387, 1974.
69. Scholz, D., Schwille, P. O., and Sigel, A., Zur metabolischen Klassifikation und Metaphylaxe der haufigsten Lithiasisformen, Urol. Ausg. A, 19, 202, 1980.
70. Kaplan, R. A., Haussler, M. R., Deftos, L. J., Bone, H., and Pak, C. Y. C., The role of 1a,25-dihydroxyvitamin D the mediation of intestinal hyperabsorption of calcium in primary hyperparathyroidism and absorptive hypercalciuria, J. Clin. Invest., 59, 756, 1977.
71. Schwille, P. O., Scholz, D., Paulus, M., Engelhardt, W., and Sigel, A., Citrate in daily and fasting urine. Results of controls, patients with recurrent idiopathic calcium urolithiasis, and primary hyperparathyroidism, Invest. Urol., 16, 457, 1979.
72. Fleisch, H. and Bisaz, S., Isolation from urine of pyrophosphate, a calcification inhibitor, Am. J. Physiol., 203, 671, 1962.
73. Robertson, W. G., Peacock, M., Heyburn, P. J., Marshall, R. W., Rutherford, A., Williams, R. E., and Clark, P. B., The significance of mild hyperoxaluria in calcium stone-formation, in Oxalate in Human Biochemistry and Clinical Pathology, Rose, G. A., Robertson, W. G., and Watts, R. W. E., Eds., Wellcome, London, 1979, 173.
74. Wrong, O., Urinary hydrogen ion excretion in patients with uric acid calculi, Proc. R. Soc. Med., 79, 313, 1966.
75. Henneman, P. H., Walloch, S., and Dempsey, E. F., The metabolic defect responsible for uric acid stone formation, J. Clin. Invest., 41, 537, 1962.
76. Coe, F. L., Strauss, A. L., Tembe, V., and LeDun, S., Uric acid saturation in calcium nephrolithiasis, Kidney Int., 17, 662, 1980.
77. Altenaler, E., Dietel, M., Dorn, G., and Montz, R., The effect of 1,25-Dihydroxycholecalciferol on the parathyroid hormone secretion of porcine parathyroid glands and human parathyroid adenomas in vitro, Acta Endocrinol., 86, 533, 1977.

78. Canterbury, J. M., Leman, S., Claflin, A. J., Henry, H., and Norman, A., Inhibition of parathyroid hormone secretion by 25-hydroxycholecalciferol and 24,25-dihydroxycholecalciferol in the dog, *J. Clin. Invest.*, 61, 1375, 1978.
79. Windeck, R., Brown, E. M., Gardner, D. G., and Auerbach, G. D., Effect of gastrointestinal hormones on isolated bovine parathyroid cells, *Endocrinology*, 103, 2020, 1978.
80. Lutwak, L., Tracer studies of intestinal calcium absorption in man, *Am. J. Clin. Nutr.*, 22, 771, 1969.
81. Blau, M., Spencer, H., Swernow, J., and Laszlo, D., Utilization and intestinal excretion of calcium in man, *Science*, 120, 1029, 1954.
82. Agnew, J. E., Kehayoglou, A. K., and Holdsworth, C. D., Comparison of three isotopic methods for the study of calcium absorption, *Gut*, 10, 590, 1969.
83. Birge, S. J., Peck, W. A., Berman, M., and Whedon, G. D., Study of calcium absorption in man: a kinetic analysis and physiologic model, *J. Clin. Invest.*, 48, 1705, 1969.
84. Raymakers, J. A., Roelofs, J. M. M., Duursma, S. A., and Visser, W. J., Measurement of three-hour calcium absorption by double isotope technique, *Neth. J. Med.*, 18, 191, 1975.
85. Wootton, R. and Reeve, J., The relative merits of various techniques for measuring radiocalcium absorption, *Clin. Sci.*, 58, 287, 1980.
86. Fordtran, J. S., Segmental perfusion techniques, *Gastroenterology*, 56, 987, 1969.
87. Norman, A. W., 1,25-dihydroxyvitamin D_3 and 24,25-dihyroxyvitamin D_3: key components of the Vitamin D endocrine system, *Contr. Nephrol.*, 18, 1, 1980.
88. Hughes, M. R., Drumbaugh, P. F., Haussler, M. R., Weredal, J. E., and Baylink, D. J., Regulation of serum 1a,25-dihydroxyvitamin D_3 by calcium and phosphate in the rat, *Science*, 190, 578, 1975.
89. Gray, R. W., Wilz, D. R., Caldas, A. E., and Lemann, J., The importance of phosphate in regulating plasma 1a,25-$(OH)_2$-vitamin D levels in humans: studies in healthy subjects, in calcium stoneformers and in patients with primary hyperparathyroidism, *J. Clin. Endocr. Metab.*, 45, 299, 1977.
90. Barilla, D. E., Zerwekh, J. E., and Pak, C. Y. C., A critical evaluation of the role of phosphate in the pathogenesis of absorptive hypercalciuria, *Min. Electr. Metab.*, 2, 302, 1979.
91. Nordin, B. E. C., *Metabolic Bone and Stone Disease*, Churchill Livingstone, London, 1973.
92. Cramer, C. F., Participation of parathyroid glands in control of calcium absorption in dogs, *Endocrinology*, 72, 192, 1963.
93. Schneider, L. E., Nowosielski, L. M., and Schedl, H. P., Insulin treatment of diabetic rats: effects on duodenal calcium absorption, *Endocrinology*, 100, 67, 1977.
94. Mirouze, J. and Monnier, I., Etude critique des techniques d'absorption du radiocalcium, *J. Urol. Nephrol.*, 12, 921, 1975.
95. Hubel, K. A., Effects of secretin and glucagon on intestinal transport of ions and water in the rat, *Proc. Soc. Exp. Biol. Med.*, 139, 656, 1972.
96. Barbezat, G. O. and Grossmann, M. I., Effect of glucagon on water and electrolyte movement in jejunum and ileum of dog, *Gastroenterology*, 762, 1971.
97. Bynum, T. E., Jacobson, E. D., and Johnson, L. R., Gastrin inhibition of intestinal abosrption in dogs, *Gastroenterology*, 61, 858, 1971.
98. Fujita, T., Effect of tetragastrin on calcium absorption from rat duodenal loop in vivo, *Endocrinol. Jpn.*, 20, 407, 1973.
99. Engelhardt, W., Grohmann, C., Schwille, P. O., and Scholz, D., Effects of vagotomy on duodenal calcium abosrption in rats, *Acta Endocrinol. Suppl.*, 225, 335, 1979.
100. Scholz, D. and Schwille, P. O., Gastrin and intestinal calcium absorption in man, *J. Min. Electr. Metab.*, 7, 76, 1982.
101. Scholz, D., Schwille, P. O., Engelhardt, W., and Morcinietz, C., Calcium metabolism following proximal gastric vagotomy in man, in *Vitamin D Basic Research and Its Clinical Application*, Norman, et al., Eds., Gruyter, Berlin, 1979, 1019.
102. Dembinski, A. B. and Johnson, L. R., Growth of pancreas and gastrointestinal mucosa in antrectomized and gastrin-treated rats, *Endocrinology*, 105, 769, 1979.
103. Scholz, D. and Schwille, P. O., Somatostatin and intestinal calcium abosrption in man, *Metabolism*, 27, (Suppl. 1), 1349, 1978.
104. Gustavsson, S., Johansson, H., Jung, B., and Lundqvist, G., Gastrointestinal motility and absorption of calcium during infusion of somatostatin in the rat, *Digestion*, 19, 170, 1979.
105. Licata, A. A., Bou, E., and Bartter, F. C., Effects of dietary protein on calcium metabolism in normal subjects and in patients with osteoporosis, 57th Annual Session of the Americal College of Physicians, Philadelphia, Pa., 1976.
106. Spencer, H., Kramer, L., Osis, D., and Norris, C., Effect of a high protein (meat) intake on calcium metabolism in man, *Am. J. Clin. Nutr.*, 31, 2167, 197.
107. Pansu, D., Chapuy, M. C., and Vign, n, G., Actions des certains sucres sur l'absorption du calcium, *Rev. Rhum.*, 38, 533, 1971.

108. **Kehayoglou, A. K., Williams, H. S., Whimster, W. F., and Holdsworth, C. D.**, Calcium absorption in the normal, bile duct ligated and cirrhotic rat, with observations on the effect of long and medium-chain triglycerides, *Gut,* 9, 597, 1968.
109. **Agnew, J. E. and Holdsworth, C. D.**, The effect of fat on calcium absorption from a mixed meal in normal subjects, patients with malabsorptive disease and patients with a partial gastrectomy, *Gut,* 12, 973, 1971.
110. **Cramer, C. F.**, Effect of Ca/P ratio and pH on calcium and phosphorus absorption from dog gut loops in vivo, *Can. J. Physiol. Pharmacol.,* 46, 171, 1968.
111. **Scholz, D. and Schwille, P. O.**, Thiazid-Wirkungen bei idiopathischer Calcium — Urolithiasis. Ergebnisse einer Doppelblindstudie über ein Jahr, *Drug Res.,* 30, 1928, 1980.
112. **Schwille, P. O. et al.**, unpublished data.
113. **Malluche, H.**, personal communication.
114. **Rose, G. A.**, unpublished data.
115. **Pak, C. Y. C.**, personal communication.

Chapter 9

HEMODIALYSIS AND CALCIUM

Syed N. Asad and Joseph M. Letteri

TABLE OF CONTENTS

I. Introduction 124

II. Classification of Bone Disease in Hemodialysis Patients 124

III. Total Body Calcium in Chronic Renal Failure and Hemodialysis Patients. . . 125

IV. Pathogenesis of Bone Disease in Hemodialysis Patients 127
    A. Parathyroid Hormone and Its Effects 127
    B. Abnormal Vitamin D Metabolism 127
    C. Role of Acidosis 127
    D. Plasma Calcitonin 128
    E. Dialysate Calcium Concentration 128
    F. Effect of Hemodialysis on Serum Calcium and Its Fractions 128
    G. Trace Elements and Other Factors 129
    H. Pathogenesis of Soft Tissue Calcification 129

V. Management of Bone Disease in Hemodialysis Patients 130

References 131

I. INTRODUCTION

Chronic maintenance hemodialysis for end-stage renal disease has evolved remarkably in the past two decades from a tentative treatment for a small group of patients in Seattle to a broadly applied therapy that sustains the lives of 100,000 patients throughout the world.[1] Most of the abnormalities associated with uremia are ameliorated by chronic maintenance hemodialysis, however, perturbations in mineral metabolism and bone disease continue as a serious consequence. With exponential increase in the number of world dialysis population, particularly in the last decade, there has been an increased incidence of bone disease and associated complications. Most of the dialysis patients who do exhibit symptoms frequently sustain fractures and suffer the complications of soft tissue calcifications.[2]

Although uremic toxins account for some of the signs and symptoms of uremic syndrome, it appears that many others are physiological responses intended to compensate for the excretory and endocrine function of the kidney. Diet, dialysis, and transplantation procedures devised to control the severity of the uremic syndrome will, in themselves, lead to further disturbances in homeostasis. This complex interplay among uremic toxins, compensatory metabolic, and endocrine changes and therapeutic procedures are important determinants of the altered calcium and phosphorus homeostasis in uremia.

The derangements in skeletal metabolism and in homeostasis control of calcium, phosphorus, magnesium, and a variety of other minerals in hemodialysis patients are characterized by impaired gastrointestinal absorption of calcium, parathyroid gland hyperactivity, resistance to biological effects of vitamin D, abnormal vitamin D metabolism, metastatic soft tissue and vascular calcification, and a steadily progressive form of bone disease.

Presently, more than 90% of the patients treated with dialysis for longer than 2 years have radiological evidence of bone disease.[3-4] The incidence and severity of bone disease, however, is quite variable. Several factors other than dialysis but including diet, geographical location, age, sex, and physical activity, are important determinants of the severity of the bone disease associated with dialysis.

II. CLASSIFICATION OF BONE DISEASE IN HEMODIALYSIS PATIENTS

The clinical and radiographic features which characterize hemodialysis bone disease and set it apart from osteodystrophy as it occurs in the majority of undialyzed patients include an increased incidence of bone pain and of fractures and a substantial reduction in the bone mass.[5]

It is clear that no simple term is adequate to classify hemodialysis bone disease which is associated with a complex interplay of factors and the variety of lesions which they may produce. Osteoporosis, osteomalacia, osteofibrosis, osteosclerosis, and hyperostasis may all occur and, yet, uremic bone is more than the sum of these lesions. We believe that a classification based on the rate of bone turnover (reabsorption/accretion) and the degree of metastatic calcification is more meaningful particularly from the treatment viewpoint. From our experience and from the experiences of others, we classify the hemodialysis bone disease into the following three groups.[6]

Group I: high turnover bone disease — This affects a vast majority of the dialysis patients and is characterized by increased levels of serum parathyroid hormone, alkaline phosphatase, and low serum calcium which rises slowly with treatment. Typically, bone histology shows increase of resorption, osteoclastic activity, fibrosis, and osteoid seams. It is characterized by decreased bone mass as measured by bone densiometry studies, total body calcium, and radiographic studies. Treatment for this group of patients has been quite successful with vitamin D administration (25-OH D or 1,25 (OH)$_2$D) and parathyroidectomy.

Group II: low turnover bone disease — This affects a small group of dialysis patients and the symptoms are characterized by spontaneous bone fractures and bone pain. The serum parathyroid level and alkaline phosphatase level tend to be normal or low and the serum calcium is generally normal or high and rises rapidly with vitamin D therapy. The bone histology reveals increased osteoid seams, decreased osteoclastic activity, and decreased bone resorption with minimal to absent fibrosis. The treatment for this group of patients has been poor in response to vitamin D therapy and parathyroidectomy.

Group III: metastatic calcification associated with high turnover bone disease — A small minority of patients, with high turnover bone disease may develop spontaneous bone fractures, bone pains, and predominant symptoms of metastatic calcification. Pruritis, ischemic syndromes, pulmonary syndromes and unusual cardiac arrhythmias, single or in combination, may all be present. Metastatic calcification may be revealed first by band keratopathy — best seen by slit-lamp examination. Accompanying conjunctivitis (uremic "red eye"), although rare, could be an early signal after treatment with vitamin D or a high-serum Ca × P product. Patients characteristically have a very high serum parathyroid hormone and alkaline phosphatase levels (> 500 IU). The bone histology shows an increase in osteoclastic activity, bone resorption, bone fibrosis, and osteoid calcification. There is a significant decrease in trabecular bone mass. Following parathyroidectomy, this group of patients shows considerable improvement in pruritis and ischemic syndromes and a decrease in bone fractures, and an increase in bone mass.

III: TOTAL BODY CALCIUM IN CHRONIC RENAL FAILURE AND HEMODIALYSIS PATIENTS

With the advent of total body neutron activation (TBNAA) and whole body counting with a sensitive system, the absolute quantity of certain minerals in the body may be accurately measured. When absolute measurements of the total body stores of such elements as calcium, potassium, sodium, phosphorus, and chloride are normalized for size, sex, age, and body habitus in the individual patients, estimates of normal, increased, and decreased stores of these elements are possible in a wide variety of metabolic conditions. When these measurements are obtained in the same patient over a long timespan, they reflect an integrated value of all the day-to-day changes in body balance of the particular element, and complement the shortterm estimates of body balance by the classical balance techniques.[7]

The total body calcium and phosphorus and calcium ratios evaluated in a series of uremic patients managed with and without chronic hemodialysis are shown in Figure 1. Total body calcium ratios were increased in the majority of the patients and the mean values were signigicantly higher than normal. Patients with elevated total body phosphorus/calcium ratios were usually those with plasma phosphorus concentration in excess of 6 mg/dℓ and a creatinine clearance of less than 20 mℓ/min.

There is a great deal of heterogeneity in the absolute total body calcium and the total body calcium ratio. In more than half of the patients, total body calcium ratio was normal.

Since total body calcium in renal disease reflects not only skeletal calcium, but also that present in soft tissues, vessels, and around joints, some indirect measurement of the bone mineral content would be useful to compare with the total body calcium stores. This could be achieved by determining the mineral content of the radius with the use of Cameron Sorenson Absorptometric technique utilizing the Norland-Cameron Absorptometer®.[8]

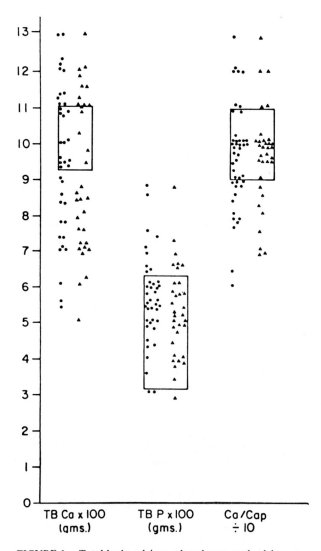

FIGURE 1. Total body calcium, phosphorus, and calcium ratio (Ca/Cap — observed Ca/ predicted calcium) in dialyzed and undialyzed patients. The rectangular areas represent the normal range (±2 SD). To derive the actual value multiply or divide the abscissa as shown (e.g., TBCa 10 × 100 = 1000 g).

In general, the bone mineral content decreases as the total body calcium ratio decreases. However, there are several subsets of dialysis patients where this generalization does not apply. In 4% of our patients, predicted bone mineral content was normal when the total body calcium ratio was below predicted normal values. This phenomenon probably reflects significant loss of calcium from skeletal sites other than the radius. In another subset of patients, the bone mineral content was low when the calcium ratio was within normal range. This subset represented approximately 13% of the dialysis patients studied. Therefore, caution is required in extrapolating observations from single isolated bone measurements to generalizations about the total body calcified tissue dynamics.[8]

IV. PATHOGENESIS OF BONE DISEASE IN HEMODIALYSIS PATIENTS

A. Parathyroid Hormone and Its Effects

The pathogenic factors responsible for secondary hyperparathyroidism originate early in the course of renal disease and are operative for many years before the patient manifests overt bone disease in the late stages of advanced renal insufficiency. A decrease in ionized calcium in the blood is probably the only important factor that increases the secretion of parathyroid hormone and leads to the consequent parathyroid hyperplasia. Major factors leading to hypocalcemia in renal failure include phosphate retention, abnormalities of vitamin D metabolism, and a reduced skeletal response to the calcemic action of parathyroid hormone.[9]

The skeletal resistance to the calcemic effect of parathyroid hormone (PTH) in renal failure may be attributed to diminished or absent production of 1,25 DHCC, has been shown by reversal in the skeletal resistance after administration of 1,25 DHCC. However, this could not be demonstrated in the in vitro skeletal tissue cultures in our laboratory.[10] One of the contributors to the resistance of bone to the calcemic effect of parathyroid hormones is the refractoriness of bone adenylate cyclase to PTH. Massry et al.[11] found that skeletal resistance to the calcemic effect of PTH in uremic patients was not improved by parathyroidectomy. Since calcium mobilization in bone has been associated with intramitochondrial mineral deposits, the recent demonstration of defective electron transport and oxidative phosphorylation of mitochondria obtained from the intestine of uremic rats suggests that a similar defect in bone cells might play a role in bone resistant to PTH.

B. Abnormal Vitamin D Metabolism

Apparent vitamin D deficiency occurs in patients with chronic renal failure. This manifests itself as decreased intestinal absorption of calcium in most patients. It is likely that decreased production of 1,25 DHCC is responsible for the apparent vitamin-D deficiency syndrome in renal failure. This thesis has received confirmation by showing in uremic patients (1) decreased or absent formation of 1,25 DHCC; (2) low or undetectable plasma levels of 1,25 DHCC; (3) a return of plasma levels of 1,25 DHCC to normal following successful renal transplantation; and (4) the correction of hypocalcemia and decreased intestinal absorption of calcium by very small quantities of 1,25 DHCC.

The absorption and the storage of vitamin D is normal in patients with renal failure. The serum 25-hydroxycholecalciferol (25 HCC) level is a chemical index of several factors involving vitamin D metabolism. It reflects dietary and active vitamin D status of the patient, rates of 25 hydroxylation by the liver, and the 1-24 hydroxylation degradation and excretion of 25 HCC and its distribution between tissue and circulating pools. The formation of 25 HCC and its disappearance are probably normal in renal failure. Measured level of 25 HCC is quite variable in uremia. We[14] have noted a significant inverse correlation between serum 25 HCC and Pi in uremia. Increased Pi may interfere with the liver hydroxylation of vitamin D. Letteri et al. described a positive correlation between 25 HCC and serum calcium in uremic patients, but did not find any correlation between levels of 25 HCC and total body calcium in uremia nor any measurable losses of 25 hydroxycholecalciferol across the dialyzer.

C. Role of Acidosis

The bone bicarbonate acts as the major cellular buffer which helps neutralize the excess of [H⁺] ions accumulated in both acute and chronic uremic acidosis. The restoration of body and bone buffers is totally incomplete despite dialysis and the bones continue to provide carbonate to maintain a steady level of extracellular [H⁺] ions and blood bicarbonate despite dietary acid loads which are incompletely excreted in uremia. Chemical, morphological characteristics of bone obtained from dialysis patients include a low bone carbonate, high bone HPO_4, increased porosity, and increased osteoid.[15] The most consistent abnormality is the depression in the level of bone carbonate which is lost predominantly as calcium carbonate and correlates with the duration of uremia. Reduced bone carbonate may affect the crystal growth and maturation in conjunction with bone resorption and apatite dissolution, thereby providing evidence that acidosis could indirectly affect the skeleton.

D. Plasma Calcitonin

The concentration of plasma calcitonin measured by radioimmunoassay is reported high in patients with renal failure as well as in dialysis patients.[16] It remains to be elucidated whether calcitonin has an actual physiological role in chronic renal failure or whether increase or immunoreactive calcitonin demonstrated[17] is due to the accumulation of metabolites of calcitonin which could otherwise be excreted by normal kidneys in healthy persons.

E. Dialysate Calcium Concentration

One of the major factors predisposing the dialysis osteopenia is the net loss of calcium across the dialyzer during dialysis with dialysate fluid containing less than 6 mg/dℓ of calcium. We, and others, have shown that dialysate calcium of 5 mg/dℓ in hemodialyzed patients is associated with a significant total body loss of 207 ± 80 mg/day.[18-20] In contrast, the use of dialysate containing 6.5 mg/dℓ calcium prevented such losses. The effect of dialysate calcium concentration on serum calcium fractions and parathyroid hormone species in plasma during hemodialysis shows variable results: immunoreactive C-terminal fragment was suppressed by increasing dialysate calcium concentration whereas the N-terminal fragments demonstrated an increase.[21] Increasing calcium concentration to 8 g/dℓ in the dialysate fluid has not been effective in suppressing parathyroid hormone and improving the bone histology in a select group of dialysis patients, and use of such a concentrate resulted in soft tissue calcification and its attended complications.[22]

F. Effect of Hemodialysis on Serum Calcium and Its Fractions

Within a few months after initiation of regular hemodialysis, the predialysis concentration of serum calcium generally reaches levels between 9.0 and 10.0 mg/dℓ. Thus, an increase in the serum calcium level generally occurs in patients who were previously hypocalcemic, while little or no change occurs in patients with initial serum calcium levels near normal. The mechanism responsible for this increase in serum calcium level is not clear.

Some of the dialysis related factors that may affect calcium hemostasis in dialysis patients include (1) concentration of calcium and magnesium in dialysate; (2) the efficiency of the dialyzer, which can affect the rate of movement of various uremic "toxins" that fall with each dialysis; (3) hemoconcentration associated with substantial loss of water during dialysis; and (4) changes in blood acid-base status.

It has been suggested that the increase in total blood calcium level toward the end of dialysis reflects an increase in protein-bound, rather than ionized fraction of calcium.[23] Table 1 shows the results of a study we performed in our center involving 17 subjects of both sexes and of different ages treated by maintenance dialysis for varying periods of time. The dialysate calcium concentration was 2.76 ± 0.24 (SD) meq/ℓ. Pre- and postdialysis blood samples were collected and analyzed for serum calcium, ionized calcium, ultrafiltrable calcium, and total proteins. All these measured parameters in-

Table 1
SERUM CALCIUM FRACTIONS AND SERUM PROTEINS

Measurement	Predialysis	Postdialysis	Corrected for hemo-concentration[a]
Total calcium (mmol/ℓ)	2.38 ± 0.29	2.76 ± 0.21[b]	2.52 ± 2
Ionized calcium (mmol/ℓ)	1.16 ± 0.12	1.28 ± 0.67[b]	1.16 ± 0.08
Protein-bound calcium (mmol/ℓ)	1.14 ± 0.21	1.44 ± 0.13[b]	1.31 ± 0.12[c]
Total protein (g/ℓ)	69 ± 7	75 ± 10[b]	69 ± 7
Blood pH	7.31 ± .03	7.43 ± 0.03	

Note: Results are expressed as mean values ± SD; probabilities (P) were assigned by using students *t* test.

[a] Percentage change in total protein is used as a correction factor for hemoconcentration.
[b] $p < 0.01$.
[c] $p < 0.05$.

creased at the end of hemodialysis. However, with a correction for hemoconcentration, only the protein-bound calcium was significantly increased. The increase in protein-bound calcium is not only affected by hemoconcentration, but also by changes in blood pH, phosphorus, and magnesium, the physical property of the protein structure itself, and parathyroid hormone.[24]

G. Trace Elements and Other Factors

Many trace elements accumulate in dialyzed patients but their role in evolution of bone disease in dialysis patients is not clear.[25] Fluoride accumulated in blood, tissues, and bone of calcium in patients receiving repeated dialysis with fluoride contaminated dialysate. Patients in a community utilizing fluoridated water for hemodialysis were reported to have a high incidence of clinically significant bone disease consisting mainly of bone pains, fractures, muscle weakness, and decreased bone mass, as opposed to extosis and osteosclerosis features in fluorosis in nonuremic patients.[26] It has been suggested that aluminum, sometimes present in a water supply, may accumulate in the patient during dialysis; the use of such water has been associated with a high evidence of bone disease.[27]

Appropriate water treatment could remove both trace elements from the water and reduce the incidence of skeletal disease and its complication. Present state-of-art would seem to dictate that water treatment should be employed where there is a substantial contamination of water with trace elements and other impurities.

Patients undergoing regular dialysis routinely receive parenteral heparin. Both clinical and experimental studies suggest that repeated heparin administration is associated with osteoporosis and may predispose to fractures in patients without renal disease.[28] Presently, it is not possible to test the hypothesis that heparin administration plays a role in the pathogenesis of bone disease in dialysis patients because it is generally not possible to carry out regular dialysis without using heparin.

H. Pathogenesis of Soft Tissue Calcification

The factors which predispose to soft tissue calcification are increased calcium (Ca) and inorganic phosphorus (Pi) product (Ca × Pi), hyperparathyroidism, hypermagnesemia, alkalosis during dialysis, and local tissue injury. The incidence of soft tissue calcification is especially high when the Ca-Pi produce is greater than 75, which is usually associated with very severe secondary hyperparathyroidism and/or uncontrolled hyperphosphatemia. However, a retrospective analysis of autopsied dialysis patients indicated that the severity of vascular calcification, present in the vast majority of the dialysis patients (79%), was not related to the duration of dialysis, patient's age, degree of parathyroid gland hyperplasia, radiographic evidence of soft tissue calcification, serum calcium and phosphorus level or Ca × Pi product, or type of severity of metabolic bone disease.[29] Therefore, the factors determining which dialysis patients will develop soft tissue calcification and how severe the lesions will be still remain obscure. Appearance and incidence of vascular calcification and whether hemodialysis has any preventative role has led to much controversy in literature.[30-31]

In earlier studies, the incidence of vascular calcification rose in proportion to the time the patients had been on hemodialysis. In patients with 3 years of dialysis, an incidence of 75% was found.[32] In patients with 9 years of dialysis, an incidence of 83% was reported by Tattler.[33] Extraosseous calcification can no longer be envisaged as the simple result of a transgression of a critical Ca and Pi product in the extracellular fluid compartments. Differences in the inorganic constituents and crystalline feature of various types of extraosseous Ca-P deposits were clearly demonstrated by Alfrey et al. which undoubtedly reflect different mechanisms.[34] Arterial and tumoral Ca-P deposits in uremic patients consisted of apatite. In contrast, uremic visceral Ca-P deposits were high in magnesium with approximately 30% of phosphorus present as pyrophosphate. Although tumoral calcification can be prevented or cured by phosphate depletion, it is unknown whether such treatment has any effect on established vascular calcification.[35] Evidently, vascular calcification is unresponsive to parathyroidectomy and renal transplantation; although vitamin D therapy has been reported to ameliorate vascular calcification.[36]

The most notable difference between adults and children on dialysis is the relative absence of vascular and nonvascular soft tissue calcification in children.[11] Although the reason for this difference is not known, it may be related to the greater amount of woven osteoid normally present in the growing skeleton.[4]

V. MANAGEMENT OF BONE DISEASE IN HEMODIALYSIS PATIENTS

The management objectives for abnormal Ca and phosphorus metabolism in dialysis patients should include: maintaining blood levels of Ca and phosphorus as normal as possible, suppressing parathyroid hormone secretion, preventing or reversing soft tissue calcification, and restoring the skeleton to normal. Of these factors, the control of serum phosphorus is probably the most important and can be accomplished by restricting the dietary intake of phosphorus to 0.8 to 1.0 g/day and by administering phosphate binders (aluminum hydroxide or aluminum carbonate). Dietary calcium should be adequate and supplemental oral Ca up to 1 g/day can be given when the serum phosphorus is controlled. The dialysate calcium concentration should be maintained at 6.0 to 6.5 mg/dℓ to prevent either loss or gain of calcium during dialysis.

After the serum phosphorus has been controlled, vitamin D sterols are indicated: in the presence of hypocalcemia unresponsive to Ca supplementation through diet or dialysis; when there is evidence of overt secondary hyperparathyroidism — particularly when serum Ca is below 11 mg/dℓ; for patients who exhibit symptoms of proximal myopathy.

Other management considerations include use of appropriate levels of dialysate magnesium (0.5 to 0.7 meq/ℓ) to avoid hypermagnesemia,[9] use of water treatment to remove flouride and aluminum, and maintenance of normal acid and base status. Finally, parathyroidectomy should be reserved for patients developing persistent hypercalcemia, metastatic calcification, and a high serum Ca and phosphorus product that is refractory to other therapy.

REFERENCES

1. Becker, E. L., Finite resources and medical triage, *Am. J. Med.*, 66, 549, 1979.
2. Goldsmith, R. S., Arnaud, C. D., and Johnson, W. J., Effect of calcium and phosphorus on patients maintained on dialysis, *Kidney Int.*, Suppl. 2, S-118, 1975.
3. Ritz, E., Krempien, G., Mehls, O., Malluche, H., Strobell, I., and Zimmerman, H., Skeletal complications of renal insufficiency and maintenance hemodialysis, *Nephron*, 10, 195, 1973.
4. Stanbury, S. W., Azotemic renal osteodystrohpy clinics, in *Endocrinology Metabolism*, MacIntyre, I., Ed., W. B. Saunders, Philadelphia, 1972, 267.
5. Ritz, E., Krempier, B., Mehls, O., and Malluchi, H., Skeletal abnormalities in chronic renal insufficiency before and during maintenance dialysis, *Kidney Int.*, 4, 116, 1973.
6. Coburn, J. W., Brickman, A. S., Sherrard, D. J., and Wing, E. G. C., Defective skeletal mineralization in uremia without relation to Vitamin D, Serum Ca or P, Am. Soc. Nephrology, 10th Ann. Meet., (Abstr.), P 3A, 1979.
7. Letteri, J. M. and Cohn, S. H., Total body neutron activation analysis in the study of mineral homeostasis in chronic renal disease, in *Calcium Metabolism in Renal Failure and Nephrolithiasis*, David, S., Ed, John Wiley & Sons, New York, 1977, chap. 7.
8. Cohn, S. H., Ellis, K. J., Caselnova, R. C., Asad, S. N., and Letteri, J. M., Correlation of radial bone mineral content with total body calcium in chronic renal failure, *J. Lab. Clin. Med.*, 86, 910, 1975.
9. Coburn, J. W., Renal Osteodystrophy, *Kidney Int.*, 17, 677, 1980.
10. Letteri, J. M. and Asad, S. N., Effect of uremic sera on parathyroid (PTH) mediated release of calcium from normal rat embryonal bone maintained in tissue culture, 4th Int. Workshop Vitamin D, (Abstr.,) Kongresshalle Berlin, February 18, 1979.
11. Massry, S. G., Coburn, J. W., Lee, D. B. N., and Kleeman, C. R., *Clinical Aspects of Metabolic Bone Disease*, Frame, B., Parfitt, A. M., and Duncan, H., Eds., Exerpta Medica, Amsterdam, 1973, 560.
12. Russell, J. E. and Alvioli, L. V., Effect of experimental chronic renal insufficiency on bone mineral and collagen maturation., *J. Clin. Invest.*, 51, 3072, 1972.
13. Haussler, M. R., *Vitamin D and Problems Related to Uremic Bone Disease*, Normal, A. W., Schaefer, K., Grigoliet, H. G., Herrath, D. V., and Ritz, E. Eds., de Gruyter, Berlin, 1975, 25.
14. Letteri, J. M., Roginsky, M., Moo, F., Scipione, R., Ellis, K., and Cohn, S., *Vitamin D and Problems Related to Uremic Bone Disease*, Normal, A. W., Schaeffer, K., Grigoliet, H. G., Herrath, D. V., and Ritz, E., Eds., de Gruyter, Berlin, 1975, 303.
15. Pellegrino, E. D., Biltz, R. M., and Letteri, J. M., Interrelationship of carbonate, phosphate, monohydrogen phosphate, calcium, magnesium and sodium in uremic bone: comparison of dialysed and nondialysed patients, *Clin. Sci. Mol. Med.*, 53, 307, 1977.
16. Shainkin-Kestenbaum, R., Lerman, R., Reuben, A., and Berlyne, G. M., Plasma calcitonin in chronic renal failure: relation to other factors of importance in bivalent ion metabolism., *Clin. Sci. Mol. Med.*, 52(6), 577, 1977.
17. Silva, O. L., Becker, K. L., Shalhoub, R. J., Snider, R. H., Biven, L. E., and Moore, C. F., Calcitonin levels in chronic renal disease, *Nephron*, 19, 12, 1977.
18. Asad, S. N., Ellis, K. J., Cohn, S. H., and Letteri, J. M., Changes in total body calcium on prolonged maintenance hemodialysis with high and low dialysate calcium, *Nephron*, 23, 223, 1979.
19. Denney, J. P., Sherrard, D. J., Nelp, W. P., Chestnut, C. H., III, and Baylink, D. J., Total body calcium and long term calcium balance in chronic renal disease, *J. Lab. Clin. Med.*, 82. 226, 1973.
20. Wing, A. J., Optimum concentration of dialysis fluid for maintenance hemodialysis, *Br. Med. J.*, 4, 145, 1968.
21. McIntosh, C. H. S., Fuch, C., Dorn, S., Quellhorst, E., Henning, H. V., Heach, R. D., and Scheler, F., Effect of dialysate calcium concentration on plasma parathyroid hormone during hemodialysis with high and low dialysate calcium, *Nephron*, 19, 88, 1977.
22. Drueke, T., Bordier, P. J., Man, N. K., Jungers, P., and Marie, P., Effect of high dialysate calcium concentration on bone remodelling, serum biochemistry and parathyroid hormone in patients with renal osteodystrophy, *Kidney Int.*, 11, 267, 1977.
23. Gosling, P., Robinson, B. H. B., and Sammons, H. G., Changes in protein bound calcium in serum of hemodialysis patients, *Clin. Sci. Mol. Med.*, 48, 521, 1975.
24. Duncan, P. H., Savory, J., and Gitelman, H. J., Effect of hemodialysis on protein bound calcium, *Kidney Int.*, 15, 283, 1979.
25. Kjellstrand, C. M., Alfrey, A. C., Eaton, J. W., Friedman, E. A., Ginn, H. E., Hull, A. R., and Odgen, D., Panel conference — toxicity of materials and medications used in dialysis *Trans. Am. Soc. Artif. Intern. Organs*, 24, 764, 1978.
26. Traves, D. R., Freeman, R. B., Kann, D. E., Ramos, L. P., and Scribner, B. H., Hemodialysis with fluoridated water, *Trans. Am. Soc. Artif. Intern. Organs*, 14, 412, 1968.

27. Ward, M. D., Feest, R. A., Ellis, H. A., Parkinson, I. S., Kerr, D. N. S., Harrington, J., and Goode, G. L., Osteomalacic dialysis osteodystrohpy: evidence for a water-bone etiological agent, probably aluminum, *Lancet*, 1, 841, 1978.
28. Griffith, G. C., Nichols, G., Jr., Asher, J. D., and Flanagan, B., Heparin osteoporosis, *JAMA*, 195, 1089, 1966.
29. Massry, S. G. and Coburn, J. W., in *Clinical Aspects of Uremia in Dialysis*, Massry, S. G. and Sellar, A. L., Eds., Charles C Thomas, Springfield, Ill., 1976, chap. 11.
30. Kuzela, D. C., Huffer, W. E., Conger, J. D., Winter, D. S., ad Hammond, W. S., Soft tissue calcification in chronic dialysis patients, *Am. J. Pathol.*, 86(2), 403, 1977.
31. Eastwood, J. B., Curtis, J. R., Gower, P. E., Roberts, A. P., and DeWardner, H. E., Maintenance hemodialysis, *Lancet*, 1, 265, 1974.
32. Parfitt, A. M., Clinical and radiographic manifestations of renal osteodystrophy, in *Calcium Metabolism in Renal Failure and Nephrolithiasis*, David, S., Ed., John Wiley & Sons, New York, 1977, chap. 4.
33. Tatler, G. L. V., Baillod, R. A., Varghese, Z., Young, W. B., Farrow, S., Wills, M. R., and Moorhead, J. B., Evolution of bone disease over ten years in 135 patients with terminal renal falure, *Br. Med. J.*, 4, 315, 1973.
34. Alfrey, A. C., Solomons, C. C., Circillio, C., and Miller, N. L., Extraosseous calcification: evidence for abnormal pyrophosphate metabolism in uremia, *J. Clin. Invest.*, 57, 692, 1976.
35. Goldsmith, R. S. and Johnson, W. J., Role of phosphate depletion and high dialysate calcium in controlling dialytic renal osteodystrophy, *Kidney Int.*, 4, 154, 1973.
36. Verberckmoes, R., Bouillon, R., and Krempien, B., Disappearance of vascular calcifications during treatment of renal osteodystrophy, *Ann. Int. Med.*, 82, 529, 1975.

Chapter 10

THE CALCIUM PARADOX: MECHANISMS AND CLINICAL RELEVANCE

T. J. C. Ruigrok

TABLE OF CONTENTS

I. Introduction 134

II. The Calcium Paradox 134
    A. Calcium Depletion 134
    B. Calcium Repletion 135
    C. Factors Influencing the Calcium Paradox 135

III. Clinical Relevance 137

IV. Concluding Comments 138

Acknowledgment 138

References 138

I. INTRODUCTION

Calcium ions are essential for excitation-contraction coupling in the myocardium. The passage of an action potential across the sarcolemma leads to a transient increase of the cytosolic calcium concentration from approximately 10^{-7} to 10^{-5} mol/ℓ and to contraction of the myofibrils. An excessive rise in cytosolic calcium, however, causes activation of calcium-dependent ATPases and impairment of the phosphorylating capacity of mitochondria, resulting in necrosis.[1] Since the extracellular calcium concentration amounts to approximately 10^{-3} mol/ℓ, the sarcolemma determines, to a high degree, the contractile and energetic state of the heart.

Perfusion of an isolated heart with a calcium-free solution results in the development of an electromechanical dissociation[2] and an increase of the sarcolemmal permeability to calcium.[3] Upon reperfusion with normal calcium-containing solution, the heart does not regain its mechanical activity, but shows a sudden and severe tissue damage.[4] This phenomenon, which is known as the calcium paradox, was only of academic interest until it was realized that calcium-free solutions are widely used in clinical practice. The following paper is devoted to a brief review of the effects of calcium depletion and repletion on the myocardium. In addition, the clinical relevance of the calcium paradox will be discussed.

II. THE CALCIUM PARADOX

A. Calcium Depletion

When an isolated heart is perfused through its coronary system with a calcium-free solution, the heart loses its mechanical activity,[5] but initially not its electrical activity.[2] The frequency of the electrical potentials decreases and after 2 to 3 min the heart shows atrioventricular conduction abnormalities and ventricular arrhythmias. After 3 to 65 min of calcium-free perfusion, ventricular fibrillation or ventricular standstill appears.[6-9]

Perfusion of isolated rat hearts with a calcium-free solution results in a marked decline of calcium binding and uptake activities of the mitochondrial fraction without any effect on the microsomal fraction.[10] The activities of sarcolemmal adenylate cyclase, Mg^{2+}-ATPase, and (Na^+-K^+)-ATPase decline,[11] and the sodium, potassium, magnesium, and calcium contents of the hearts diminish.[11,12] During calcium-free perfusion, rat hearts release only small amounts of creatine kinase (CK) into the coronary effluent.[13] Rabbit and guinea pig hearts, on the contrary, release substantial amounts of CK from the onset of exposure to a calcium-free solution.[9] If rat hearts are made anoxic, omission of calcium from the perfusate exacerbates the release of CK.[14]

No ultrastructural changes are perceptible during the first 3 min of calcium-free perfusion.[15] Prolonged perfusion with calcium-free solution leads to a number of ultrastructural changes, such as separation of the intercalated discs, dilation of the tubular system, swelling of mitochondria and sarcoplasmic reticulum, contraction of myofibrils, and complete separation of the cells.[7,8,13,15,16] The most conspicuous modification brought about by calcium depletion is a separation of the external lamina from the surface coat of the sarcolemma, resulting in the formation of fluid-filled blebs.[3,7,12,13,17,18] This separation of superficial layers of the sarcolemma is irreversible and correlates with an increase of the sarcolemmal permeability to calcium[3] (see also Reference 19). Carbohydrate-calcium-carbohydrate bridges may be involved in the cohesion of membrane superficial layers. It has been suggested that calcium-free perfusion disrupts these bridges with the consequent formation of blebs.[17] The structure of the lipid bilayer appears also to be affected by calcium depletion.[13,17] The formation of blebs and the increased permeability to calcium are prevented when a minimum of 5×10^{-5} mol/ℓ calcium is present in the extracellular fluid.[17]

B. Calcium Repletion

It can be concluded that calcium depletion ultimately leads to severe electrophysiological, contractile, ultrastructural, and biochemical alterations. Omission of calcium from the perfusion fluid for only 3 to 5 min is sufficient to induce the calcium paradox in rat heart on reintroduction of calcium.[16,20] The most conspicuous changes brought about by this short calcium-free period are electromechanical dissociation,[2] atrioventricular conduction abnormalities and ventricular arrhythmias,[6,9,21] and a separation of the external lamina from the surface coat of the sarcolemma.[3,17,18] The calcium paradox is characterized by an influx of calcium into the cells, exhaustion of tissue high-energy phosphates, the rapid onset of myocardial contracture, massive release of cell constituents, and excessive ultrastructural damage.

Readmission of calcium to a rat heart, after an adequate calcium-free perfusion, results in ventricular tachycardia followed by ventricular fibrillation and finally arrest, while the electrical activity of the atria is unaffected.[9,21] After a brief burst of uncoordinated mechanical activity the heart develops contracture.[8,9,22] The heart loses its normal color and acquires a pale and mottled appearance. This phenomenon, whereby the heart maintains its red color and electrical activity when the perfusion fluid is changed from normal to calcium-free, but loses these properties on return to normal, was first described by Zimmerman and Hülsmann and termed the calcium paradox.[4] The bleaching of the heart is accompanied by a massive release of myoglobin (which colors the effluent a yellowish-brown) and enzymes such as CK and lactate dehydrogenase (LDH).[4,8,9,13,18,22-24]

The activities of sarcolemmal adenylate cyclase, Ca^{2+}-ATPase, Mg^{2+}-ATPase, and (Na^+-K^+)-ATPase show a sharp decrease, the tissue levels of sodium and calcium increase,[8,11,25] and those of potassium and magnesium decrease.[11,25] Microsomal calcium binding and uptake activities decrease during reperfusion with calcium-containing solution, while the mitochondrial calcium binding and uptake activities show a marked increase.[10] It is possible that upon reintroduction of calcium in the perfusion fluid the mitochondria are altered in such a way that they serve as an adaptive mechanism to handle the massive influx of calcium from the extracellular space, as well as to compensate for the loss of calcium pump mechanism of the sarcoplasmic reticulum.[10]

Readmission of calcium following a short period of calcium depletion causes massive ultrastructural damage characterized by severe contracture with considerable disruption of the myofilaments and swelling of mitochondria with formation of electron-dense particles. Disruption of the sarcolemma and intercalated discs, and even complete separation of myocardial cells have been observed.[8,13,15,16,18,24,26-28]

The calcium paradox is accompanied with a sudden and severe decline of myocardial high-energy phosphates, as has been shown with phosphorus nuclear magnetic resonance spectroscopy[27] and freeze-clamp methods.[29] After 30 sec of calcium repletion, myocardial creatine phosphate (CP) and ATP levels are decreased by 65% and 45%, respectively. In the same period, there is an increase in creatine (15%), ADP (85%), and AMP (2800%). During continued reperfusion with calcium the concentration of all compounds decreases gradually. The effluent fluid contains large amounts of creatine and AMP, and relatively minor amounts of CP, ATP, and ADP.[29] A similar decline of high-energy phosphates occurs in rabbit, guinea pig, and mouse hearts.[9] Occurrence of the calcium paradox in rabbit hearts, however, has been the subject of controversy.[9,30]

C. Factors Influencing the Calcium Paradox

Occurrence of the calcium paradox can be influenced by pH, composition, and temperature of the perfusion fluids and by the energetic state of the heart. When the pH of the calcium-free solution is lowered, the calcium-free period has to be extended in order to evoke the calcium paradox. Inversely, a higher pH decreases the length of the calcium-free period, necessary to induce the calcium paradox. It has been suggested that a high H⁺ concentration hampers the calcium movements across the sarcolemma and at subcellular level.[31] Reducing the sodium concentration of the calcium-free solution augments the recovery of calcium-deprived hearts and prevents the ultrastructural damage upon reintroduction of calcium in the perfusion fluid. Reduced loss of calcium during the period of calcium-deprivation is the presumed mechanism of this protective effect.[15,25]

In the original calcium paradox experiments,[4] isolated rat hearts were continuously perfused with oxygenated glucose-containing solutions. This enabled the hearts to maintain their high-energy stores until the end of the calcium-free period at a high level by oxidation of glucose. In energy-depleted hearts (i.e., during anoxic perfusion in the absence of glucose), the calcium paradox does not occur unless the hearts are reoxygenated and the electron transport system is reactivated. A well-functioning electron transport system is not a condition for the occurrence of the calcium paradox. In the absence of electron transport the calcium paradox does occur, provided that ATP is present. This situation can be obtained either by anoxic perfusion with glucose or by perfusion with oxygenated glucose-containing solutions in the presence of potassium cyanide.[20]

Accumulation of calcium by mitochondria is an important defense mechanism by which cells try to control an increased influx of calcium from the extracellular space, e.g., owing to an increased permeability of the sarcolemma.[32] The mitochondrial accumulation of calcium together with phosphate can be supported by electron transport or ATP.[33] During this process electron-dense granules of calcium phosphate are deposited in the mitochondrial matrix. Mitochondria most likely play a key role in the origin of the calcium paradox, since the conditions for the calcium paradox are the same as for the accumulation of calcium by mitochondria: availability of either electron transport or ATP.[20,33]

When perfusion of isolated rat hearts with a calcium-free solution is performed at 4°C, both contractility and electrical activity disappear. Readmission of calcium to these hearts permits rapid recovery of cardiac function upon rewarming to 37°C.[8] After 10 to 12 min of calcium-free perfusion between 37 and 30°C, a progressive inhibition of the calcium paradox injury is observed upon reperfusion with calcium-containing solution.[18,27] When calcium-free perfusion is performed below 30°C, reintroduction of calcium in the perfusion fluid causes negligible damage to the myocardium. By extending the calcium-free perfusion period, however, it is possible to predispose all hearts to the calcium paradox, even when calcium-free perfusion is performed at 20°C. At 20°C, calcium-free perfusion has to be continued for 80 min to induce massive enzyme release upon reperfusion with calcium-containing solution.[34] Whether or not hypothermia during the reperfusion phase prevents the calcium paradox has been the subject of controversy.[18,34]

Presence of propranolol or the calcium antagonists verapamil or D600 in the reperfusion medium does not prevent the calcium paradox.[25] A similar observation is achieved when verapamil is present during both the calcium-free and reperfusion period.[25,35] Verapamil does not reduce the massive enzyme release, but reduces the initial rate at which myocardial CP and ATP stores are depleted and decreases the shortening of the mean sarcomere length during reperfusion with calcium-containing solution.[35] This indicates that inhibition of calcium influx via the slow channels does not protect heart muscle against the deleterious effects of readmitting calcium after a period of calcium depletion.

In most methods developed for preparing suspensions of isolated adult heart cells, the cells are exposed for some time to a calcium-free solution. As a result, the majority of these suspensions are intolerant to calcium. It has been suggested that occurrence of the calcium paradox is responsible for degeneration of the cells upon exposure to physiological levels of calcium.[36,37] Dimethylsulfoxide (DMSO) appears to protect isolated heart cells against calcium-induced degeneration.[37] It has been established that DMSO does not protect isolated rat hearts against the effects of calcium-free perfusion which predispose the myocardium to the calcium paradox. However, the calcium paradox damage can be reduced by adding DMSO to the reperfusion medium.[21] It is possible that DMSO, with its reported ability to protect against cellular and tissue edema,[38] protects the sarcolemma against stretching and rupturing, and in so doing, reduces the massive enzyme release during the reperfusion phase. Another explanation is that DMSO influences energy-dependent transmembrane fluxes of calcium, which would lead to exhaustion of tissue high-energy phosphates[27,29] and myocardial contracture.[35,39]

III. CLINICAL RELEVANCE

The clinical relevance of the calcium paradox is well recognized at present. The use of calcium-free cardioplegic solutions[40,41] during ischemic cardiac arrest may be hazardous by increasing the sarcolemmal permeability to calcium and, in so doing, making the heart susceptible to the calcium paradox.[42-44] Ischemia in itself may also lead to an increased permeability of the sarcolemma to calcium. In contrast with the calcium-free situation where the endogenous stores of high-energy phosphates are well maintained,[29] energy depletion is responsible for the impairment of sarcolemmal integrity in ischemic myocardium. Verapamil is able to protect the heart against the effects of ischemia,[45] but not against the alterations brought about by calcium depletion.[35] The biochemical and ultrastructural changes which may be induced by reperfusing ischemic myocardium[46,47] or reoxygenating anoxic myocardium[48,49] are identical to those noted during the calcium paradox. Strong evidence has been provided that not the mere readmission of oxygen, but reactivation of the electron transport system and a consequent accumulation of calcium by the mitochondria play a crucial role in the paradoxical extension of ischemic damage upon reperfusion of the coronary system.[24,50-53]

1.	Ischemia (aortic cross-clamping)	+		Reactivation of electron transport system (reperfusion with blood)
2.	Calcium depletion (calcium-free cardioplegia)	+		Readmission of calcium (reperfusion with blood)
	↓			↓
	Increase of sarcolemmal permeability to calcium			Energy- and calcium-dependent cell damage

When ischemic myocardium is reperfused with calcium-free blood, the reperfusion damage is prevented.[54] Likewise, no calcium paradox is observed when, after a short calcium-free perfusion, calcium is readmitted to energy-depleted hearts.[20]

It has been suggested, on the basis of an experimental study in which myocardial preservation was investigated using calcium-free and calcium-containing cardioplegic solutions, that the calcium paradox does not as readily occur in an in vivo as in an in vitro heart preparation.[55] Coronary perfusion with calcium-free Tyrode solution (for 10 min at 37°C) during cardiopulmonary bypass in dogs, however, results in irreversible myocardial damage upon reperfusion with blood: development of a systolic arrest and release of large amounts of CK into the circulation.[56] Using the same experimental set-up, the effect of the Bretschneider procedure on the recovery of dog hearts in the post-ischemic phase after 90 min of aortic occlusion was studied.[57] In one series of experiments, Bretschneider's "Kardioplegische Lösung-HTP" was used; in a second series 5×10^{-5} mol/ℓ CaCl$_2$ (which has been shown to prevent the calcium paradox)[17] was added to this calcium-free solution. The coronary system was perfused with cardioplegic solution (6°C) for 6 min at the onset, and for 1 min after 30 and 60 min of ischemic arrest. Functional recovery of the heart and plasma CK activity during the post-ischemic phase was not significantly different in the two groups of dogs, indicating that low temperature, mode of administration, and possibly composition of the Bretschneider solution prevents the calcium paradox. This solution is characterized by a low sodium concentration (15 mmol/ℓ) and a low pH (6.8 at 37°C), which reduces the potential hazard of the absence of calcium.[15,25,31] Perfusion of isolated rat heart with the Bretschneider solution at 20°C, on the other hand, may lead to considerable reperfusion damage, although no full calcium paradox will occur.[58] This reperfusion damage can be prevented by addition of 5×10^{-5} mol/ℓ CaCl$_2$ (only 6 mg/ℓ!) to the Bretschneider solution. It is concluded, therefore, that excessive or continuous infusion of calcium-free cardioplegic solutions in clinical practice can be hazardous and should therefore be avoided.

The calcium paradox is not limited to the heart but may also occur after exposure of kidneys to calcium-free solutions for preservation or transplantation purposes.[59] The use of calcium-free solutions in preparing veins for peripheral vascular and coronary bypass surgery may predispose the intima to the calcium paradox and lead to early graft failure.[60]

IV. CONCLUDING COMMENTS

Calcium depletion and calcium repletion can be extremely harmful to the myocardium. The damaging effect of calcium depletion has been attributed to a separation of the external lamina from the surface coat of the sarcolemma. This separation of superficial layers correlates with an increase of the sarcolemmal permeability to calcium. As a consequence, reintroduction of calcium in the extracellular fluid results in an influx of calcium into the cells and a sudden and severe tissue damage: the calcium paradox.

Calcium-free solutions are widely used in a number of clinical situations. Since the calcium paradox is not limited to the heart, it is strongly recommended not to expose human tissue to solutions containing less than 5×10^{-5} mol/ℓ calcium, which is an absolute requirement for maintenance of normal cell membrane structure.

ACKNOWLEDGMENT

The author is grateful to Professor Dr. F. L. Meijler for his constructive criticism.

REFERENCES

1. **Fleckenstein, A.,** Specific inhibitors and promoters of calcium action in the excitation-contraction coupling of heart muscle and their role in the prevention or production of myocardial lesions, in *Calcium and the Heart,* Harris, P. and Opie, L. H., Eds., Academic Press, London, 1971, 135.
2. **Mines, G. R.,** On functional analysis by the action of electrolytes, *J. Physiol. (London),* 46, 188, 1913.
3. **Frank, J. S., Langer, G. A., Nudd, L. M., and Seraydarian, K.,** The myocardial cell surface, its histochemistry, and the effect of sialic acid and calcium removal on its structure and cellular ionic exchange, *Circ. Res.,* 41, 702, 1977.

4. Zimmerman, A. N. E. and Hülsmann, W. C., Pardoxical influence of calcium ions on the permeability of the cell membranes of the isolated rat heart, *Nature (London)*, 211, 646, 1966.
5. Ringer, S., A further contribution regarding the influence of the different constituents of the blood on the contraction of the heart, *J. Physiol. (London)*, 4, 29, 1883.
6. Weiss, D. L., Surawicz, B., and Rubenstein, I., Myocardial lesions of calcium deficiency causing irreversible myocardial failure, *Am. J. Pathol.*, 48, 653, 1966.
7. Muir, A. R., The effects of divalent cations on the ultrastructure of the perfused rat heart, *J. Anat.*, 101, 239, 1967.
8. Holland, C. E. and Olson, R. E., Prevention by hypothermia of paradoxical calcium necrosis in cardiac muscle, *J. Mol. Cell. Cardiol.*, 7, 917, 1975.
9. Hearse, D. J., Humphrey, S. M., Boink, A. B. T. J., and Ruigrok, T. J. C., The calcium paradox: metabolic, electrophysiological, contractile and ultrastructural characteristics in four species, *Eur. J. Cardiol.*, 7, 241, 1978.
10. Lee, S. L. and Dhalla, N. S., Subcellular calcium transport in failing hearts due to calcium deficiency and overload, *Am. J. Physiol.*, 231, 1159, 1976.
11. Dhalla, N. S., Tomlinson, C. W., Singh, J. N., Lee, S. L., McNamara, D. B., Harrow, J. A. C., and Yates, J. C., Role of sarcolemmal changes in cardiac pathophysiology, in *Recent Advances in Studies on Cardiac Structure and Metabolism*, Vol. 9, Roy, P.-E. and Dhalla, N. S., Eds., University Park Press, Baltimore, 1976, 377.
12. Tomlinson, C. W., Yates, J. C., and Dhalla, N. S., Relationship among changes in intracellular calcium stores, ultrastructure, and contractility of myocardium, in *Recent Advances in Studies on Cardiac Structure and Metabolism*, Vol. 4, Dhalla, N. S., Ed., University Park Press, Baltimore, 1974, 331.
13. Ashraf, M., Correlative studies on sarcolemmal ultrastructure, permeability, and loss of intracellular enzymes in the isolated heart perfused with calcium-free medium, *Am. J. Pathol.*, 97, 411, 1979.
14. Nayler, W. G., Grau, A., and Slade, A., A protective effect of verapamil on hypoxic heart muscle, *Cardiovasc. Res.*, 10, 650, 1976.
15. Yates, J. C. and Dhalla, N. S., Structural and functional changes associated with failure and recovery of hearts after perfusion with Ca^{2+}-free medium, *J. Mol. Cell. Cardiol.*, 7, 91, 1975.
16. Zimmerman, A. N. E., Daems, W., Hülsmann, W. C., Snijder, J., Wisse, E., and Durrer, D., Morphological changes of heart muscle caused by successive perfusion with calcium-free and calcium-containing solutions (calcium paradox), *Cardiovasc. Res.*, 1, 201, 1967.
17. Crevey, B. J., Langer, G. A., and Frank, J. S., Role of Ca^{2+} in maintenance of rabbit myocardial cell membrane structural and functional integrity, *J. Mol. Cell. Cardiol.*, 10, 1081, 1978.
18. Hearse, D. J., Humphrey, S. M., and Bullock, G. R., The oxygen paradox and the calcium paradox: two facets of the same problem?, *J. Mol. Cell. Cardiol.*, 10, 641, 1978.
19. Isenberg, G. and Klöckner, U., Glycocalyx is not required for slow inward calcium current in isolated rat heart myocytes, *Nature (london)*, 284, 358, 1980.
20. Ruigrok, T. J. C., Boink, A. B. T. J., Spies, F., Blok, F. J., Maas, A. H. J., and Zimmerman, A. N. E., Energy dependence of the calcium paradox, *J. Mol. Cell. Cardiol.*, 10, 991, 1978.
21. Ruigrok, T. J. C., De Moes, D., Slade, A. M., and Nayler, W. G., The effect of dimethylsulfoxide on the calcium paradox, *Am. J. Pathol.*, 103, 390, 1981.
22. Ruigrok, T. J. C., Burgersdijk, F. J. A., and Zimmerman, A. N. E., The calcium paradox: a reaffirmation, *Eur. J. Cardiol.*, 3, 59, 1975.
23. De Leiris, J. and Feuvray, D., Factors affecting the release of lactate dehydrogenase from isolated rat heart after calcium and magnesium free perfusions, *Cardiovasc. Res.*, 7, 383, 1973.
24. Ganote, C. E., Worstell, J., and Kaltenbach, J. P., Oxygen-induced enzyme release after irreversible myocardial injury; effects of cyanide in perfused rat hearts, *Am. J. Pathol.*, 84, 327, 1976.
25. Alto, L. E. and Dhalla, N. S., Myocardial cation contents during induction of calcium paradox, *Am. J. Physiol.*, 237, H713, 1979.
26. Muir, A. R., A calcium-induced contracture of cardiac muscle cells, *J. Anat.*, 102, 148, 1968.
27. Bulkley, B. H., Nunnally, R. L., and Hollis, D. P., "Calcium paradox" and the effect of varied temperature on its development; a phosphorus nuclear magnetic resonance and morphologic study, *Lab. Invest.*, 39, 133, 1978.
28. Singal, P. K., Matsukubo, M. P., and Dhalla, N. S.,Calcium-related changes in the ultrastructure of mammalian myocardium, *Br. J. Exp. Pathol.*, 60, 96, 1979.
29. Boink, A. B. T. J., Ruigrok, T. J. C., Maas, A. H. J., and Zimmerman, A. N. E., Changes in high-energy phosphate compounds of isolated rat hearts during Ca^{2+}-free perfusion and reperfusion with Ca^{2+}, *J. Mol. Cell. Cardiol.*, 8, 973, 1976.
30. Lee, Y. C. P. and Visscher, M. B., Perfusate cations and contracture and Ca, Cr, PCr, and ATP in rabbit myocardium, *Am. J. Physiol.*, 219, 1637, 1970.
31. Bielecki, K., The influence of changes in pH of the perfusion fluid on the occurrence of the calcium paradox in the isolated rat heart, *Cardiovasc. Res.*, 3, 268, 1969.

32. Carafoli, E., Mitochondrial uptake of calcium ions and the regulation of cell function, in *Biochemical Society Symposia,* Vol. 39, Biochemical Society, London, 1974, 89.
33. Brierley, G. P., Murer, E., and Bachmann, E., Studies on ion transport. III. The accumulation of calcium and inorganic phosphate by heart mitochondria, *Arch. Biochem. Biophys.,* 105, 89, 1964.
34. Boink, A. B. T. J., Ruigrok, T. J. C., De Moes, D., Maas, A. H. J., and Zimmerman, A. N. E., The effect of hypothermia on the occurrence of the calcium paradox, *Pflugers Arch.,* 385, 105, 1980.
35. Ruigrok, T. J. C., Boink, A. B. T. J., Slade, A., Zimmerman, A. N. E., Meijler, F. L., and Nayler, W. G., The effect of verapamil on the calcium paradox, *Am. J. Pathol.,* 98, 769, 1980.
36. Farmer, B. B., Harris, R. A., Jolly, W. W., Hathaway, D. R., Katzberg, A., Watanabe, A. M., Whitlow, A. L., and Besch, H. R., Isolation and characterization of adult rat heart cells, *Arch. Biochem. Biophys.,* 179, 545, 1977.
37. Clark, M. G., Gannon, B. J., Bodkin, N., Patten, G. S., and Berry, M. N., An improved procedure for the high-yield preparation of intact beating heart cells from the adult rat; biochemical and morphologic study, *J. Mol. Cell. Cardiol.,* 10, 1101, 1978.
38. Görög, P., and Kovács, I. B., Effect of dimethyl sulfoxide (DMSO) on various experimental inflammations, *Curr. Ther. Res. Clin. Exp.,* 10, 486, 1968.
39. Hearse, D. J., Garlick, P. B., and Humphrey, S. M., Ischemic contracture of the myocardium: mechanisms and prevention, *Am. J. Cardiol.,* 39, 986, 1977.
40. Kirsch, U., Rodewald, G., and Kalmár, P., Induced ischemic arrest; clinical experience with cardioplegia in open-heart surgery, *J. Thorac. Cardiovasc. Surg.,* 63, 121, 1972.
41. Bretschneider, H. J., Hübner, G., Knoll, D., Lohr, B., Nordbeck, H., and Spieckermann, P. G., Myocardial resistance and tolerance to ischemia: physiological and biochemical basis, *J. Cardiovasc. Surg.,* 16, 241, 1975.
42. Tyers, G. F. O., Metabolic arrest of the ischemic heart, *Ann. Thorac. Surg.,* 20, 91, 1975.
43. Jynge, P., Hearse, D. J., and Braimbridge, M. V., Myocardial protection during ischemic cardiac arrest; a possible hazard with calcium-free cardioplegic infusates, *J. Thorac. Cardiovasc. Surg.,* 73, 848, 1977.
44. Jynge, P., Hearse, D. J., De Leiris, J., Feuvray, D., and Braimbridge, M. V., Protection of the ischemic myocardium; ultrastructural, enzymatic, and functional assessment of the efficacy of various cardioplegic infusates, *J. Thorac. Cardiovasc. Surg.,* 76, 2, 1978.
45. Robb-Nicholson, C., Currie, W. D., and Wechsler, A. S., Effects of verapamil on myocardial tolerance to ischemic arrest; comparison to potassium arrest, *Circulation* 58, (Suppl. 1), 119, 1978.
46. Whalen, D. A., Hamilton, D. G., Ganote, C. E., and Jennings, R. B., Effect of a transient period of ischemia on myocardial cells. I. Effects on cell volume regulation, *Am. J. Pathol.,* 74, 381, 1974.
47. Kloner, R. A., Ganote, C. E., Whalen, D. A., and Jennings, R. B., Effect of a transient period of ischemia on myocardial cells. II. Fine structure during the first few minutes of reflow, *Am. J. Pathol.,* 74, 399, 1974.
48. Hearse, D. J., Humphrey, S. M., and Chain, E. B., Abrupt reoxygenation of the anoxic potassium-arrested perfused rat heart: a study of myocardial enzyme release, *J. Mol. Cell. Cardiol.,* 5, 395, 1973.
49. Ganote, C. E., Seabra-Gomes, R., Nayler, W. G., and Jennings, R. B., Irreversible myocardial injury in anoxic perfused rat hearts, *Am. J. Pathol.,* 80, 419, 1975.
50. Shen, A. C. and Jennings, R. B., Myocardial calcium and magnesium in acute ischemic injury, *Am. J. Pathol.,* 67, 417, 1972.
51. Shen, A. C. and Jennings, R. B., Kinetics of calcium accumulation in acute myocardial ischemic injury, *Am. J. Pathol.,* 67, 441, 1972.
52. Hearse, D. J., Reperfusion of the ischemic myocardium, *J. Mol. Cell. Cardiol.,* 9, 605, 1977.
53. Peng, C. F., Kane, J. J., Murphy, M. L., and Straub, K. D., Abnormal mitochondrial oxidative phosphorylation of ischemic myocardium reversed by Ca^{2+}-chelating agents, *J. Mol. Cell. Cardiol.,* 9, 897, 1977.
54. Ashraf, M., White, F., and Bloor, C. M., Ultrastructural influences of reperfusing dog myocardium with calcium-free blood after coronary artery occlusion, *Am. J. Pathol.,* 90, 423, 1978.
55. Engelmap, R. M., Auvil, J., O'Donoghue, M. J., and Levitsky, S., The significance of multidose cardioplegia and hypothermia in myocardial preservation during ischemic arrest, *J. Thorac. Cardiovasc. Surg.,* 75, 555, 1978.
56. Ruigrok, T. J. C., Hitchcock, J. F., and Zimmerman, A. N. E., Calcium paradox in dog heart during cardiopulmonary bypass, *J. Mol. Cell. Cardiol.,* 10, (Suppl. 1), 89, 1978.
57. Ruigrok, T. J. C., Sneek, J. H. J., Van Dort, J. T. M., and Zimmerman, A. N. E., Cardioplegia in the presence or absence of calcium, *J. Mol. Cell. Cardiol.,* 12, (Suppl. 1), 139, 1980.
58. Ruigrok, T. J. C. and De Moes, D., Calcium paradox in rat heart caused by successive perfusion with Bretschneider's histidine-buffered cardioplegic solution and a calcium-containing solution, submitted, 1981.

59. Nozick, J. H., Zimmerman, A. N. E., Poll, P., and Mankowitz, B. J., The kidney and the calcium paradox, *J. Surg. Res.*, 11, 60, 1971.
60. Nozick, J. H., Farnsworth, P., Montefusco, C. M., Parsonnet, V., Ruigrok, T. J. C., and Zimmerman, A. N. E., Autogenous vein graft thrombosis following exposure to calcium-free solutions (calcium paradox), *J. Cardiovasc. Surg.*, 22, 166, 1981.

Chapter 11

CALCIUM AND CANCER

Leopold J. Anghileri

TABLE OF CONTENTS

I.	Introduction	144
II.	Ionic Environment During Carcinogenesis	144
III.	Calcium and Cell Injury	145
	A. Role of Mitochondria	147
	B. Calcium and Metabolism of the Cell	149
IV.	Calcium and Oncogenic Viruses	150
V.	Cytostatics and Calcium	150
VI.	Conclusions	150
References		151

I. INTRODUCTION

Most of the scientific literature concerning calcium and cancer deals with disturbances of calcium homeostasis and abnormalities in bone metabolism provoked by the presence of the tumor.[1] Besides few early attempts to correlate calcium content of tissues with the neoplastic development,[2-4] no further efforts have been done to determine the possible role of intracellular Ca^{2+} concentration changes in carcinogenesis. One of the reasons for this lack of information on the effects of intracellular environment changes has been the unavailability of analytical methods capable to determine, without interferences, the extremely low ionic concentrations involved in that process. This problem, however, has been in many cases solved by the development, during the last two decades, of a suitable sophisticated analytical technology (atomic absorption spectrometry, ion-selective electrodes, isotopic analysis, metallochromic indicator dyes, fluorescent probes, etc.)

To fully estimate the possible implications of calcium in the neoplastic phenomena, we must consider that Ca^{2+} plays a unique and critical role in a wide variety of physiological processes of the cell. To a number of authors, the modifications at the cell membrane level are the principal factors triggering the malignant transformation of the cell.[5-9] In this review, as a chronologically primary event, the interaction plasma membrane-carcinogen and its subsequent effects, especially in what concerns the ionic environment, will be discussed in the light of the available experimental evidence.

II. IONIC ENVIRONMENT DURING CARCINOGENESIS

The follow-up of chemically induced experimental tumors has shown a relationship between ionic concentration changes in the cell and the preneoplastic and neoplastic states.[10,11] The most striking observation in tumor has been a very significant increase of extracellular cations (calcium and sodium) while the intracellular ones (magnesium and potassium) were decreased. On concentration basis these changes were more important for calcium (augmented up to 5.4-fold) and sodium (augmented up to 2.4-fold). This finding appears to indicate a cell membrane impairment as a concomitant phenomenon of malignancy. Coincidentally, a primary alteration of the cell membranes by interaction with the carcinogen can be inferred from the observation that amino azo dye carcinogens bind to the protein of cell membranes.[12] On the other hand, chemical carcinogens show an increased interaction with the plasma membrane which is related to their lipophilic properties determined by specific molecular characteristics.[13] It should be noted in this connection that permeability differences between cancer and normal cells have been suggested on the basis of the modified action of thyroxine on the two types of cell,[14] as well as on the different release of intracellular material.[15,16]

Even in cases where there is no detectable cell membrane-carcinogen interaction, changes produced in the ionic environment can be important in the functional integrity of the cell (including that of cell membranes). The experimental evidence indicates that the fluid state of lipids is very important in the regulation of the biological properties of the membrane, such as nutrient and passive ion transport.[17,18] On the other hand, calcium is known to change the fluidity of cell membrane lipids,[19] a property which points out that the interaction of calcium-cell membrane lipid might explain the role of Ca^{2+} in the functional regulation of a great many characteristics of biological membranes.[20] In connection with membrane changes and the stabilizing role of divalent cations (calcium and magnesium), the analyses of these cations bound to tumor phospholipids have shown a consistent low binding with respect to that of normal tissue (Table 1).[21-23] In addition to this, a more recent experimental work on tumor growth

Table 1
CALCIUM, MAGNESIUM, AND PHOSPHOLIPID IN DIFFERENT TYPES OF HUMAN INTRACRANIAL TUMORS[22]

	Astrocytoma (n = 7)	Meningioma (n = 13)	Glioblastoma (n = 8)	Normal brain (n = 6)
Total calcium	168.3 ± 22.8	2,103 ± 1,154	383.5 ± 179.5	125.8 ± 52.8
Total magnesium	56.9 ± 5.0	125.8 ± 24.8	52.4 ± 7.4	65.0 ± 10.9
Complexed calcium	15.4 ± 4.8	9.2 ± 2.0	34.4 ± 3.6	75.8 ± 48.8
Complexed magnesium	4.6 ± 0.9	4.0 ± 0.9	10.0 ± 2.5	34.4 ± 3.1
Complexed phospholipid as phosphorus	22.9 ± 8.5	20.1 ± 3.9	76.5 ± 23.7	77.3 ± 13.9

Note: All the values as µg/g wet tissue and mean value ± SD.

Table 2
DIVALENT CATION-BINDING CHANGES DURING DS SARCOMA GROWTH INHIBITION BY CALCITONIN

	Group calcitonin	Group control
Number of tumors	23	16
Tumor weight (g)	9.7 ± 1.7 ($p<0.01$)	14.6 ± 1.6
Lipid calcium (µg/g tumor)	12.8 ± 1.8 ($p<0.01$)	6.3 ± 0.6
Lipid magnesium (µg/g tumor)	3.1 ± 0.3 ($p<0.01$)	1.8 ± 0.2

Note: All the values as mean value ± SEM.

From Anghileri, L. J. and Delbrück, H., *Experientia,* 35, 1664, 1979. With permission.

inhibition by the hypocalcemic effect of calcitonin,[24] has shown a significant growth inhibition only when the concentration of divalent cations bound to the phospholipid was increased with respect to the controls (Table 2). These results seem to indicate a relationship between malignancy and cell membrane characteristics (among them permeability impairment), which is in agreement with the concept that the divalent cations control the stability of the membrane by the formation of a ternary complex between the anionic groups of a protein and a phospholipid molecule.

III. CALCIUM AND CELL INJURY

The extracellular fluid is very rich in Ca^{2+} ($10^{-3}M$) while intracellular Ca^{2+} concentration is much lower ($10^{-6}M$). The electrical potential across the plasma membrane tends to drive Ca^{2+} into the cell, and this large electrochemical gradient is maintained by the relative impermeability of the plasma membrane to Ca^{2+}, and by an active extrusion (calcium pump).[25] Disruption of the integrity of the plasma membrane provokes an influx of Ca^{2+} across the plasma membrane and down a steep concentration gradient. This influx of Ca^{2+} can produce two types of effect on the cell.

1. A massive internal calcification which overwhelms the mitochondrial calcium-buffer system and leads to cell death by necrosis. Coincidentally, the early stages of chemical carcinogenesis are known to coincide with notable cell death.[26] However, it has been demonstrated that mitochondrial calcification is not necessarily followed by cell death and that intramitochondrial calcification may be reversible[27,28] and this characteristic might be related to the second type of effect.

Table 3
STIMULUS DURATION-DETERMINED PATTERN OF THE EFFECTS OF INCREASED INTRACELLULAR CALCIUM CONCENTRATION DURING HEPATO-INTOXICATION BY THIOACETAMIDE

Short lasting
    Acute intoxication
        (A) Toxic cell death (necrosis) — irreversible
        (B) Buffered calcium overload and cell membrane repair — reversible

Long lasting
    Chronic intoxication
        (A) Toxic cell death (necrosis) — irreversible
        (B) Cell dedifferentiation (preneoplastic state) →
        → Cancer cell (neoplastic state) — irreversible

2. Less drastic, with a regeneration process taking place in which the damaged cells become dedifferentiated and more primitive,[29] with a slow shifting toward the glycolytic state, and with succeeding generations of cell becoming more and more autonomous from tissue level controls. This stage of cell transformation presents the characteristics of a preneoplastic state.[31]

The pattern and the magnitude of these changes are dependent on the type of cell and the duration of Ca^{2+} concentration change (Table 3).

Among the structural changes of the cell caused by this modification of the ionic environment there are microtubule depolymerization,[31] closing of intracellular junctions,[32] and alterations of the lipid-protein membrane structure with the subsequent change of membrane potential and transport processes leading to changes in other ion fluxes.[33] A suitable model to study these effects on the cell as a function of time during which Ca^{2+} concentration is changed is provided by the hepatotoxicity of thioacetamide. A single dose of this compound provokes a reversible action characterized by a 34-fold increase of calcium in liver at 24 hr after administration, and returning to normal after 8 days.[34] Some cells died, but the damaged membranes of most cells are soon repaired (reversible change). In contrast to this, a long lasting chronic administration of a lower dose develops a cirrhotic condition which very often is accompanied by cholangiocarcinoma development.[11] Studies performed with the hepatocarcinogen DL-ethionine have shown that the acute toxicity is reflected by an extensive polyribosome disaggregation which is not accompanied by the polysome-membrane separation that characterizes neoplasia.[35] On the contrary, chronic toxicity leading to hepatocarcinogenesis, which is not accompanied by disaggregation, is characterized by the detachment of polysomes. This is an indication that polysomal disaggregation is not the essential mechanism of carcinogenesis but the integrity (structural and functional) of the intracellular membrane seems to be essential for preventing the neoplastic development.[36,37] In connection to this, it has been suggested that the primary carcinogenic event is the destruction of the proteins which hold the polysomes and because only membrane-bound polysomes are capable of synthesizing the polysome-anchoring proteins, this lack of attachment proteins assure the perpetuation of the defect.[38]

In the case of thioacetamide the induction of a tumor appears to support the proposed role of Ca^{2+} concentration changes when the effect is properly prolonged in order to have an irreversible change leading to carcinogenesis. This effect of ionic changes triggering the cellular events directing to carcinogenesis has been suggested in a more general way by other authors.[39-41] On the other hand, the calcium dependence of toxic cell death (which generally accompanies carcinogenesis)[26] has been recently demonstrated.[42] Some authors have suggested that the loss of regulation of many syntheses of enzymes observed in tumor cells,[43] may be due to the loss of ability of the inducing and repressing compounds to enter the cell as compared to the normal cells.[44] Cancer is a phenomenon characterized by a lack of intercellular communication[45,46] in which the cells are more autonomous than their normal counterparts, and they do not respond to the signals which coordinate the functions of the neighboring cells. Cell division appears to be regulated by persistent periodic, rhythmic, or cyclic variations of concentration of some key molecules. Biochemical oscillations, both extra- and intracellular have been suggested to play a fundamental role in this interdependency of cell functions and biochemical signals.[47] In normal cells, differences in phase between oscillating concentrations in neighboring cells (asynchrony) keep a harmonic rate of cell division. Contrary to this, a breakdown in intercellular communication among tumor cells provokes a synchrony which is characteristic to neoplastic growth.[48]

A. Role of Mitochondria

Warburg's theory[49] suggests that tumor cells arise from normal cells in two phases: the first is an irreversible damage to the respiratory mechanism followed by a second phase of replacement of the lost respiratory energy by glycolytic energy, and as a consequence of this change of energy source the highly developed normal somatic cells are converted into more primitive, randomly growing cancer cells. Coincidentally, the mitochondria (which constitutes the respiratory mechanism of the cell) and the glycolytic system are principal cellular oscillators[50,51] which by means of intracellular oscillations control not only their own metabolism but also that of neighboring cells. There is evidence that metabolic oscillations are coupled to the mitochondrial oscillations and that the balance between aerobic respiration and glycolysis is controlled by this coupling.[48]

The alteration of the mitochondrial mechanism for oxidative phosphorylation by a carcinogen has been related to damage of the mitochondrial membrane.[5] Mitochondria has been found to be more sensitive than the rest of the cell to the toxic effects of many carcinogenic substances.[52] The impairment of the membrane permits the release of the mitochondrial genome of circular DNA into the extramitochondrial environment,[53] and this genetic material may eventually alter the nuclear genome of the cell in a similar way as when an appropriate exogenous oncogenic virus invades a host cell to transform it into a cancer cell.[54,55] In addition to this, direct damage to the mitochondria will result in a slow enhancement of glycolysis (energetic compensation) leading toward the neoplastic state, at the same time that the mitochondrial damage is being passed on to the progeny if the mitochondrial DNA has been affected. As a consequence, it is possible that in eucaryotic cells the two distinct genetic systems, the nuclear DNA and the cytoplasmic DNA are involved in the pathways to carcinogenesis. A damage to the nucleus affecting the nuclear genome controlling the cell membrane may produce a slow accumulation of Ca^{2+} in the cell with the consequent impairment of its ability to form permeable junctions with other cells. Since the mitochondrial membrane could also be damaged, a concomitant irreversible damage of cell respiration could take place.

It is noteworthy that some authors have suggested that mitochondrial DNA represents a second genetic system of the cell and, therefore, mutability of mitochondrial DNA is an important possibility which may induce pathological changes in the cell.[56] On the other hand, the experimental evidence indicates that mitochondrial DNA plays an important role in the specification of cell membrane characteristics, and on this fact is based the hypothesis according to which carcinogens — and possibly certain oncogenic viruses — affect primarily the mitochondrial genetic apparatus which consequently can cause various degrees of heritable defects in the plasma membrane.[57] The modification of plasma membrane permeability toward different ions, provoking

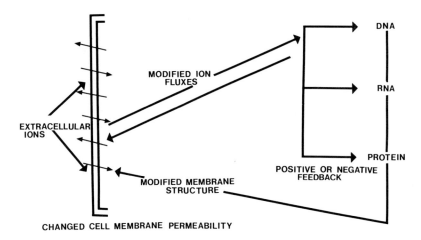

FIGURE 1. Probable relationships between changes in ionic permeability of the cell membrane and gene expression affecting its characteristics.

a change of intracellular concentrations, is capable to act on feedback mechanisms altering the translation of RNA into protein,[58,59] and likely the transcription of DNA into RNA.[60,61] It is obvious that these feedback interactions between plasma membrane permeability and gene expression might be capable to modify the state of differentiation of the cell (Figure 1).

The experimental evidence indicates that mitochondria of tumor cells show alterations in the outer and inner membranes which are an inherent characteristic of the neoplastic state.[62,63] Changes in the genetic apparatus of the mitochondria are probably the cause of these alterations. A normal synthesis of mitochondrial protein, related to a proper functioning of the mitochondrial genetic apparatus, is needed for the maintenance of the cell respiratory chain, "sine qua non" condition for normal cell division.[64] On the other hand, the abundant literature indicates that mitochondrial alterations (structural as well as biochemical) may be a primary stage in oncogenesis,[65] that short periods of feeding carcinogenic azo dyes decrease the yield of mitochondria from liver,[66,67] that the mass of mitochondria isolated from hepatoma is less than that obtained from normal liver,[68,69] that there is a correlation in animals fed with amino azo dyes between tumor incidence and the minimum swelling (response to hypotonic media) of liver mitochondria,[70] and that some precancerous enzymic modifications can be related to structural alterations of the mitochondria provoked by the carcinogen.[71]

On the basis of experiences on induced morphological differentiation in malignant mouse neuroblastoma by means of substances which are known to be antivitamin K agents and which release Ca^{2+} from mammalian mitochondria, the hypothesis that malignancy is caused by disturbed vitamin K physiology of the cell has been proposed.[72] In this case, the release of Ca^{2+} from mitochondria, with a consequent increase of the cytoplasmic concentration, is the factor leading to cell differentiation. However, other experiments indicate that the role of vitamin K in tumor growth is related to its important role in a postribosomal step for carboxylation of glutamyl residues to gamma-carboxyglutamic acid residues, and proteins containing this residue have very strong Ca^{2+} - and phospholipid-binding properties which make possible their involvement in the alteration of certain surface properties of tumor cells by impairment of vitamin K physiology.[73] Concerning these effects it is noteworthy that vitamin A, a strong antitumorigenic agent, also provokes modifications of the cancer cell surface and induces cell differentiation.[74,75] On the other hand, vitamin A is also involved in calcium metabolism and in energy metabolism of the mitochondria.

B. Calcium and Metabolism of the Cell

The importance of the ionic environment in the various cell compartments is pointed out by the concept of cell metabolism controlled by the ionic concentrations.[76] This concept emphasizes the importance of Ca^{2+} concentration on enzyme systems involved in energy production and macromolecular biosynthesis. Together with Mg^{2+}, Ca^{2+} plays a major role in the regulation of many basic cell functions. In this respect, it is interesting to note that some characteristics of the altered behavior of tumor cells have been related to modifications of their calcium and magnesium metabolism.[77-80] One of the most striking characteristics of the cancer cell, its high aerobic glycolysis, is strongly stimulated by an increase of the Ca/Mg ratio.[81]

The specific properties of the mitochondrial Ca^{2+}-transport system play a key role in the rapid regulation of the Ca/Mg ratio outside the mitochondria, where changes in this ratio are able to alter the activity of Ca^{2+}- and Mg^{2+}-sensitive enzymes. It should be emphasized that the fact that the principal regulating factor is not simply the absolute concentration of ionic calcium in the environment, but rather the Ca/Mg ratio in the vicinity of the ion-sensitive enzyme.[82] In tumor cells the mitochondria show the characteristic to retain for long periods the cytoplasmic Ca^{2+} without releasing it back to the cytoplasm. In terms of the ionic control theory, this behavior which provokes the sequestration of cytoplasmic Ca^{2+} modifies the Ca/Mg ratio in the cytoplasm and consequently leads to changes in the rate of metabolism sensitive to that ratio. Coincidentally, a high accumulation of Ca^{2+} in mitochondria has been observed during carcinogenesis and in tumors.[10,83-85] These results, however, present a fundamental point of uncertainty which is general to all the information on intracellular location of ions and small molecules obtained by experimental methods that alter the conditions in the living cell. Homogenization by mechanical disruption, osmotic shock, or nitrogen pressure shock can produce translocations of ions and small molecules which in the intact cell are prevented by their compartmental distribution. This problem has been already pointed out by several authors.[86,87] As a consequence, the results of subcellular distribution after compartmental disruptions must be cautiously considered, particularly in the case of small ions as Ca^{2+} because artifactual interactions are almost unavoidable.

In the case of cell injury, a membrane alteration permits an increased net flux of Ca^{2+} into the cell forcing mitochondria to maintain the intracellular Ca^{2+} homeostasis by sequestering the excess of Ca^{2+}.[88,89] This massive loading with Ca^{2+} will damage the mitochondria and the oxidative phosphorylation mechanism, a phenomenon which will become irreversible when the abnormal Ca^{2+} influx takes place for a prolonged period of time. This sequence of events is clearly shown by acute hepatotoxicity. The toxic effect of carbon tetrachloride (which as thioacetamide is a carcinogen) takes place by its conversion to a highly reactive free radical (CCl_3^-) by action of the oxidases of the endoplasmic reticulum. Subsequently the cell membrane is impaired by destruction of its lipid component and it becomes permeable to the extracellular Ca^{2+}.[90] The high levels of extracellular Ca^{2+} entering the cytoplasm are sequestered by the mitochondria, and as a consequence of this uptake there is an impairment of the mitochondrial mechanism of oxidative phosphorylation.[34,91-93] In acute intoxication due to a single dose this liver transformation by internal calcification of the cells, is a relatively controlled and transient phenomenon (the importance of cell damage and the duration of the healing process is dose-dependent) but it may progress (after repeated administration of the toxic substance) until the concentration of mitochondrial Ca^{2+} makes impossible the functioning of mitochondria. In absence of the mitochondrial buffer system the concentration of Ca^{2+} in the cytosol may increase to values which provoke the cell death (necrotic calcification).[26] On the other hand, cells which are submitted to a chronic exposure to the toxic agent, and which do not die by necrosis, are slowly shifted toward the glycolytic state in order to survive the impairment of the respiratory system. This shift toward a primitive state provoked by a metabolic disorder implicates that the succeeding generations of cells become more and more autonomous from tissue-level controls, lacking a response to the autonomic nervous system as well as to the immune system.[29] When they have reached genetic autonomy, because of their lack of intracellular communication and the fact that they carry altered mitochondrial or nuclear DNA, they will persist indefinitely as tumor cells.

IV. CALCIUM AND ONCOGENIC VIRUSES

Since some virus-host cell interactions are used as a cancer model,[94] we can consider the host cell transformation resulting from that interaction as additional evidence of the involvement of Ca^{2+} in carcinogenesis. It is a well-known fact that viruses change the fluidity of the host cell membrane provoking ion fluxes and leak out of small molecules across it.[95,96] It has been suggested that the ion fluxes are produced to provide the ionic conditions necessary to the virus developmental program,[97] and which are to change the intracellular medium to become more like the external one.[98] In addition to this, there is experimental evidence that calcium bound to the virus outside the cell is used to provide the free energy (by release being transported down a concentration gradient) necessary to favor virus penetration and disassembly when it enters a cell and releases both RNA and protein.[99,100]

V. CYTOSTATICS AND CALCIUM

An experimental fact is the known interference of several groups of antibiotics (ionophores, tetracyclines, polymyxines, aminoglycosides, etc.) with calcium ions. Their action being reflected by increasing or decreasing calcium-dependent effects. In most cases this action seems to be related to a competition between the antibiotics and calcium for membrane binding sites involved in the transport of the ion.[101] There is experimental evidence that the antibiotics might interfere by binding competition with the calcium-phospholipid interaction at cellular membranes.[102,103] Several very effective antitumor compounds present a remarkable inhibitory effect on the cellular respiratory mechanisms, while the antitumor effect is generally accompanied by a considerable toxicity apparently related to changes in calcium metabolism.

Adriamycin and daunomycin have shown effects both in calcium transport by a biphasic model of membrane and in mitochondria of tumor cells.[104] The interaction cytostatic-phospholipid, taking place by binding through the amino sugars to be membrane phospholipids may be the cause of Ca^{2+}-flux impairment at the mitochondrial membrane level.[104-106] On the other hand, adriamycin and daunomycin show some chelating properties: adriamycin provokes hypocalcemia in experimental animals,[107] the antiblastic activity of daunomycin is enhanced in presence of low levels of Ca^{2+},[108] while there is a reduction of in vivo antitumoral activity by simultaneous administration of Ca^{2+}.[109] These effects appear to strengthen the hypothesis of cytostatic activity due, at least partially, to the chelating characteristics of these anthracycline antibiotics. These properties may also be responsible for the membrane associated actions of adriamycin, altering surface properties and promoting dissociation of DNA-membrane complexes.[110] Vincristine, another antimitotic agent widely used in chemotherapy of tumors, is also capable to modify Ca^{2+} transport.[111]

Finally, it is interesting to point out that other drugs, such as desmethylimipramine, capable to inhibit cell respiration can increase the survival time of tumor-bearing animals.[112]

VI. CONCLUSIONS

The bibliography reviewed in this chapter seems to indicate a close relationship between cancer and calcium-dependent biological processes of the cell. This calcium dependency, especially in what concerns cell membrane physiology, can be inferred from the results of experimental carcinogenesis.

It is evident that the understanding of the biological mechanisms regulating the physicochemical characteristics of the membrane and the interrelationships between the various ion fluxes across the cell membrane, might provide a more clear and precise interpretation of the events triggering the neoplastic transformation of the cell. Unfortunately, the lack of sophistication of the rather difficult bioinorganic and physicochemical approaches, does not appeal to a more general use of them in cancer research. We must hope, that for the sake of the cancer phenomenon explanation, in the future more experimental work will be done on this line of cell biology investigation.

REFERENCES

1. Mundy, G. R., Calcium and cancer, Life Sci., 23, 1735, 1978.
2. Kishi, S., Fujiwara, T., and Nakahara, W., Comparison of chemical composition between hepatoma and normal liver tissues: sodium, potassium, calcium, magnesium, iron, iodine and chloride, including sodium chloride, Gann, 31, 1, 1937.
3. Hickie, R. A. and Kalant, H., Calcium and magnesium content of rat liver and Morris hepatoma 5123tc, Cancer Res., 27, 1053, 1967.
4. Anghileri, L. J., Cellular lipoproteins, calcium and magnesium in Novikoff hepatoma, liver of tumor-bearing and liver of normal rats, Oncology, 29, 152, 1974.
5. Hadler, H. I., Daniel, B. G., and Pratt, R. D., The induction of ATP-energized mitochondrial volume changes by carcinogenesis with N-hydroxy-N-acetylaminofluorenes when combined with showdomycin — a unitary hypothesis for carcinogenesis, J. Antibiot., 24, 405, 1971.
6. Hoelzl-Wallach, D. F., Cellular membranes and tumor behavior: a new hypothesis, Proc. Natl. Acad. Sci. USA, 61, 868, 1968.
7. Kalckar, H. M., Galactose metabolism and cell "sociology", Science, 150, 305, 1965.
8. Montagnier, L. and Torpier, G., Membrane changes in virus transformed cells, Bull. Cancer, 63, 123, 1976.
9. Smith, A. E. and Kenyon, D. H., A unifying concept of carcinogenesis and its therapeutic implication, Oncology, 27, 459, 1973.
10. Anghileri, L. J., Heidbreder, M., Weiler, G., and Dermietzel, R., Liver tumors induced by 4-dimethylaminoazobenzene: experimental basis for a chemical carcinogenesis concept, Arch. Geschwultsforsch, 46, 639, 1976.
11. Anghileri, L. J., Heidbreder, M., Weiler, G., and Dermietzel, R., Hepatocarcinogenesis by thioacetamide: correlations of histological and biochemical changes and possible role of cell injury, Exp. Cell Biol., 45, 34, 1977.
12. Arcos, J. C., Gosch, H. H., and Zickafoose, D., Fine structural alterations in cell particles during chemical carcinogenesis. III. Selective action of hepatic carcinogens other than 3'-methyl-4-dimethylaminoazobenzene on different types of mitochondrial swelling. Effect of stimulated liver growth, J. Biophys. Biochem. Cytol., 10, 23, 1961.
13. Litman, G. W. and Litman, R. T., Interaction of chemical carcinogens with plasma membranes: the effect of dimethylaminoazobenzene on erythrocyte osmotic fragility, Biochem. Biophys. Res. Commun., 60, 865, 1974.
14. Le Breton, E. and Moulé, Y., Biochemistry and physiology of the cancer cell, in The Cell, Vol. 5, Brache, J. and Mirsky, A. E., Eds., Academic Press, New York, 1961, 497.
15. Holmberg, G., On the in vivo release of cytoplasmic enzymes from ascites tumor cells as compared with strain L cells, Cancer Res., 21, 1386, 1961.
16. Anghileri, L. J. and Dermietzel, R., Cell coat in tumor cells — effect of trypsin and EDTA: a biochemical and morphological study, Oncology, 33, 17, 1976.
17. Amatruda, J. M. and Finch, E. D., Modulation of hexose uptake and insulin action by cell membrane fluidity, J. Biol. Chem., 254, 2619, 1979.

18. Jain, M. K. and White, H. B., in *Advances in Lipid Research*, Vol. 15, Paoletti, R. and Kritchevsky, D., Eds., Academic Press, New York 1977, 1.
19. Gordon, L. M., Sauerheber, R. D., and Esgate, J. A., Spin label studies on rat liver and heart plasma membranes: effects of temperature, calcium and lanthanum on membrane fluidity, *J. Supramol. Struct.*, 9, 299, 1978.
20. Nicolson, G. L., Poste, G., and Ji, T. H., Dynamic Aspects of Cell Surface Organization in Cell Surface Reviews, Poste, G. and Nicolson, G., Eds., North-Holland Publ., Amsterdam, 1977, 1.
21. Anghileri, L. J. and Dermietzel, R., Calcium-phosphate-phospholipid complexes in experimental tumors. Their possible relationship with tumor calcification, *Z. Krebsforsch.*, 79, 148, 1973.
22. Anghileri, L. J., Ca^{2+}- and Mg^{2+}-Phospholipid Complexes and Their Possible Role in Ionic Permeability Changes During Carcinogenesis, Symposium on Nutrition and Cancer, American College of Nutrition, Houston, Texas, June 9, 1977.
23. Anghileri, L. J., Stavrou, D., and Weidenbach, W., Phospholipids and calcification in human intracranial tumors, *Arch. Geschwultsforsch.*, 47, 330, 1977.
24. Anghileri, L. J. and Delbrück, H., Divalent cation-phospholipid complexes and tumor growth inhibition, *Experientia*, 34, 1664, 1979.
25. Rasmussen, H., Cell communication, calcium ion, and cyclic adenosine monophosphate, *Science*, 170, 404, 1970.
26. Price, J. M., Harman, J. W., Miller, E. C., and Miller, J. A., Progressive microscopic alterations in the liver of rats fed hepatic carcinogens 3′-methyl-4-dimethylaminoazobenzene and 4′-fluoro-4-dimethylaminoazobenzene, *Cancer Res.*, 12, 192, 1952.
27. Carafoli, E. and Tiozzo, R., A study of energy-linked calcium transport in liver mitochondria during CCl_4 intoxication, *Ex. Mol. Pathol.*, 9, 131, 1968.
28. Sell, D. E. and Reynolds, E. S., Liver parenchymal cell injury, *J. Cell Biol.*, 41, 736, 1969.
29. Gold, P. and Freeman, S. O., Specific carcinoembryonic antigens of the human digestive system, *J. Exp. Med.*, 122, 467, 1965.
30. Anghileri, L. J., Cell membrane ionic permeability, calcium ion, mitochondria, and carcinogenesis, *Arch. Geschwultsforsch.*, 48, 497, 1978.
31. Schliwa, M., The role of divalent ions in the regulation of microtubule assembly. In vivo studies on microtubules of the heliozoan axopodium using the ionophore A23187, *J. Cell Biol.*, 70, 527, 1976.
32. Rose, B. and Loewenstein, W. R., Permeability of cell junction depends on local cytoplasmic calcium activity, *Nature (London)*, 254, 250, 1975.
33. Lew, V. L. and Ferreira, H. G., Variable calcium sensitivity of a potassium selective channel in intact red-cell membranes, *Nature (London)*, 263, 336, 1977.
34. Anghileri, L. J., Metabolism of calcium and magnesium in liver during acute thioacetamide intoxication, *Int. J. Clin. Pharmacol.*, 14, 101, 1976.
35. Gabriel L., Canuto, R. A., Gravela, E., Garceo, R., and Feo, F., Alterations of liver ergastoplasmic membranes during DL-ethionine hepatocarcinogenesis — aminopyrine-demethylase activity and ribosome-membrane interaction, *Life Sci.*, 15, 2119, 1974.
36. Apffel, C. A., Nonimmunological host defenses. A review, *Cancer Res.*, 36, 1527, 1976.
37. Apffel, C. A., The endoplasmic reticulum membrane system and malignant neoplasia, *Progr. Exp. Tumor Res.*, 22, 317, 1978.
38. Apffel, C. A. and Ira, T., Microsomal aspects of carcinogenesis and neoplasia, *Med. Hypothesis*, 5, 23, 1979.
39. Cone, D. D., Unified theory on the basic mechanisms of normal mitotic control and oncogenesis, *J. Theoret. Biol.*, 30, 151, 1971.
40. Mitchell, R. S., Elgas, R. J., and Balk, S. D., Proliferation of Rous Sarcoma virus-infected, but not of normal, chicken fibroblasts in oxygen-enriched environments: preliminary report, *Proc. Natl. Acad. Sci. USA*, 73, 1265, 1976.
41. Whitfield, J. F., MacManus, J. P., Rixon, R. H., Boynton, A. L., Yondale, T., and Swierenga, S., The positive control of cell proliferation by the interplay of calcium ions and cyclic nucleotides: a review, *In Vitro*, 12, 1, 1976.
42. Schanne, F. A. X., Kane, A. B., Young, E. E., and Farber, J. L., Calcium dependence of toxic cell death: a final common pathway, *Science*, 206, 700, 1979.
43. Pitot, H. C., Some biochemical essentials of malignancy, *Cancer Res.*, 23, 1474, 1963.
44. Pardee, A. B., Cell division and a hypothesis of cancer, *Natl. Cancer Inst. Monogr.*, 14, 7, 1964.
45. Loewenstein, W. R., Some reflections on growth and differentiation, *Perspect. Biol. Med.*, 11, 260, 1968.
46. Loewenstein, W. R., Intercellular communications, *Sci. Am.*, 2, 79, 1970.
47. Burton, A. C., Cellular communication, contact inhibition, cell blocks and cancer: the impact of the work of W. R. Loewenstein, *Perspect. Biol. Med.* 14, 301, 1971.
48. Smith, A. E. and Kenyon, D. H., An unifying concept of carcinogenesis and its therapeutic implications, *Oncology*, 27, 459, 1973.

49. Warburg, O., On the origin of cancer cell, *Science,* 123, 309, 1956.
50. Chance, B. and Yoshioka, T., Sustained oscillations of ionic constituents of mitochondria, *Arch. Biochem. Biophys.,* 117, 451, 1966.
51. Parker, K., Relation of structure to energy coupling in rat liver mitochondria, *Fed. Proc. Fed. Am. Soc. Exp. Biol.,* 29, 1533, 1970.
52. Graffi, A., Zelluläre Speicherung cancerogener Kohlenwasserstoffe, *Z. Krebsforsch.,* 49, 477, 1939.
53. Borst, P. A., Mitochondrial nucleic acids, *Ann. Rev. Biochem.,* 41, 333, 1972.
54. Dulbecco, R., Cell transformation by viruses, *Science,* 166, 962, 1969.
55. Baluda, M. A. and Nayak, D. P., DNA complementary to viral RNA in leukemia cells induced by avian myeloblastosis virus, *Proc. Natl. Acad. Sci. USA,* 66, 329, 1970.
56. Nass, M. M. K., Mitochondrial DNA: advances, problems, and goals, *Science,* 165, 25, 1969.
57. Wilkie, D. T., Egilson, V., and Evans, I. H., Mitochondria in oncogensis, *Lancet,* 1, 697, 1975.
58. Douzon, P. and Maurel, P., Ionic regulation in genetic translation systems, *Proc. Natl. Acad. Sci. USA,* 74, 1013, 1977.
59. Shinohara, T. and Piatigorsky, J., Regulation of protein synthesis, intracellular electrolytes and cataract formation in vitro, *Nature (London),* 270, 406, 1977.
60. Herman, R. C. and Moyer, R. W., In vivo repair of bacteriophage T5 DNA: as assay for viral growth control, *Virology,* 66, 393, 1975.
61. Lacombe, C. and Laigle, A., A circular dichroism study of DNA-basic peptide association in the absence or in the presence of Ca^{2+}, *Nucleic Acid Res.,* 4, 1783, 1977.
62. Feo, F., Bonelli, G., and Matli, A., Effect of trypsin on morphological integrity and some functional activities of mitochondria from normal liver and AH-130 Yoshida ascites hepatoma, *Life Sci.,* 9, 1235, 1970.
63. White, M. T., Arya, D. V., and Tewari, K. K., Biochemical properties of neoplastic cell mitochondria, *J. Natl. Cancer Inst.,* 53, 553, 1974.
64. Mittwoch, V., Kirk, D. J., and Wilkie, D., Effects of mitochondrial inhibition by chloramphenicol on the mitotic cycle of human cell cultures, *J. Med. Gen.,* 11, 260, 1974.
65. White, M. T. and Tewari, K. K., Structural and functional changes in Novikoff hepatoma mitochondria, *Cancer Res.,* 33, 1645, 1973.
66. Price, J. M., Miller, E. C., Miller, J. A., and Weber, G. M., Studies on the intracellular composition of livers from rats fed various aminoazo dyes. II. 3'-methyl-2'-methyl-, and 2-methyl-4-dimethylaminoazobenzene and 4'-fluoro-4-dimethylaminoazobenzene, *Cancer Res.,* 10, 18, 1950.
67. Striebich, M. J., Shelton, E., and Schneider, W. C., Quantitative morphological studies on the livers and liver homogenates of rats fed 2-methyl-or 3'-methyl-4-dimethylaminoazobenzene, *Cancer Res.,* 13, 279, 1953.
68. Schneider, W. C., Intracellular distribution of enzymes. II. The distribution of succinic dehydrogenase, cytochrome oxidase, adenosinetriphosphatase, and phosphorus compounds in normal rat liver and in rat hepatoma, *Cancer Res.,* 6, 685, 1946.
69. Schneider, W. C. and Hogeborm, G. H., Intracellular distribution of enzymes. VI. The distribution of succino-oxidase and cytochrome-oxidase activities in normal mouse liver and liver hepatoma, *J. Natl. Cancer Inst.,* 10, 969, 1950.
70. Arcos, J. C., Griffith, G. W., and Cunningham, R. W., Fine structural alterations in cell particles during chemical carcinogenesis. II. Further evidence for their involvement in the mechanism of carcinogenesis. The swelling of rat liver mitochondria during feeding of aminoazo dyes, *J. Biophys. Biochem. Cytol.,* 7, 49, 1960.
71. Clerici, E. and Cudkowicz, G., Certain effects of p-dimethylaminoazobenzene, deficient diet and hypotonic media on mitochondrial enzymes in experimental rat liver carcinogenesis, *J. Natl. Cancer Inst.,* 16, 1459, 1956.
72. Egilsson, V., Cancer and Vitamin K, *Lancet,* 2, 254, 1977.
73. Hilgard, P., Cancer and vitamin K, *Lancet,* 2, 403, 1977.
74. Brandes, D., Radiation response of L1210 leukemia cells treated with vitamin A-alcohol, *J. Natl. Cancer Inst.,* 52, 945, 1974.
75. Hogan-Ryan, A. and Fenelly, J. J., Neuraminidase like effect of vitamin A on cell surface, *Eur. J. Cancer,* 14, 113, 1978.
76. Bygrave, F. L., in *Control Mechanisms in Cancer,* Criss, W. E., Ono, T., and Sabine, J. R., Eds., Raven Press, New York, 1976, 411.
77. Rubin, H., Control role of magnesium in coordinate control of metabolism and growth in animal cells, *Proc. Natl. Acad. Sci. USA,* 72, 3551, 1975.
78. Boynton, A. L. and Whitfield, J. F., Different calcium requirements for proliferation of conditionally and unconditionally tumorigenic mouse cells, *Proc. Natl. Acad. Sci. USA,* 73, 1651, 1976.
79. Moore, L. and Postan, I., Energy-dependent calcium uptake by fibroblast microsomes, *Ann. N.Y. Acad. Sci.,* 307, 177, 1978.

80. Whitfield, J. F., Boynton, A. L., MacManus, J. P., Sikorska, M. and Tsang, B. K., The regulation of cell proliferation by calcium and cyclic AMP, *Mol. Cell. Biochem.*, 27, 155, 1979.
81. Bossi, D., Cittadini, A., Wolf, F., Milani, A., Magalini, S., and Terranova, T., Intracellular calcium and magnesium content and aerobic lactate production in intact Ehrlich ascites tumor cells, *FEBS Lett.*, 104, 6, 1979.
82. Meh, J. and Bygrave, F. L., The role of mitochondria in modifying calcium-sensitive cytoplasmic metabolic activities. Modification of piruvate kinase activity, *Biochem. J.*, 128, 415, 1972.
83. Anghileri, L. J. and Heidbreder, M., Cell membrane ionic permeability and mitochondria changes during 4-dimethylaminoazobenzene carcinogenesis, *Arch. Geschwulstforsch.*, 46, 389, 1976.
84. Williams, J. F., Cook, P. C., Matthaei, K. I., and Halley, J. B. W., Pyridine and adenine nucleotide ratios and futile substrate cycling in regulation of energy metabolism and proposed hyperthermic regression of neoplasm, in *Control Mechanisms in Cancer*, Criss, W. E., Ono, T., and Sabine, J. R., Eds., Raven Press, New York, 1976, 425.
85. Lanone, K. F., Hemington, J. G., Ohnishi, T., Morris, H. P., and Williamson, J. R., Defects of anion and electron transport in Morris hepatoma mitochondria, in *Hormones and Cancer*, McKerns, E., Ed., Academic Press, New York, 1974, 131.
86. Kay, E. R. M., Smellie, R. M. S., Humphrey, G. F., et al., A comparison of cell nuclei isolated from rabbit tissues by aqueous and non-aqueous procedures, *Biochem. J.*, 62, 160, 1956.
87. Slater, E. C., Sarcosomes (muscle mitochondria), mitochondria and other cytoplasmic inclusions, *Symp. Soc. Exp. Biol.*, 10, 110. 1957.
88. Lehninger, A. L., in *The Mitochondrion*, Academic Press, New York, 1964, 175.
89. Borle, A. B., Calcium metabolism at the cellular level, *Fed. Proc. Fed. Am. Soc. Exp. Biol.*, 32, 1944, 1973.
90. Thiers, R. E., Reynolds, E. S., and Vallee, B. L., The effect of carbon tetrachloride poisoning on subcellular metal distribution in rat liver, *J. Biol. Chem.*, 235, 2130, 1960.
91. Reynolds, E. S., Thiers, R. E., and Vallee, B. L., Mitochondrial function and metal content in carbon tetrachloride poisoning, *J. Biol. Chem.*, 237, 3546, 1962.
92. Reynolds, E. S., Liver parenchymal cell injury. Initial alterations of the cell following poisoning with carbon tetrachloride, *J. Cell Biol.*, 19, 139, 1963.
93. Moore, L., Davenport, G. R., and Landon, E. J., Calcium uptake of a rat liver microsomal subcellular fraction in response to in vivo administration of carbon tetrachloride, *J. Biol. Chem.*, 251, 1197, 1976.
94. Cairns, J., The cancer problem, *Sci. Am.*, Nov., 64, 1975.
95. Levanon, A., Kohn, A., and Inbar, M., Increase in lipid fluidity of cellular membranes induced by adsorption of RNA and DNA virions, *J. Virol.*, 22, 353, 1977.
96. Peterhaus, E., Brouse, E., and Wyler, R., ATP production and Ca^{2+} uptake by mitochondria in chick embryo fibroblasts during replication of Semliki Forest virus, *Experientia*, 33, 827, 1977.
97. Herrlich, P., Rahmsdorf, H. J., and Schweiger, M., Regulation of macromolecular synthesis by membrane changes, *Adv. Biosci.*, 12, 523, 1973.
98. Ponta, H., Altendorf, K. H., Schweiger, M., Hirsch-Kauffman, M., et al., *E. coli* membranes become permeable to ions following T7-virus infection, *Mol. Gen. Genet.*, 149, 145, 1976.
99. Butler, P. J. G. and Durham, A. C. H., Tobacco mosaic virus protein aggregation and the virus assembly, *Adv. Protein Chem.*, 31, 187, 1977.
100. Durham, A. C. H., The role of small ions, especially calcium, in virus disassembly, takeover, and transformation, *Biomedicine*, 28, 307, 1978.
101. Corrado, A. P., Prado, W. A., and Pimenta de Morais, I., Competitive antagonism between calcium and aminoglycoside antibiotics in skeletal and smooth muscles, in *Concepts of Membranes in Regulation and Excitation*, Rocha, E., Silva, M., and Suarez-Kurtz, G., Eds., Raven Press, N.Y., 1975, 201.
102. Schacht, J., Inhibition by neomycin of polyphosphoinositide turnover in subcellular fractions of guinea pig cerebral cortex in vitro, *J. Neurochem.*, 27, 1119, 1976.
103. Lodhi, S., Weiner, N. D., and Schacht, J., Interactions of neomycin and calcium in synaptosomal membranes and polyphosphoinositide monolayers, *Biochim. Biophys. Acta.*, 426, 781, 1976.
104. Anghileri, L. J., Ca^{2+}-transport inhibition by the antitumor agents adriamycin and daunomycin, *Arzneim. Forsch. Drug Res.*, 27, 1177, 1977.
105. Revis, N. W. and Marusic, N., Effects of doxorubicin and its aglycone metabolite on calcium sequestration by rabbit heart, liver and kidney mitochondria, *Life Sci.*, 25, 1055, 1979.
106. Duarte-Karim, M., Ruysschaert, D. M., and Hildebrand, O., Affinity of adriamycin to phospholipids — a possible explanation for cardiac mitochondrial lesions, *Biochem. Biophys. Res. Commun.*, 71, 658, 1976.

107. **Young, D. M., Mettler, F. P., and Fioravanti, J. L.**, Adriamycin, Verapamil and calcium metabolism, *Proceedings of AACR,* AACR Abstr. No. 357, p. 90, 1976.
108. **Bossa, R., Cegani, M., Dubini, F., et al.**, Interferences between daunomycin and calcium, *Progress in Chemotherapy,* Vol. 3, G.K. Daikos, Athens, 1974.
109. **Bossa, R., Galatulas, I., and Mantovani, E.**, Interference by calcium with antileukemic effect of daunomycin in mice, *Int. Res. Commun. Syst.,* 2, 1508, 1974.
110. **Murphree, S. A., Cunningham, L. S., Hwang, K. M., et al.**, Membrane associated actions of adriamycin, *Proceedings of AACR,* AACR Abstr. No. 164, 1976.
111. **Yasin, R., Highes, B. P., and Parker, J. A.**, The effect of vincristine on the calcium transport and phospholipid composition of rat skeletal muscle microsomes, *Lab. Invest.,* 29, 207, 1973.
112. **Bossa, R., Fiorini, G., Galatulas, I., et al.**, Activity of desmethylimipramine on brain and Ehrlich ascites tumor respiration and glycolysis, *Naunyn-Schmiedebergs Arch. Pharmakol.,* 269, 453, 1971.

Chapter 12

CALCIUM IN PSYCHIATRIC ILLNESS

John Scott Carman

TABLE OF CONTENTS

I. Affective Illness 158
    A. Why Calcium? 158
    B. Pharmacological Evidence Implicating Calcium 159
    C. Calcitropic Agents: Modulators of Effect and Activity............. 160
        1. Calcium Lactate 160
        2. Parathormone 160
        3. Dihydrotachysterol 160
        4. Calcitonin 161
            a. Synthetic Salmon Calcitonin (SCT) 161
            b. Divergent Effects of SCT on Calcium in CSF vs. Serum 161
            c. Opposite Effects of DHT and SCT on Periodicity 161
            d. Adverse Effects of Calcitonin: Depression 162
            e. Adverse Effects of Calcitonin: Anorexia 162
            f. Calcitonin: Aborts Episodes of Malignant Hyperpyrexia and Reduces Serum CPK 162
            g. Calcitonin: Reduction of Prolactin 163
            h. Mechanism of Action of Calcitonin — Lack of Clear Effect on Monoamines? 163
            i. Is Calcitonin a Neuroactive Peptide With Direct Effects? 163
            j. Differences in CSF Human Calcitonin (HCT) by Behavioral State 163

II. Anxiety 163

III. Calcium and Schizophrenia 164

IV. Calcium and Neuroleptics... 164

V. Calcium and Pain... 164
    A. Calcitonin and Pain... 165

VI. Alcoholism 165

VII. Nonspecific Influences 165

References... 165

I. AFFECTIVE ILLNESS

A. Why Calcium?

The observation that lithium salts normalize periodic extremes of motion and emotion has awakened interest in possible etiologic relevance or psychopharmacologic use of other cations. The following considerations led us to focus particularly on calcium:

1. Psychiatric symptoms are frequent in primary disorders of calcium metabolism.
    a. Anergy, lethargy, and depression, progressing to catatonic stupor, but often without evidence of intellectual dysfunction, have been observed in more than 40% of patients with hyperparathyroidism[1-2] vs. a lesser percent in Cushing's (35%) or Addison's (15%) disease or in hypothyroid myxedema (30%).[1-33]

    A direct proportionality between the magnitude of hypercalcemia and the severity of psychiatric symptoms has been suggested.[2,18] Similarly, increased calcium secondary to vitamin D intoxication has been associated with mood disturbance.[26-27]
    b. On the other hand, hypocalcemic states appear unusually able to mimic "functional" excitement, irritability, paranoia, and most important and unique — cyclic mood disturbance.[34-57]
2. Behavioral symptoms analagous to affect disturbance in man follow experimental manipulations of CNS calcium in animals.
    a. Less than 10% reductions of CSF calcium by ventricular infusion of citrate, EDTA, or artificial CSF low in calcium produce arousal, excitement, hyperactivity, and irritability.[58-63]
    b. Increased ventricular calcium produced sedation and catatonic behavior in cats and goats.[64-67]
3. Primary mood shifts — secondary calcium shifts.
    a. Only two studies out of twelve[68-79] found significant group differences between normal and depressed patients on CSF calcium[68] or CSF calcium/magnesium ratio.[69]
    b. However, mean 10% reductions (though not always significant) in CSF and/or serum calcium have been reported during antidepressant response to ECT,[78-79] lithium, tricyclic antidepressants, MAOI, and sleep deprivation.
    c. However, in longitudinal studies of serum constituents in rapidly cycling manic depressive or periodic catatonic patients, *serum* calcium *increased* about as much as *CSF* calcium *decreased* — in precise coincidence with the "switch" into mania or psychotic agitation or excitement (Table 1).[90-93]

    In normal or depressed humans[94-95] even *major increases* in serum calcium are accompanied only by *minimal increases* in CSF calcium — only 10% as large as those in serum. Nevertheless, in such "rapid cycling" manic depressive patients, periodic serum calcium elevations appear to trigger more enduring and opposite reductions in CSF calcium and, which, in turn, may account for periodic episodes of psychotic agitation.

    Decreased serum cortisol, already observed during recurrent switches into mania,[96-97] could explain these phasic increases in serum calcium, but its effects on CSF calcium are unknown and its interactions with calcitonin and PTH are uncertain.[98] On the other hand, however, the observed calcium increase in CSF of depressed patients might be primary — by increasing hypothalamic CRF release,[99] it might produce the lack of dexamethasone suppressability of plasma cortisol seen in those same patients.[100] Thus, as we learn more and more pieces of the puzzle of the etiology of affective illness, we continue to find nothing which could not have been caused by a central disorder of calcium regulation or distribution.[101]

Table 1
NONSPECIFIC FACTORS INFLUENCING CALCIUM CONCENTRATION IN SERUM AND CSF

Factors	Influence on Ca
↑ Stress[225-226]	↓ Serum total + ionized; negative balance refractory to vitamin D
↑ Activity[74,226-228]	↓ Serum total; positive balance; possible ↓ CSF
Erect posture[229]	↑ Serum total ionized
↓ Sleep[230-231]	↑ Serum total and ↓ CSF
Circadian rhythm[35,232-233]	Serum lowest at 3 a.m., peak at 7 p.m., CSF similar; thus, if a patient underwent a 180° phase shift, 7 a.m. specimens might actually represent 7 p.m. value for the organism, and consequently appear elevated
Dietary changes[234-238]	
↑ Ca	↑ Serum total + ionized overall; and amplitude of circadian flux in serum total Ca; while Thyr-PTH-X or calcium restriction markedly exaggerates that flux
↑ P, ↑ Mg, ↑ Prot., ↓ Vit. D, ↓ Calories	↓ Total serum

B. Pharmacological Evidence Implicating Calcium

1. Progressive restriction of dietary calcium mitigated and finally abolished both the rhythmic rises in serum calcium and the periodic psychotic episodes in one patient.[92]
2. Lithium, which reduces the efficency of alimentary calcium absorption,[102] may exert its prophylactic effect on recurrent agitation by mimicking the effects of dietary restriction. Indeed, in rapidly cycling bipolar patients, whether or not lithium abolished behavioral cycling appeared invariably associated with whether or not lithium abolished this serum calcium cycle.[101]

   Lithium treatment is associated with a transient increase in serum total calcium,[103-107] a more enduring hypermagnesemia,[106-108] hypocalciuria,[102,109-115] altered RBC/plasma ratio of Ca or Mg,[116] and increase in ionized calcium significantly correlated with the serum lithium level.[117] Birch[118,119] has suggested — on the basis of lithium's physiochemical similarities to calcium and magnesium ions — that lithium competes with those ions for their usual sites in cells and bone, thereby causing demineralization.[105,120,121] To the extent that divalent cation displacement is germane to lithium's mode of action might predict response.[104] Increases in serum calcium and magnesium were noted early in treatment in those depressed patients who subsequently responded to lithium,[104] while neither ion was altered significantly in the nonresponders. Those patients who became temporarily hypomanic early in the course of lithium treatment exhibited much greater and highly significant increases in serum calcium but no significant change in magnesium. Similiar large acute hypercalcemic response to lithium administration occurs in rats which are surgically rendered unable to produce calcitonin.[122] It is tempting to speculate that periodic calcitonin insufficiency might characterize those patients with a liability to lithium-induced hypercalcemia and hypomania. Moreover, in rapidly cycling patients, the efficacy of lithium in antimanic prophylaxis seemed linked to its ability to disrupt previously rhythmic increases in serum calcium and phosphorus.[101] On the other hand, the hypermagnesemic shift on lithium may augur development of a neurotoxic reaction to that drug.[116,123]

The recent observation that chronic lithium pretreatment in depressed patients may diminish the hypercalcemic effect of *parathormone* without altering its hypophosphatemic action[124] suggests that lithium may obstruct the osteogeneic, but not the nephrogenic, action of parathormone. It might do so by displacing (labile) bone calcium, making it less available to increased resorption, or by inhibition of vitamin D activation, thus, accounting for the decreased gut calcium absorption reported on lithium.[102] Recently, several groups have reported functional[125-127] or structural hyperparathyroidism in patients long maintained on lithium.[106,126,127] Such glandular hyperplasia may represent a compensatory response to lithium's block of the hypercalcemic effect of parathormone.[124]

Interestingly, lithium appears to reverse the nocturnal refractoriness of bone to PTH[108] — perhaps explaining its osteopenic effects. Also interesting is the observation that despite the similar changes in blood chemistry noted in unipolar and bipolar patients, only the bipolar group exhibited osteopenia on lithium.[105]

Conversely, lithium reabsorption is not affected by PTH.[128] However, more than 25% reduction in serum and RBC lithium concentrations ($p < 0.005$, paired-t) were noted after calcitonin in manic patients who were coadministered constant doses of lithium throughout three active and three placebo days.[129] One plausible explanation of the decreases in blood lithium may be sequestration in CSF, bone, or mitochondria, i.e., lithium behaving like calcium under the influence of CT.[130] However, diminished alimentary absorption of new lithium during the hours between SCT administration and blood collection appears a more plausible explanation, since the reductions following SCT are very nearly mimicked by withholding all lithium during the same interval, and since both *giving* SCT and *withholding* lithium produce no additive lowering of RBC or plasma lithium levels.

C. Calcitropic Agents: Modulators of Effect and Activity

To explore the causal relevance of these ionic shifts to the attendant behavioral alterations, we performed double-blind, placebo-controlled trials in similar patients of agents which might enhance or inhibit the serum and CSF calcium findings.

1. Calcium Lactate

Six patients received 750 mg calcium lactate or identical placebo capsules by mouth at 10 p.m. Two week-long series of nightly calcium treatments were arranged in random order with two week-long series of nightly placebo.

Calcium produced significant increases in agitation (mania) ratings by a mean 30% of baseline ($p < 0.05$, paired-t) in all six patients. On the other hand, when three periodically psychotic patients on routine hospital diet were given single nonblind oral calcium lactate loads of 3 g, no switches were induced during stable periods where one would not have been predicted.[101]

2. Parathormone

Richter[55] reported that idiopathic hypoparathyroidism in animals and man may cause dramatic cycles in activity and mood. However, while acute PTH infusion increased serum calcium, it had minimal effect on CSF calcium.[131] Similarly, open parathormone infusions to seven depressed patients had no effect on mood, despite induction of modest acute increases in serum calcium.[124]

3. Dihydrotachysterol

Dihydrotachysterol (DHT) 0.2 mg was administered orally three times daily to eight otherwise medication-free patients for approximately 1 month, preceded and followed by placebo periods of about equal length. DHT, a synthetic analogue of vitamin D, increased — within normal range — serum total calcium, decreased CSF calcium in the only patient measured, and markedly increased ratings of manic symptoms. Motor activity increased, in four patients wearing wristband monitors,[132] by as much as 50% of placebo levels.[101]

4. Calcitonin
a. Synthetic Salmon Calcitonin (SCT)

Like human calcitonin, SCT is a peptide produced in neural crest-derived tissues[133] and capable of diminishing serum calcium and phosphorus. In the two trials of this hormone, 15 of 17 manic patients received a random-coded sequence of six daily injections (three active and three placebo). Activity counts decreased after SCT by 40% of baseline (placebo) levels in five patients studied by electronic wristband activity monitors.[132] A substantial and significant reduction in manic symptoms occurred after active but not placebo SCT, beginning as early as 2 hr after injection and lasting as long as 36 hr (though usually less than 24).[129,134] This effect was discernable even in patients previously thought to have received maximal antimanic effect from their concurrent (stable) doses of neuroleptics or lithium. There was no significant change in psychotic symptoms or anxiety for these or another 11 schizophrenic patients receiving the hormone.

Important issues remain to be explored, including: whether repeated administration produced tachyphylaxis, cumulative effects, or, on cessation, rebound symptoms. Both studies employed the same arbitrarily chosen dose (140 MRC Units), and effects of higher or lower doses need to be explored. No major differences were observed between the behavioral and biochemical effects of SCT or placebo given every 3 days at 7 p.m. (in the first SCT trial) or at 7 a.m. daily (in the second SCT trial).

b. Divergent Effects of SCT on Calcium in CSF vs. Serum

In the first SCT trial, the size of the hypocalcemic response to calcitonin was significantly greater in seven baseline depressed or stuporous patients (with a mean decrease of 0.44 ± 0.03 meq/ℓ) than in the eight manic or psychotically agitated patients (with a mean decrease of 0.11 ± 0.03 meq/ℓ; $p < 0.002$, student-t). In the second trial, all patients studied were at least somewhat manic or agitated and the mean calcium decrease for that group as a whole was *also* 0.11 meq/ℓ.

SCT increased CSF calcium by 10% in three patients who underwent LPs during the first trial ($p < 0.05$, paired-t).[134] Similarly, CSF calcium and CSF PTH increased in five monkeys from 2 to 10 hr after injection of SCT.[135]

Thus, DHT and SCT seem uniquely able to induce opposite but equal changes in serum vs. CSF calcium — mimicking or reversing the equal but opposite shifts observed to occur spontaneously during periodic onsets of mania. The opposite behavioral effects of pharmacologic trials of diametrically opposite calcium regulatory hormones suggest that state-related shifts in CSF and serum calcium may bear some causal relevance to cyclic mood flux. In the second trial, the reduction in BPRS "manic" composite score correlated positively with the reduction in serum calcium induced by salmon calcitonin. Moreover, our data already suggest that the feedback mechanisms regulating CSF calcium in response to changes in peripheral calcium may be unusually reactive and tend to overshoot in manic patients. This is further suggested by the reduction of urinary calcium excretion following oral calcium loads in manic patients.[239]

c. Opposite Effects of DHT and SCT on Periodicity

On placebo, three of the patients participating in both the DHT and first SCT trials had exhibited periodic spontaneous exacerbations. On DHT, these episodes grew more frequent, while intervening periods of lucidity became shorter and cycle length decreased by about 50%. When 20 consecutive injections of active SCT were given every third evening to each of three rapidly cycling patients, anticipated increases in serum calcium and phosphorus were reduced in amplitude, and the onsets of expected and formerly quite regular agitated episodes were delayed or aborted. Consequently, total period length (interval between consecutive onsets of agitation) increased by about 70% on calcitonin.[136]

d. Adverse Effects of Calcitonin: Depression

Depression increased on SCT vs. placebo in 10 of 12 patients in the first trial ($p =$ ns; paired-t) including two schizophrenic patients in whom intense dysphoria culminated in serious suicide attempts. Unusually high baseline serum levels of endogenous human CT, earlier measured in those two patients, may have signaled some idiosyncratic vulnerability to SCT-induced depression. Although depression increased in 14 of 18 patients in the second trial, there were no suicide attempts. Taken together and analyzed by sign test, both trials suggest a depressogenic effect of SCT in 24 of 30 patients ($p < 0.01$).

The five most verbal patients were asked to rate the severity and describe the quality of subjective mood effects produced by SCT. Although blind to the sequence of injections, each was able to correctly identify every day on which he or she received active calcitonin by a feeling of generalized dysphoria, heaviness, lethargy, and anergic hopelessness, which one patient compared to her experience following an infusion of physostigmine earlier in her hospitalization.

e. Adverse Effects of Calcitonin: Anorexia

Nausea became troublesome in four patients and severe vomiting resulted in the only patient aborted from the second trial. Anorexia produced a modest (2 to 4%) but significant weight loss in 26 of 30 patients in the course of both trials.

The development of nausea or vomiting did not influence behavioral rating changes, and the finding of anorexia and weight loss supports the recent demonstration, in monkeys and rats, of a *marked* decline in oral ingestive behaviors, following considerably higher doses by weight.[137] This anorexia seems unlikely to be due to the recently described hyperglycemic effect of SCT,[138] since no consistent changes in blood glucose were noted after SCT in this group. Alternatively, SCT might enhance the hypothalamic satiety mechanism[139-140] by changing brain calcium concentrations or by some direct CNS effect of the peptide. On the other hand, decreased ingestive behaviors might simply represent one aspect of the nonspecific reduction in activity reported following SCT[101] injection in rats. A further possible contributor to this weight loss may be CT-induced diuresis — an effect noted in animals[141] and roughly half of our patients.

f. Calcitonin: Aborts Episodes of Malignant Hyperpyrexia and Reduces Serum CPK

Episodes of malignant catatonic hyperpyrexia in one patient were lengthened and intensified by DHT and aborted by SCT.[142] Certainly, the temperature reduction in this patient might have resulted from increased CSF calcium.[143-144] Massive increases in CPK leakage from mouse skeletal muscle may be triggered by physiologic increases in extracellular calcium,[145] suggesting that transient serum calcium spikes might augment leakage of muscle enzymes in sensitive pyschotic patients. Indeed, serum calcium increases preceded serum CPK increases in periodic psychotic exacerbations by about 4 days,[101] and mean serum CPK increased by 80% of baseline in eight patients on DHT. Finally, mean serum CPK decreased by more than 40% after SCT injections in 17 of 18 patients, from both SCT trials, who exhibited baseline serum CPK greater than 100 IU/ℓ.

g. Calcitonin: Reduction of Prolactin

SCT antagonized serum prolactin elevations by about 40% in 12 of 14 patients kept on constant neuroleptics during both trials[129] and reduced CSF prolactin by about 40% in four of four otherwise unmedicated monkeys.[146] The fact that two hypercalcemic agents — PTH and DHT[101] — have opposite effects to those of SCT, not only on serum calcium but on serum prolactin, suggests that the SCT effect on prolactin may be calcium-mediated. Indeed, ambient calcium is a very potent regulator of prolactin secretion by fish in vivo and by rat pituitaries in vitro.[147] Furthermore, since PRL is a potent hypercalcemic hormone, its reduction by SCT might provide a central mechanism of calcitonin induced hypocalcemia, as proposed by Stekolnikov and Abdukaramov.[148]

The minimum conclusion from all of this could be that behavioral effects of SCT do not represent merely some potentiation of coadministered lithium or neuroleptics. If anything, antimanic effects of SCT emerge *despite* a reduction in serum and RBC lithium levels on constant lithium dose and *despite* reductions in serum prolactin baseline elevations of which are taken as a barometer of postsynaptic dopamine blockade, which has been felt to be integral to the therapeutic effects of neuroleptics (see below).

h. Mechanism of Action of Calcitonin — Lack of Clear Effect on Monoamines?

The relationship of these calcium and calcitonin findings to the extensively investigated alterations of monoaminergic function in mood remains unclear. Since both alpha- and beta-noradrenergic receptors stimulate calcitonin secretion, independent of changes in serum calcium,[149] and since mania has been characterized both as a state of noradrenergic excess and as a state of reduced or altered CSF calcitonin (below), it seems quite unlikely that the effects of SCT on behavior (above) or levels in psychopathologic states of HCT might be explained by changes in norepinephrine. Conversely, our own preliminary data suggest that SCT had no effect on static levels of eight catecholamine metabolites in the CSF of five monkeys.[135]

i. Is Calcitonin a Neuroactive Peptide With Direct Effects?

We recently demonstrated that SCT gains ingress to CSF of primates from 2 to 30 hr after subcutaneous administration, with peak levels measurable from 12 to 18 hr after injection.[135] This agrees with the earlier observation of Stekolnikov and Abdukarimov.[148] SCT has direct neuronal effects independent of systemic or even local extracellular concentrations of calcium.[101] Since the time course of SCT ingress into CSF more closely matches the behavioral effects of SCT than the time course of SCT-induced increase of CSF calcium (4 to 10 hr), its behavioral effects may represent direct effects of the peptide itself.

j. Differences in CSF Human Calcitonin (HCT) by Behavioral State

Lumbar CSF was collected from 90 subjects at 9 a.m., after 9 hr of bedrest and fasting, and at least 2 weeks free of medication. CSF was analyzed for HCT by a sensitive and specific radioimmunoassay with a detection limit of 20 pg/mℓ and a coefficient of variation for lab error of less than 10%. By this assay, 26 (63%) of 41 normal volunteers and 16 (64%) of 25 depressed patients had measurable CSF HCT. On the other hand, only 2 (14%) of 14 manics had detectable CSF HCT ($p < 0.05$, $x2$).[239] Thus, the possibility must be considered that, in at least some cases of mania, a relative deficiency of or insensitivity to the endogenous calcitonin peptide *may* have etiologic significance.

II. ANXIETY

Several years ago, Pitts and McClure observed that serum lactate was higher after exertion in some patients with anxiety neurosis,[150] that intravenous lactate could provoke acute anxiety symptoms in these patients,[151] and that calcium coadministration could prevent that effect of lactate.[152] It was subsequently demonstrated that transient hypocalcemic episodes were not and could not be responsible for the original lactate effect.[153] Moreover, SCT-induced hypocalcemia failed to alter anxiety in manic or schizophrenic patients. Of 43 patients seen in follow-up after intestinal bypass surgery, those 19 with prominent anxiety and depression symptoms had lower serum 25-OH-vitamin D than 24 patients without psychiatric symptoms.[48]

III. CALCIUM AND SCHIZOPHRENIA

While there are no significant group differences between schizophrenic and control CSF calcium,[154-156] increased CSF calcium and decreased serum calcium have been described during spontaneous remissions of acute schizophrenic episodes within the same patient.[157-158]

IV. CALCIUM AND NEUROLEPTICS

Neuroleptics appear likewise to reduce serum[159-161] and increase CSF calcium.[239] It has been noted that the reduction of serum total calcium levels in patients on neuroleptics reaches a maximum just prior to the onset of dystonic reations.[157] Moreover, lower baseline serum calcium may indicate a greater vulnerability to acute EPS[157] or other acute toxicity.[161] The mild hypocalcemia induced by these agents may result from impairment of the first step of vitamin D activation by the liver,[161] from increased uptake of calcium into mitochondria,[162-163] or from hypothermia.[161]

Finally, acute neuroleptic toxicity may be treatable with calcium[160-161] or vitamin D.[161] On the other hand, calcitonin, like neuroleptics, reduces serum and increases CSF calcium; does not influence dopamine metabolite levels in CSF; but reduces serum prolactin in patients on neuroleptics (PRL),[165] and CSF PRL in medicated free monkeys.

Calcitonin ameliorated symptoms of tardive dyskinesia in a double-blind, placebo-controlled trial in three patients.[239] Other calcium effects of neuroleptics may be integral to their therapeutic or side effects. For example, trifluroperazine,[166] pimozide, haloperidol, and chlorpromazine[167] all quite specifically inhibit calcium binding to the calcium-dependent protein activator of cyclic nucleotide phosphodiesterase, whereas antidepressant and antianxiety agents have much weaker effects and pentobarbital, morphine, amphetamine, LSD, histamine, or dopamine have none. Moreover, if increases in CSF calcium, secondary to neuroleptics, adequately nullify the reduction in CSF calcium seen in acute psychosis, it may enhance receptor affinity of an endogenous pyschotogen.[168]

V. CALCIUM AND PAIN

Decreased serum calcium should decrease pain receptor sensitivity peripherally.[169] Similarly, decreasing calcium uptake by or binding to brain by lanthanum,[171] various opiates,[184] and beta-endorphin[185] has been associated with analgesia. Furthermore, naloxone antagonizes such brain calcium depletion as well as the analgesia.[186] Conversely, concurrent administration of calcium or calcium ionophores — but not Ba, Mg, Sr, or Zn — by a variety of routes antagonized morphine analgesia[187-200] and delayed or prevented development of tolerance.[201] Reduction or chelation of calcium enhanced morphines analgesic potency.[193] Quite opposite, their initial calcium-depleting effects, chronic opiates, increased calcium content of synaptosomes.[202-207]

A. Calcitonin and Pain

When given to patients with Paget's disease, calcitonin decreases bone pain out of proportion to its normalization of the radiologic or biochemical signs of that illness.[208] Interestingly, it has been suggested that CT was produced in endorphin-producing cells from a common peptide precursor.[209] Similarly, administered parenterally or intraventricularly calcitonin increased pain threshold on standard tests of nociception.[210-211] This effect was antagonized by calcium but not by naloxone.[195-196] On the other hand, CT also reduces local inflammatory response[212] by inhibiting prostaglandin synthesis — an effect not blocked by calcium.[212] Our own preliminary experience with salmon calcitonin has shown it a mild analgesic in three patients with chronic intractable pain. However, it has displayed equivocal value as treatment of opiate withdrawal, actually worsening the intensity of the withdrawal symptoms in the two patients and precipitating abstinence in a third, despite maintenance continuing of that patient on a constant daily dose of methadone.

VI. ALCOHOLISM

Alcoholic livers fail to activate vitamin D by 25-Ohation. Such failure is only partially explained by discernible cirrhotic disease. It may be a function of the effects of ethanol on calcium regulators, i.e., it increases both CT[217] and PTH release[218] (although, presumably, the latter is in response to the decreased serum calcium caused by CT). Despite these hormonal changes, there are neither any discernible alterations in bone density nor bone mineral composition in chronic alcoholics.[219] One author suggested that the decrease in active D was simply dietary in origin.[216]

Nearly 30% reduction in brain calcium has been reported on the first day of ethanol ingestion. This depletion can be prevented by pretreatment with naxolone, which has no effect by itself on brain calcium.[220] Another report[221] localized this calcium loss to synaptosomes and suggested that over 2 weeks of alcohol consumption was required to reduce calcium binding, whereas acute administration of ethanol over a wide range of concentrations or acute withdrawal of ethanol after chronic pretreatment, actually increased synaptosomal transport of calcium.[221-223]

On the other hand, a high calcium diet may adversely influence ethanol toxicity — since increased calcium in liver mitochondria can inhibit acetaldehyde oxidation by the liver[224] and increased intraventricular calcium enhances ethanol narcosis and intoxication (whereas chelating agents antagonize ethanol narcosis and intoxication).

VII. NONSPECIFIC INFLUENCES

Table 1 lists several nonspecific factors which might have contributed to or mitigated the alterations of CSF and serum calcium reported (above) in primary psychiatric disorders.

REFERENCES

1. **Whybrow, P. C. and Hurwitz, T.**, Psychological disturbances associated with endocrine disease and hormone therapy, in *Hormones, Behavior and Therapy,* Sachar, E. J., Ed., Raven Press, New York, 1976, 125.
2. **Peterson, P.**, Psychiatric disorders in primary hyperparathyroidism, *J. Clin. Endocrinol. Metab.,* 28, 1491, 1968.

3. Coffey, R. J. and Lee, T. C., Acute hyperparathyroidism, *Am. Surg.*, 36(5), 257, 1970.
4. Bruck, J. and Katschnig, H., On a depressive syndrome in primary hyperparathyroidism, *Wien Z. Nervenheilkd, Deren Grenzgeb.*, 27, 107, 1969.
5. Tiano, S., Wijsenbeek, H., and Munitz, H., Mental changes due to parathyroid adenoma, *Harefuah*, 79, 220, 1970.
6. Marks, C., Hyperparathyroidism and its clinical effects, *Am. J. Surg.*, 116, 40, 1968.
7. Fruensgaard, K., Mental changes in primary hyperparathyroidism. A survey and report of a fatal case, *Ugeskr. Laeg.*, 132, 2391, 70.
8. Cogan, M. G., Covey, C. M., Arieff, A. I., Wisniewski, A., Clark, O. H., Lazarowitz, V., and Leach, W., Central nervous system manifestations of hyperparathyroidism, *Am. J. Med.*, 65, 963, 1978.
9. Kosinski, K., Roth, S. I., and Chapman, E. H., Primary hyperparathyroidism with 31 years of hypercalcemia, *JAMA*, 236, 590, 1976.
10. Noble, P., Depressive illness and hyperparathyroidism, *Proc. R. Soc. Med.*, 67, 1066, 1974.
11. Low, J. C., Schaaf, M., Earll, J. M., Piechocki, J. T., and Li, T. K., Ionic calcium determination in primary hyperparathyroidism, *JAMA*, 223, 152, 1973.
12. Cooper, A. F. and Schapira, K., Case report: depression, catatonic stupor, and EEG changes in hyperparathyroidism, *Psychol. Med.*, 3, 509, 1973.
13. Flanagan, T. A., Goodwin, D. W., and Alderson, P., Psychiatric illness in a large family with familial hyperparathyroidism, *Br. J. Psychiat.*, 117, 693, 1970.
14. Sigstad, H. and Jacobsen, C. D., Tertiary hyperparathyroidism. Report of a case, *Acta Med. Scand.*, 188, 337, 1970.
15. Hecht, A., Gershberg, H., and St. Paul H., Primary hyperparathyroidism. Laboratory and clinical data in 73 cases, *JAMA*, 233, 519, 1975.
16. Gatewood, J. W., Organ C. H., Jr., and Mead, B. T., Mental changes associated with hyperparathyroidism, *Am. J. Psychiatr.*, 132, 129, 1975.
17. Bontoux, D., Barbier, J., Sudre, Y., Alcalay, M., and Devries, R. X., Primary hyperparathyroidism. Our experience with 30 cases, *Rev. Rhum. Mal. Osteoartic*, 44, 155, 1977.
18. Evaldsson, U., Ertekin, C., Ingvar, D. H., and Waldenström, J. G., Encephalopathia hypercalcemica. A clinical and electroencephalographic study in myeloma and other disorders, *J. Chron. Dis.*, 22, 431, 1969.
19. Reinfrank, R. F., Primary hyperparathyroidism with depression, *Arch. Intern. Med.*, 108, 606, 1961.
20. Hockaday, T. D. R., Keynes, W. M., and McKenzie, J. K., Catatonic stupor in an elderly female with hyperparathyroidism, *Br. Med. J.*, 1, 85, 1966.
21. Mandel, M. M., Recurrent psychotic depression associated with hypercalcemia and parathyroid adenoma, *Am. J. Psychiatr.*, 117, 234, 1960.
22. Nielson, H., Familial occurrency, gastrointestinal symptoms, and mental disturbances in hyperparathyroidism, *Acta Med. Scand.*, 151, 359, 1955.
23. Rookus, P. and Speelman, J. J., Psychiatric disturbances in hyperparathyroidism, *Psychiatr. Neurol. Neurochir.*, 64, 46, 1961.
24. Karpati, G. and Frame, B., Neuropsychiatric disorders in primary hyperparathyroidism, *Arch. Neurol.*, 10, 387, 1964.
25. Hecht, A. and Gershbert, H., Primary hyperparathyroidism: laboratory and clinical data in 73 cases, *JAMA*, 233, 519, 1975.
26. Anderson, D. C., Cooper, A. F., and Naylor, G. J., Vitamin-D intolication with hypernatremia potassium and water depletion, and mental depression, *Br. Med. J.*, 4, 744, 1968.
27. Lynch, H. T., Lemon, H. M., Henn, M. J., Ellington, R. J., and Grissom, R. L., Vitamin-D intoxicated patient with hypoparathyroidism, *Arch. Int. Med.*, 114, 375, 1964.
28. Eitinger, L., Hyperparathyroidism with mental changes, *Nord, Med. Ark.*, 14, 1581, 1942.
29. Cope, O., The story of hyperparathyroidism at the Massachusetts General Hospital, *N. Engl. J. Med.*, 274, 1174, 1966.
30. Frank, M., Nathan, P., and Lacebnik, J., Clinical experience with hyperparathyroidism in sixty patients, fifty-one of them having urolithiasis, *Urol. Int.*, 23, 315, 1968.
31. Watson, L., Clinical aspects of hyperparathyroidism, *Proc. R. Soc. Med.*, 61, 1123, 1968.
32. Bergeron R., Murphy, R., and Warrne, K., Actue pancreatitis, actue hyperparathyroidism, and low magnesium syndrome: a case report, *Lahey Clin. Found. Bul.*, 12, 181, 1961.
33. Potts, J. and Roberts, B., Clinical significance of magnesium deficiency and its relation to parathyroid disease, *Am. J. Med. Sci.*, 235, 206, 1958.
34. Gertner, J. M., Hodsman, A. B., and Neuberger, J. N., 1-Alpha-hydroxycholecalciferol in the treatment of hypocalcaemic psychosis, *Clin. Endocrinol. (Oxford)*, 5, 539, 1976.
35. Mikkelsen, E. J. and Reider, A. A., Post-parathyroidectomy psychosis: clinical and research implications, *J. Clin. Psychiat.*, 40, 352, 1979.

36. Dahlmann, W. and Prill, A., Chronic secondary hypoparathyroidism in connection with a partially reversible psycho-organic syndrome (transl.), *Z. Neruol.*, 206, 223, 1974.
37. Christie-Brown, J. R., The psychiatric aspects of disturbed calcium metabolism. Mood changes following parathyroidectomy, *Proc. R. Soc. Med.*, 61, 1121, 1968.
38. Hay, G. G., Jolley, D. J., and Jones, R. G., A case of the Capgras syndrome in association with pseudo-hypoparathyroidism, *Acta Psychiatr. Scand.*, 50, 73, 1974.
39. Fourman, P., Rawnsley, K., Davis, R. H., Jones, K. H., and Morgan, D. B., Effect of calcium on mental symptoms in partial parathyroid insufficiency, *Lancet*, 2, 914, 1967.
40. Christiaens, L., Fontaine, G., Farriaux, J. P., and Biserte, G., Chrinic psuedohypoparathyroidism apropos of 3 familial cases, *Acta Poediatr. Belg.*, 21, 5, 1967.
41. Boursier, B., Nervous depression due to magnesium deficiency, crenotherapy by sulfur, calcium and magnesium water, *Presse Therm. Clim.*, 104, 174, 1967.
42. Nyland, H. and Skre, H., Cerebral calcinosis with late onset encephalopathy. Unusual type of pseudo-pseudohypoparathyreoidism, *Acta Neurol. Scand.*, 56, 309, 1977.
43. Kocian, J., Honzak, R., and Horackova, E., Attempt to influence neurotic symptoms in malabsorption syndromes with calcium, *Cesk. Gastroenterol. Vyz.*, 23, 78, 1969.
44. Katz, S. H. and Foulks, E. F., Mineral metabolism and behavior: abnormalities of calcium homeostasis, *Am. J. Psys. Anthropol.*, 32, 299, 1970.
45. Popea, A. and Demetresco, T., Tetany and psychosis, *Rev. Stiint. V. Adamachi*, 19, 531, 1930.
46. Knospe, H., Tetany psychosis, *Nibatsschr. Psychiat. Neurol*, 99, 503, 1938.
47. Emerson, K., Walsh, F. B., and Howard, J. E., Idiopathic hypoparathyroidism: a report of two cases, *Case Rep.*, 1256.
48. Bach, P. and Hey, H., Depression or asthenia related to metabolic disturbances in obese patients after intestine bypass surgery, *Acta Psychiatr. Scand.*, 59, 462, 1979.
49. Ritter, G., Prill, A., and Villes, E., Neuropsychiatric aspects of so-called psuedohypoparathyroidism, *Dtsch. Z. Nervenheilkd.*, 197, 1, 1970.
50. Dinkle, L., Clinical aspects of cerebral nuclear calcinosis, *Muench. Med. Wochenschr.*, 113, 1346, 1971.
51. Chopra, I. J. and Nugent, C. A., Concurrence of features of pseudohypoparathyroidism, pseudo-pseudohypoparathyroidism and basal-cell nevus syndrome, *Am. J. Med. Sci.*, 260, 171, 1970.
52. Dudley, A. W., Jr. and Hawkins, H., Mineralization of the central nervous system in pseudopseudohypoparathyroidism (PPH), *J. Neurol. Neurosurg. Psychiatr.*, 33(2), 147, 1970.
53. Denko, J. D., and Kaelbling, R., The psychiatric aspects of hypoparathyroidism, *Acta Psychiatr. Scand.*, 38, 1, 1962.
54. Hay, G. G., Jolley, D. J., and Jones, R. G., A case of the Capgras syndrome in association with pseudo-hypoparathyroidism, *Acta Psychiatr. Scand.*, 50, 73, 1974.
55. Richter, C. P., Honeyman, W. M., and Hunter, H., Behavior and mood cycles apparently related to parathyroid insufficiency, *J. Neurol. Neurosurg. Psychiatr.*, 3, 19, 1940.
56. Greene, J. A. and Swanson, L. W., Psychosis in hypoparathyroidism: with a report of five cases, *J. Ment. Sci.*, 86, 1233, 1940.
57. Snowden, J. A., Mackie, A. C., and Pearce, J. B., Hypocalcemic myopathy with paranoid psychosis, *J. Neurol. Neurosurg. Psychiatr.*, 9, 48, 1976.
58. Weed, L. H. and Wedgeforth, R., Experimental irrigation of the subarachnoid space, *J. Pharmacol. Exp. Ther.*, 13, 317, 1919.
59. Huggins, C. B. and Hastings, A. B., Effect of calcium and citrate injections into cerebrospinal fluid, *Proc. Soc. Exp. Biol. Med.*, 30, 459, 1933.
60. Mullin, F. J., Hastings, A. B., and Lees, W. M., Neuromuscular response to variations in calcium and potassium concentration in the cerebrospinal fluid, *Am. J. Physiol.*, 121, 719, 1938.
61. Koenigstein, H., Shifting of cations in the cerebrospinal fluid as a cause of pruritus, *J. Invest. Dermatol.*, 17, 99, 1951.
62. Pappenheimer, J. R., Heisay, S. R., and Jordon, E. F., Perfusion of the cerebral ventricular system in unanesthetized goats, *Am. J. Psychol.*, 203, 763, 1962.
63. Veale, W. L. and Myers, R. D., Emotional behavior, arousal and sleep produced by sodium and calcium ions perfused within the hypothalamus of the cat, *Physiol. Behav.*, 7, 601, 1971.
64. Stern, K. and Chvoles, G. J., Effective de l'injection intraventriculaire des ions Ca et K, *C. R. Soc. Biol.*, 112, 568, 1953.
65. Cloetta, M., Fischer, H., and Van Der Loeff, M. R., Die biochemie von Schlaf und Erregung mit besondere Berucksichtigung der Bedeutung der Kation, *Arch. Exp. Pathol. PharmaKol.*, 174, 589, 1934.
66. Marguardt, P. and Riemschneider, H. H. T., Uber die wirking von intrazisternal injeziertem calcium, *Arch. Int Pharmacodyn.*, 85, 273, 1951.
67. Feldberg, W. and Sherwood, S. L., Effects of calcium and potassium injected into the cerebral ventricles of the cat, *J. Physiol.*, 139, 408, 1957.

68. Hakim, A. J., Bomb, B. S., Pandey, S. K., and Singh, S. V., A study of cerebrospinal fluid calcium and magnesium in depression, *J. Assoc. Physicians India*, 23, 311, 1975.
69. Harris, W. H. and Beauchemin, J. A., CSF Ca, Mg, and their ratio in psychoses of organic and functional origin, *Yale J. Biol. Med.*, 29, 117, 1956.
70. Breyer, E. and Quadbeck, G., The cerebrospinal fluid content of magnesium and other cations in central nervous system diseases, *Dtsch. Z. Nervenheilkd.*, 187, 595, 1965.
71. Bjorum, N., Plenge, P., and Rafaeisen, O. J., Electrolytes in CSF in endogenous depression, *Acta Psychiatr. Scand.*, 48, 533, 1972.
72. Bech, P., Kirkegaard, C., Bock, E., Johanessen, M., and Rafaelsen, O. J., Hormones, electrolytes and cerebrospinal fluid proteins in manic-melancholic patients, *Neuropsychobiology*, 4, 99, 1978.
73. Jimerson, D. C., Post, R. M., Carman, J. S., van Kammen, D. P., Wood, J. H., Goodwin, F. K., and Bunney, W. E., CSF Calcium: clinical correlates in affective illness and schizophrenia, *Biol. Psychiatr.*, 14, 37, 1979.
74. Ueno, Y., Aoki, N., Yabuki, R., and Kuraishi, F., Electrolyte metabolism in blood and cerebrospinal fluid in psychoses, *Folia Psychiatr. Neurol. Jpn.*, 15, 304, 1961.
75. Scholberg, H. A. and Goodall, E., The phosphorus and calcium content of the blood plasma and cerebrospinal fluid in the psychoses, *J. Ment. Sci.*, 72, 51, 1926.
76. Weston, P. G. and Howard, M. Q., The determination of Na, K, Ca, and Mg in the blood and spinal fluid of patients suffering from manic-depressive insanity, *Arch. Neurol. Psychiatr.*, 8, 179, 1922.
77. Katzenelbogen, S., *The CSF and the blood*, Johns Hopkins Press, Baltimore, 1935.
78. Henry, G. W. and Ebeling, W. E., Blood calcium and phosphorus in the personality disorders: effect of U-V irradiation, *Arch. Neurol. Psychiatr.*, 16, 48, 1925.
79. Faragala, F. F. and Flach, F. F., Studies of mineral metabolism in mental depression. I. The effects of imipramine and electric convulsive therapy on calcium balance and kinetics, *J. Nerv. Ment. Dis.*, 151(2), 120, 1970.
80. Carman, J. S., Post, R. M., Goodwin, F. K., and Bunney, W. E., Calcium and the electroconvulsive therapy of severe depressive illness, *Biol. Psychiatr.*, 12, 5, 1977.
81. Mellerup, E. T., Bech, P., Sorensen, T., Fuglsang-Frederiksen, A., and Rafaelsen, O. J., Calcium and electroconvulsive therapy of depressed patients, *Biol. Psychiatr.*, 14, 711, 1979.
82. Carman, J. S. and Wyatt, R. J., Alterations in CSF and serum total calcium with changes in psychiatric state, in *Neuroregulators and Psychiatric Disorders*, Usdin, E., Hamburg, D. A., and Barchas, J. D., Eds., Oxford University Press, New York, 1977, 488.
83. Bucci, L. and Johnson, E. E., Determination of ionized calcium in mentally ill patients, *Psychosomatics*, 7(4), 228, 1966.
84. Malleson, A., Frizel, D., and Marks, V., Ionized and total plasma calcium and magnesium before and after modified E. C. T., *Br. J. Psychiatr.*, 114(510), 631, 1968.
85. Hakim, A. H., Bomb, B. S., Garg, A. R., and Singh, S. V., A comparative study of serum calcium and magnesium in cases of endogenous depression, reactive depression schizophrenia and conversion reaction, *J. Assoc. Physicians India*, 23(8), 513, 1975.
86. Naylor, G. J., Fleming, L. W., Stewart, W. K., McNamee, H. B., and Le Poidevin, D., Plasma magnesium and calcium levels in depressive psychosis, *Br. J. Psychiatr.*, 120(559), 683, 1972.
87. Bjorum, N., Electrolytes in blood in endogenous depression, *Acta Psychiatr. Scand.*, 48(1), 59, 1972.
88. Frizel, D., Coppen, A., and Marks, V., Plasma magnesium and calcium in depression, *Br. J. Psychiatr.*, 115(529), 1375, 1969.
89. Gibbons, J. L., Electrolytes and depressive illness, *Postgrad. Med. J.*, 39, 19, 1963.
90. Carman, J. S., Post, R. M., Runkle, D. C., Bunney, W. E., and Wyatt, R. J., Increased serum calcium and phosphorus with the switch into manic or excited psychotic states, *Br. J. Psychiatr.*, 135, 55, 1979.
91. Henry, G. W. and Ebeling, W. E., Blood Calcium and phosphorus in the personality disorders: effect of U-V irradiation, *Arch. Neurol. Psychiatr.*, 16, 48, 1925.
92. Speijer, N., Treatment of a periodical psychosis (degenerative psychosis) based upon hematological and biochemical deviations from the normal, *Folia Psychiatr. Neurol. Neurochur. Neerl.*, 53, 718, 1950.
93. Fischback, R., [Changes in calcium metabolism in depression during medication of thymoleptics], *Arzneim. Forsch.*, 21(1), 27, 1971.
94. Herbert, F. K., The total and diffusible calcium of serum and the calcium of cerebrospinal fluid in human cases of hypocalcemia and hypocalcemia and hypercalcemia, *Biochem. J.*, 27, 1978, 1933.
95. Goldstein, D. A., Romoff, M., Bogin, E., and Massry, S. G., Relationship between the concentrations of calcium and phosphorus in blood and cerebrospinal fluid, *J. Clin. Endocr. Metab.*, 49(1), 58, 1979.

96. Gredèn, J., DeVigne, J., Albala, A., Tarika, J., Buttenheim, M., Eiser, A., and Carroll, B., Serial Dexamethasone Suppression Tests (DST) among rapid cycling bipolar patients, paper presented at the annual meeting of the Society of Biological Psychiatry, Boston, Mass., 1980.
97. Rubinow, D. and Post, R., Correlation of plasma cortisol and mood in a rapid cycling manic-depressed patient, paper presented at the annual meeting of the Society of Biological Psychiatry, Boston, Mass., 1980.
98. Lanuza, D. M. and Marotta, S. F., Circadian and basal interrelationships of plasma cortisol and cations in women, *Aerosp. Med.*, 864, 1974.
99. Carroll, B., Curtis, G., and Mendels, J., Neuroendocrine regulation in depression. I. Limbic system-adrenocortical dysfunction, *Arch. Gen. Psychiatr.*, 33, 1039, 1976a.
100. Carroll, B., Curtis, G., and Mendels, J., Neuroendocrine regulation in depression. II. Discrimination of depressed from nondepressed patients, *Arch. Gen. Psychiatr.*, 33, 1051, 1976b.
101. Carman, J. S. and Wyatt, R. J., Calcium: bivalent cation in the bivalent pyschoses, A. E. Bennett Award paper for clinical research, *Biol. Psychiatr.*, 14, 295, 1979.
102. Nielson, J. L., Pedersen, E. B., Amidsen, A., and Darling, S., Reduced renal calcium excretion during lithium therapy, *Psychopharmacology*, 54, 101, 1977.
103. Aronoff, M. S., Evens, R. G., and Durell, J., Effect of lithium salts on electrolyte metabolism, *J. Psychiatr. Res.*, 8, 139, 1971.
104. Carman, J. S., Post, R. M., Teplitz, T. A., and Goodwin, F. K., Letter: divalent cations in predicting antidepressant response to lithium, *Lancet*, 2, 1454, 1974.
105. Baastrup, P. C., Christiansen, C., and Transbol, I., Calcium metabolism in lithium-treated patients. Relation to uni-bipolar dichotomy, *Acta Psychiatr. Scand.*, 49, 601, 1973.
106. Christiansen, C., Baastrup, P. C., and Transbol, I., Lithium, hypercalcemia, hypermagnesemia, and hyperparathyroidism: letter, *Lancet*, 2, 969, 1976.
107. Srinivasan, V., Parvathi, Devi, S., and Venkoba Roa A., Lithium effects on serum magnesium and calcium in manic depressive psychosis, in *Lithium In Medical Practice*, Johnson, F. N. and Johnson, S, Eds., MTP Press, Lancaster, 1978, W3BR447.
108. Mellerup, E. T., Lauritsen, B., Dam, H., and Rafaelson, O. J., Lithium effects on diurnal rhythm of calcium, magnesium, and phosphate metabolism in manic-melancholic disorder, *Acta. Psychiatr. Scand.*, 53, 360, 1976.
109. Crammer, J., Letter: lithium, calcium and mental illness, *Lancet*, 1, 215, 1975.
110. Tupin, J. P., Schlagenhauf, G. K., and Creson, D. L., Lithium effects on electrolyte excretion, *Am. J. Psychiatr.*, 126, 285, 1975.
111. Lovett Doust J. W., Huszka, L., Influence of some psychoactive drugs on mental metabolism in man, *Int. Pharmacopsychiatr.*, 8, 159, 1973.
112. Mellerup, E. T., Plenge, P., Ziegler, R., and Rafaelson, O. J., Lithium effects on calcium metabolism in rats, *Int. Pharmacol. Berl.*, 5, 258, 1970.
113. Bjorum, N., Hornum, I, Mellerup, E. T., Plenge, P. K., and Rafaelson, O. J., Lithium calcium and phosphorus, *Lancet*, 1, 1243, 1975.
114. Andreoli, V. M., Villani, F., and Brambilla, G., Calcium and magnesium excretion induced by lithium carbonate, *Psychopharmacol. Berl.*, 25, 77, 1972.
115. Radaelsen, O. J. and Mellerup, E. T., Mechanism of action of lithium salts. Biochemical aspects, *Psychiatr. Neurol. Neurochir.*, 76, 523, 1973.
116. Elizur, A., Graff, E., Steiner, M., and Davidson, S., Intra/extra red blood cell lithium and electrolyte distributions as correlates of neurotoxic reactions during lithium therapy, in *The Impact of Biology on Modern Psychiatry*, Gershon, E. S., Ed., Plenum Press, New York, 1977, 134.
117. Toffaletti, J., McComb, R. B., and Bowers, G. N., Increase in dialyzable calcium associated with treatment with lithium, *Clin. Chem.*, 25, 1806, 1979.
118. Birch, N. M., Letter: the role of magnesium and calcium in the pharmacology of lithium, *Biol. Psychiatr.*, 7, 269, 1973.
119. Birch, N. J., Greenfield, A. A., and Hullen, R. P., Lithium therapy and alkaline earth metal metabolism, a biochemical screening study, *Psychol. Med.*, 7, 613, 1977.
120. Birch, N. J., Lithium accumulation in bone after oral administration in rat and in man, *Clin. Sci. Mol. Med.*, 46, 409, 1974.
121. Christiansen, C., Baastrup, P. C., and Transbol, I., Osteopenia and dysregualation of divalent cations in lithium-treatment patients, *Neuropsychobiology*, 1, 344, 1975.
122. Plenge, P. and Mellerup, E. T., Lithium effects of serum Calcium, magnesium and phosphorus in rats, *Psychopharmacologia*, 49, 187, 1976.
123. Zaplet, A. M. and Groh, J., The effect of lithium on potassium, calcium, sodium and magnesium levels in serum and erythrocytes in manic depressive psychoses, *Acta Nerv. Sup.*, 20, 295, 1978.
124. Gerner, R. H., Post, R. M., Spiegel, A. M., and Murphy, D. L., Effects of parathormone and lithium treatment on calcium and mood in depressed patients, *Biol. Psychiatr.*, 12, 145, 1977.

125. Nielsen, J. L., Christensen, M. S., Pedersen, E. B., Darling, S., and Amdisen, A., Parathyroid hormone in serum during lithium therapy, *Scand. J. Clin.*, 37, 369, 1977.
126. Garfinkle, P. E., Ezrin, C., and Stancer, H. C., Hypothyroidism and hyperparathyroidism associated with lithium, *Lancet*, 2, 331, 1973.
127. Christenson, T. A. T., Lithium hypercalcemia and hyperparathyroidism, *Lancet*, 2, 144.
128. Steele, T. H. and Dudgeon, L., Reabsorption of lithium phosphate by the rat kidney: role of the parathyroids, *Kid. Int.*, 5, 203, 1974.
129. Carman, J., Crews, E., Bancroft, A., Wyatte, E., Cooper, B., Ledbetter, J., Daniels, J., Munson, S., and Bond, W., Calcium and calcium-requalting hormones in the biphasic periodic psychoses, *J. Oper. Psychiatr.*, 11, 5, 1980.
130. Rasmussen, J. and Pechet, M., Calcitonin, thyrocalcitonin, in *Pharmacology of the Endocrine System and Related Drugs*, Rasmussen, H., Ed., Oxford University Press, New York, 1972, 237.
131. Merritt, H. H. and Bauer, B., Equalibrium between cerebrospinal fluid and blood plasma: calcium content of serum cerebrospinal fluid and acqeous humor at different levels of parathyroid activity, *J. Biol. Chem.*, 90, 233, 1931.
132. Gershon, M. D. and Nunez, E. A., Histochemical radioautographical studies of serotonin and parafollicular cells in the thyroid gland of the prehibernating bat, *Endocrinology*, 86, 160, 1970.
133. Colburn, T. R., Smith, B. M., and Guarini, J. J., An ambulatory activity monitor with solid state memory, *ISA Trans.*, 15, 149, 1976.
134. Carman, J. S. and Wyatt, R. M., Use of calcitonin is psychotic agitation or mania, *Arch. Gen. Psychiat.*, 36, 72, 1979.
135. Carman, J. S. Perlow, M. J., Potter, D., Kansal, P. C., Karoum, F., Wyatt, R. J., Sequential changes of calcium, calcitropic hormones and catecholamines in monkey CSF following subcutaneous salmon calcitonin, 1980, submitted.
136. Carman, J. S. and Wyatt, R. J., Calcium: pacesetting the periodic psychoses, *Am. J. Psychiatr.*, 136, 1035, 1979.
137. Perlow, M. J., Freed, W. J., Carman, J. S., and Wyatt, R. J., Calcitonin reduces feeding in man, monkey and rat, *Pharmacol. Biochem. Behav.*, 12, 609, 1980.
138. Gattereau, A., Bielmaim, P., Durivage, J., and Larochelle, P., Hyperglycemic effect of synthetic salmon calcitonin, *Lancet*, 2, 1076, 1977.
139. Seone, J. R. and Baile, C. A., Ionic changes in cerebrospinal fluid and feeding, drinking temperature of sheep, *Psysiol. Behav.*, 10, 915, 1973.
140. Myers, R. D., Bender, S. A., Kristie, M. K., and Brophy, P. D., Feeding produced in the satiated rat by elevating the concentration of calcium in the brain, *Science*, 176, 1124, 1972.
141. Simonnet, G., Maura, A. M., and Bagdiantz, A., Calcitonin diuretic effect in the rabbit, *Horm. Metab. Res.*, 10, 347-452, 1979.
142. Carman, J. S. and Wyatt, R. J., Calcitonin and malignant catatronia, *Lancet*, 2, 1124, 1977.
143. Myers, R. D. and Buckman, J. E., Deep hyperthermia induced in the golden hamster by altering cerebralcalcium levels, *Am. J. Physiol.*, 223, 1313, 1972.
144. Myers, R. D., Gisolvi, C. V., and Mora, F., Ca levels in the brain underlie temp control during exercise in the primate, *Nature (London)*, 266, 178, 1977.
145. Soybel, D., Morgan, J., and Cohen, I., Calcium augmentation of enzyme leakage from mouse skeletal muscle and its possible site of action, *Res. Comm. Chem. Pathol. Pharmacol.*, 20, 317, 1978.
146. Isaac R., Merceron, R. E., and Caillens, G., Effect of parathormone on plasma prolactin in man, *J. Clin. Endocn. Metab.*, 48, 18, 1978.
147. Tashjian, A. H., Lomedico, M. E., and Maina, D., Role of calcium in the thyrotropin-releasing hormone-stimulated release of prolactin from pituitary cells in culture, *Biochem. Biophys. Res. Commun.*, 81, 798, 1978.
148. Stekolnikov, L. I. and Abdukaramov, A., Physiochemical study of cerebrospinal fluid under thyrocalcitonin action, *Biofizika*, 14, 921, 1969.
149. Metz, S. A., Deflos, L. J., and Baylink, D. J., Neuroendocrine modulation of calcitonin and parathyroid hormone in man, *J. Clin. Endocr. Metab.*, 47, 151-159, 1978.
150. Pitts, F. N., Jr., McClure, J. N., Jr., Lactate metabolism in anxiety neurosis, *N. Engl. J. Med.*, 277(25), 1329, 1967.
151. Pitts, F. N., Jr., The biochemistry of anxiety, *Sci. Am.*, 220(2), 69, 1969.
152. Pitts, F. N., Jr., Biochemical factors in anxiety neurosis, *Behav. Sci.*, 16(1), 82, 1971.
153. Ackerman, S. H. and Sachar, E. J., The lactate theory of anxiety: a review and reevaluation, *Psychosom. Med.*, 36(1), 69, 1974.
154. Pandey, S. K., Devpura, J. C., Bedi, H. K., and Babel, C. S., An estimation of magnesium and calcium in serum and C.S.F. in schizophrenia, *J. Assoc. Physicians India*, 21(2), 203, 1973.
155. Ueno, Y., Aoki, N., Yabuki, T., and Kuraishi, F., Electrolytes in blood and in psychoses, *Folia Psychiatr. Neurol. Jpn.*, 15, 304, 1961.

156. Jimerson, D. C., Post, R. M., Carman, J. S., van Kammen, D. P., Wood, J. H., Goodwin, F. K., and Bunney, W. E., CSF calcium: clinical correlates in affective illness and schizophrenia, *Biol. Psychiatr.*, 14(1), 37, 1979.
157. Carman, J. S. and Wyatt, R. J., Alterations in CSF and serum total calcium with changes in psychiatric state, in *Neuroregulators and Psychiatric Disorders,* Usdin, E., Hamburg, D. A., and Barchas, J. S., Eds., Oxford University Press, New York, 1977, 488.
158. Carman, J. S., Post, R. M., Runkle, D. C., Bunney, W. E., Wyatt, R. J., Increased serum calcium and phosphorus with the "switch" into manic or excited psychotic states, *Br. J. Psychiatr.*, 135, 55, 1979.
159. Alexander, P. E., van Kammen, D. P., and Bunney, W. E., Serum calcium and magnesium in schizophrenia: relationship to clinical phenomena and neuroleptic treatment, *Br. J. Psychiatr.*, 133, 143, 1978.
160. Alexander, P. E., van Kammen, D. P., and Bunney, W. L., Serum calcium and magnesium levels in schizophrenia, *Arch. Gen. Psychiatr.,* 36, 1372, 1979.
161. Martin, M. E., Means, R. A., Solomon, J. D., and Bergeim, O., Effect of chlorpromazine on serum calcium and magnesium in the rat, presented at National Mental Health Center, Saint Elizabeth's Hospital, Washington, D.C.
162. Batra, S., The effects of drugs on calcium uptake and calcium release by mitochondria and sarcoplasmic reticulum of frog skeletal muscle, *Biochem. Pharmacol.,* 23, 89, 1974.
163. Seeman, P., Staiman, A., Lee, T., and Chau-Wong, M., The membrane actions of tranquilizers in relation to neuroleptic-induced parkinsonism and tardive dyskinesia, *Adv. Biochem. Psychopharmacol.,* 9(0), 137, 1974.
164. Carman, J. S., Perlow, M. J., Potter, D., Kansal, P. C., Karoum, F., and Whatt, R. J., Sequential changes of calcium, calcitropic hormones and calecholeamines in monkey CSF following subcutaneous salmon calcitonin, 1980. submitted.
165. Carman, J. S. and Wyatt, R. J., Reduction of serum prolactin following subcutaneous salmon calcitonin, *Lancet,* 2, 1267, 1977.
166. Levin, R. M. and Weiss, B., Specificity of the binding of trifluoperazine to the calcium-dependent activator of phosphodiesterase and to a series of other calcium-binding proteins, *Biochim. Biophys. Acta,* 540, 197, 1978.
167. Levin, R. M. and Weiss, B., Selective binding of antipsychotics and other psychoactive agents to the calcium-dependent activator of cyclic nucleotide phosphodiesterase, *J. Pharmacol. Exp. Ther.*, 454, 1979.
168. Proctor, C. D., Cho, J. B., Ashley, L. G., Potts, J. L., Eaton, H. E., Jr., McGriff, J. E., Douglas, J. G., and Amoroso, C. P., Factors affecting an influence of blood plasma from schizophrenics on an action of 3,4-dimethoxyphenylethylamine, *Arch. Int. Pharmacodyn. The.,* 172(1), 106, 1968.
169. Lembeck, F. and Juan, J., Influence of changes calcium and potassium concentration on the algesic effect of bradykinesis and acetylcholine, *Nauyn-Schmiedebergs, Arch. Pharmacol.*, 299, 289, 1977.
170. Harris, R. A., Loh, H. H., and Way, E. L., Antinociceptive effects of lanthanum and cerium in nontolerant and morphine tolerant-dependent animals, *J. Pharmacol. Exp. Ther.,* 196(2), 288, 1976.
171. Harris, R. A., Loh, H. H., and Way, E. L., Analgetic effects of lanthanum: cross-tolerance with morphine, *Brain Res.,* 100, 221, 1975.
172. Cardenas, H. L. and Ross, D. L., Calcium depletion of synaptosomes after morphine treatment, *Br. J. Pharmacol.,* 57, 521, 1976.
173. Ross, D. H., Calcium content and finding in synaptosomes during morphine treatment, *Neurochem. Res.,* 2, 581, 1977.
174. Ross, D. H. and Cardenas, H. L., Levorphanol inhibition of calcium binding to synaptic membranes in vitro, *Life Sci.,* 20, 1455, 1977.
175. Cardenas, H. L. and Ross, D. H., Morphine induced calcium depletion in discrete regions of rat brain, *J. Neurochem.,* 24, 487, 1975.
176. Mule, S. J., Inhibition of phospholipid-facilitated calcium transport by central nervous system-acting drugs, *Biochem. Pharmacol.,* 18, 339, 1969.
177. Ross, D. H., Medina, M. A., and Cardenas, H. L., Morphine and ethanol: selective depletion of regional brain calcium, *Science,* 186, 63, 1974.
178. Ross, D. H., Tolerance to morphine-induced calcium depletion: characterization with reserpine and protein synthesis inhibitors, *Br. J. Pharmacol.,* 55, 431, 1975.
179. Kaku, T., Kaneto, H., and Koida, M., Proceedings: concomitant changes of synaptosomal Mg^{++}-Ca^{++}-ATPase activity with rise and fall of morphine tolerance and dependence, *Jpn. J. Pharmacol.,* 24(0), 123, 1974.
180. Harris, R. A., Yamamoto, H., Loh, H., and Way, E. L., Alterations in brain calcium localization during the development of morphine tolerance and dependence, in *Opiates and Endogenous Opioid Peptides,* Kasterlitz, H. W., Ed., Amsterdam, 1976.

172   The Role of Calcium in Biological Systems

181. **Sanghvi, I. S. and Gerson, S.**, Commentary: brain calcium and morphine action, *Biochem. Pharmacol.*, 26(13), 1183, 1977.
182. **Yamamoto, H., Harris, R. A., Loh, H. H., and Way, E. L.**, Effects of acute and chronic morphine treatments on calcium localization and binding in brain, *J. Pharmacol. Exp. Ther.*, 205(2), 255, 1978.
183. **Guerrero-Munoz, F., Cerreta, K. V., Guerrero, M. L., and Way, E. L.**, Effect of morphine on synaptosomal Ca^{++} uptake, *J. Pharmacol. Exp. Ther.*, 209(1), 132, 1979.
184. **Harris, R. A., Yamamoto, H., Loh, H. H., and Way, E. L.**, Discrete changes in brain calcium with morphine analgesia, tolerance-dependence, and abstinence, *Life Sci.*, 20(3), 501, 1977.
185. **Guerrero-Munoz, F., Guerrero, M. D. L., Way, E. L., and Li, C. H.**, Effect of B-Endorphin on calcium uptake in the brain, *Science*, 206, 89, 1979.
186. **Ross, D. H., Lynn, S. C., and Cardenas, H. L.**, Selective control of calcium levels by naloxone, *Life Sci.*, 18, 789, 1975.
187. **Ross, D. H. and Cardenas, H. L.**, Nerve cell calcium as a messenger for opiate and endorphin actions, in *Neurochemical Mechanisms of Opiates and Endorphins,* Loh, H. H. and Ross, D. H., Eds., Raven Press, New York, 1979, 301.
188. **Kakunaga, R., Kaneto, H., and Hano, K.**, Pharmacologic studies on analgesics. VII. Significance of the calcium ion in morphine analgesia, *J. Pharmacol. Exp. Ther.*, 153, 134, 1966.
189. **Hamandas, K., Sawynok, J., Sawynok, J., and Sutak, M.**, Antagonism of morphine action on brain acetylcholine release by methylxanthines and calcium, *Eur. J. Pharmacol.*, 49, 309, 1978.
190. **Sanfacon, G. and Labrecque, G.**, Acetylcholine antirelease effect of morphine and its modification by calcium, *Psychopharmacology,* 55, 151, 1977.
191. **Widman, M., Rosin, D., and Dewey, W. L.**, Effects of divalent cations, lanthanum, cation chelators and an ionophore on acetylcholine antinociception, *J. Pharmacol. Exp. Ther.*, 205, 311, 1978.
192. **Williams, E. Y., Rickman, E. E., and Elder, Z. B.**, The quest of therapy for heroin addiction. Experience with calcium gluconate, *J. Natl. Med. Assoc.*, 64(3), 205, 1972.
193. **Harris, R. A., Loh, H. H., and Way, E. L.**, Effects of divalent cation chelators and an ionophore on morphine analgesia and tolerance, *J. Pharmacol. Exp. Ther.*, 195, 488, 1976.
194. **Hitzemann, R. J., Hitzemann, B. A., and Loh, H. H.**, On the specificity of trypsin to inhibit norepiniphrine transport in nerve ending particles, *Life Sci.*, 24, 323, 1975.
195. **Satoh, M., Amano, H., Nakazawa, R., and Takagi, H.**, Inhibition by calcium of analgesia induced by intracisternal injection of porcine calcitonin, in Mice *Res. Commun. Chem. Pathol. and Pharmacol.*, 26, 213, 1979.
196. **Braga, P., Ferri, S., Santagostine, A., Olgiati, V. R., and Pecile, A.**, Lack of opiate receptor involvement in centrally induced calcitonin analgesia, *Life Sci.*, 22, 971, 1978.
197. **Kakunaga, R., Kaneto, H., and Hano, K.**, Pharmacologic studies on analgesics. VII. Significance of the calcium ion in morphine analgesia, *J. Pharmacol. Exp. Ther.*, 153, 134, 1966.
198. **Elliott, H. W., Kokka, N., and Way, E. L.**, Influence of calcium-deficit on morphine inhibition of QO_2 of rat cerebral cortex slices, *Proc. Soc. Exp. Biol. Med.*, 113, 1049, 1963.
199. **Kaneto, H.**, Inorganic ions: the role of calcium, in *Narcotic Drugs: Biochemical Pharmacology,* Clovet, D. H., Ed., Plenum Press, New York, 1971, 300.
200. **Sanghvi, I. S. and Gershon, S.**, Morphine dependent rats: blockade of precipitated abstinence by calcium, *Life Sci.*, 18(6), 649, 1976.
201. **Bhargava, H. N.**, The effects of divalent ions on morphine analgesia and abstinence syndrome in morphine-tolerant and -dependent mice, *Psychopharmacology,* 57(2), 223, 1978.
202. **Harris, R. A., Yamamoto, H., Loh, H. H., and Way, E. L.**, Alterations in brain calcium localization during the development of morphine tolerance and dependence, in *Opiates and Endogenous Opioid Peptides,* Kasterlitz, H. W., Ed., North-Holland, Amsterdam, 1976, 361.
203. **Harris, R. A., Yamamoto, H., Loh, H. H., and Way, E. L.**, Discrete changes in brain calcium with morphine analgesia, tolerance-dependence and abstinence, *Life Sci.*, 20, 501, 1977.
204. **Ross, D. H.**, Calcium content and binding in synaptosomal subfraction during chronic morphine treatment, *Neurochem. Res.*, 2, 51, 1977.
205. **Ross, D. H.**, Effects of opiate drugs on the metabolism of calcium in synaptic tissue, in *Proceedings of the Drug Abuse Conference,* Weiss, G. B., Ed., Plenum Press, New York, 1978, 241.
206. **Ross, D. H. and Cardenas, H. L.**, Levorphanol inhibition of calcium binding to synaptic membranes, *Life Sci.*, 20, 1455, 1977.
207. **Yamamoto, H., Harris, R. A., Loh, H. H., and Way, E. L.**, Effects of acute and chronic morphine treatments on calcium localization and binding in brain, *J. Pharmacol. Exp. Ther.*, 205, 255, 1978.
208. **Chen, J. R., Rhee, R. S. C., Wallach, S., Avramides, A., and Flores, A.**, Neurologic disturbances in Paget disease of bone: response to calcitonin, *Neurology,* 29, 448, 1979.
209. **Weber, E. and Woigt, K. H.**, Calcitonin is not contained within the common precursor to corticotropin and endorphin in the rat, *Biochem. Biophys. Res. Commun.*, 89, 360, 1979.
210. **Komendatova, M. V. and Milogradova, G. P.**, The effect of thyrocalcitonin on the algesthesia in animals, *Farmakol. I. Toksikol.*, 39, 413, 1976.

211. Pecile, A., Ferri, S., Braga, P. C., and Olgiati, V. R., Effects of intracerebroventricular calcitonin in the conscious rabbit, *Experientia,* 31, 331, 1975.
212. Ceserani, R., Colombo, M., Olgiati, V. R., and Pecile, A., Calcitonin and prostaglandin system, *Life Sci.,* 25, 1851, 1979.
213. Dechavanne, M., Barbier, Y., Prost, G., Pehlivanian, E., and Tolot, F., (Calcium 47 absorption in alcoholic cirrhosis. Effect of 25 hydroxycholecalciferol), *Nouv. Presse. Med.,* 3(41-43), 2549, 1974.
214. Luisier, M., Vodoz, J. F., Donath, A., Courvoisier, B., and Garcia, B., (25-hydroxy vitamin D deficiency with reduction of intestinal calcium absorption and bone density in chronic alcoholism), *Schweiz. Med. Wochenschr.,* 107(43), 1529, 1977.
215. Vodoz, J. F., Luisier, M., Donath, A., Courvoisier, B., and Garcia, B., (Decrease of intestinal absorption of 47-calcium in chronic alcoholism), *Schweiz. Med. Wochenschr.,* 107(43), 1525, 1977.
216. Verbanck, M., Verbanck, J., Brauman, J., and Mullier, J. P., Bone histology and 25-OH vitamin D plasma levels in alcoholics without cirrhosis, *Calcif. Tiss. Res.,* 22, 538, 1977.
217. Shah, J. H., Bowser, E. N., Hargis, G. K., Wongsurawat, N., Banerjee, P., Henderson, W. J., and Williams, G. A., Effect of ethanol on parathyroid hormone secretion in the rat, *Metabolism,* 27(3), 257, 1978.
218. Mezey, E., Potter, J. J., and Merchant, C. R., Effect of ethanol feeding on bone composition in the rat, *Am. J. Clin. Nutr.,* 32(1), 25, 1979.
219. Blum, K., Futterman, S., Wallace, J. E., and Schwertner, H. A., Naloxone-induced inhibition of ethanol dependence in mice, *Nature (London),* 265(5589), 49, 1977.
220. Ross, D. H., Adaptative changes in Ca^{++}-membrane interactions following chronic ethanol exposure, *Adv. Exp. Med. Biol.,* 85A, 459, 1977.
221. Sun, A. Y., Seaman, R. N., and Middleton, C. C., Effects of acute and chronic alcohol administration on brain membrane transport systems, *Adv. Exp. Med. Biol.,* 85A, 123, 1977.
222. Michaelis, E. K. and Myers, S. E., Calcium binding to brain synaptosomes. Effects of chronic ethanol intake, *Biochem. Pharmacol.,* 28, 2081, 1979.
223. Cederbaum, A. I., The effect of calcium on the oxidation of acetaldehyde by rat liver mitochondria, *Life Sci.,* 22, 111, 1977.
224. Ereckson, C. K., Tyler, T. D., and Harris, R. A., Ethanol: modification of acute intoxication by divalent cations, *Science,* 199, 1219, 1978.
225. McCance, N., Calcium balance in states of emotional stress, *Q. J. Med.,* 16, 33, 1947.
226. Hofmann, P., Schwille, P. O., and Thun, R., Hypocalcemia during restraint stress in rats — indication that gastric ulcer prophylaxis by exogenous calcium interferes with calcitonin release, *Res. Exp. Med.,* 175, 159, 1979.
227. Jubiz, W., Canterbury, J. M., and Reiss, E., Circadian rhythm in serum parathyroid hormone concentration in human subjects: correlation with serum calcium, phosphorus, albumin, and growth hormone levels, *J. Clin. Invest.,* 51, 2040, 1972.
228. Renoe, B. W., McDonald, J. M., and Ladenson, J. H., Influence of posture on free calcium and related variables, *Clin. Chem.,* 25(10), 1766, 1979.
229. Bojanovsky, J., Koch, W., and Tolle, R., Elektrolytveranderungen unter antidepressiver therapie, *Arch. Psychiatr. Nevenkr.,* 218, 379, 1974.
230. Gerner, R. H., Post, R. M., Gillin, J. C., and Bunney, W. E., Biological and behavioral effects of one night's sleep deprivation in depressed patients, *J. Psychiatr. Res.,* 15, 21, 1979.
231. March, G. L. and McKeown, B. A., Diurnal variations in plasma calcium parathormone and calcitonin levels and crop gland activity in pigeon, *Endocr. Exp.,* 11, 263, 1977.
232. Staub, J. F., Perault-Staub, A. M., and Milhaud, F., Endogenous nature of circadian rhythms in calcium metabolism, *Metabolism,* 237, 311, 1979.
233. Potts, J. T. and Deftos, L. F., Parathyroid hormone, thyrocalcitonin, vitamin D, bone and bone mineral metabolism, in *Diseases of Metabolism,* Bondy, P. K., Ed., W. B. Saunders, Philadelphia, 1969, 251.
234. McBean, L. D. and Speckmann, E. W., A recognition of the interrelationship of calcium with various dietary components, *Am. J. Clin. Nutr.,* 27, 603, 1974.
235. Vipperman, P. E., Peo, E. R., and Cunningham, R. J., Effect of dietary calcium and phosphorus level upon calcium, phosphorus and nitrogen balance in swine, *J. Anim. Sci.,* 38, (4), 758, 1974.
236. Allen, L. H., Bartlett, R. S., and Block, G. D., Reduction of renal calcium reabsorption in man by consumption of dietary protein, *J. Nutr.,* 109, 1345, 1979.
237. Licata, A. A., Bou, E., Bartter, F. C., and Cox, J., Effects of dietary protein on urinary calcium in normal subjects and in patients with nephrolithiasis, *Metabolism,* 28(9), 895, 1979.
238. Anand, C. R. and Linkswiler, H. M., Effect of protein intake on calcium balance of young men given 500 mg calcium daily, *J. Nutr.,* 104, 695, 1974.
239. Carman, J. C., unpublished data.

Chapter 13

CALCIUM DEFICIENCY IN PLANTS

E. W. Simon

TABLE OF CONTENTS

I.	Introduction	176
II.	Transport of Calcium Within Plants	177
III.	The Symptoms of Calcium Deficiency	179
	A. Water-Soaked and Necrotic Lesions	179
	1. Stems	179
	2. Leaves	180
	3. Roots	180
	4. Fruits	180
	B. Cracking of Fruit	181
IV.	Consequences of Calcium Deficiency at the Molecular Level	182
	A. Cell Bursting	182
	B. Enhanced Membrane Permeability	184
	C. Discussion	185
V.	Enhancing Calcium Levels in Plants	186
References		187

I. INTRODUCTION

Fruit and vegetable crops may suffer from a wide variety of disorders. Most of them can be ascribed to the activity of microorganisms, but in some the tissue appears to be suffering from a disorder in its own physiology. The key role of calcium in many of these physiological disorders first emerged when it was shown in 1936 that the incidence of bitter pit in apples was inversely related to the calcium content of the fruit.[1] A few years later the same was found to be true for blossom-end rot of tomatoes.[2] Since that period, calcium deficiency has proved to be the central factor in a variety of other physiological disorders and, by 1975, Shear[3] was able to list over 30 disorders in fruits and vegetables which could be attributed to calcium deficiency; several further disorders can be added to the list.[4-6] In his review, Shear[3] was able to reduce a large body of information about associated factors to a simple generalization: moisture conditions, light intensity, and the composition and molarity of the nutrient solution each seem to influence the severity of the disorders by their effect on the calcium content of the tissue in question. It should be mentioned, however, that there still remain indications that magnesium level, for instance, and certain environmental conditions may have a more direct influence on the disorders.[7,8]

An important feature of calcium deficiency in crop plants is its localization, symptoms appearing typically in bulky fruits, or heading vegetables like lettuce and Brussels sprouts. The concentration of calcium in these tissues is below that in the rest of the plant, indicating that the deficiency is not so much a failure to take up enough calcium from the soil, as its uneven distribution through the plant. Thus, apples showing symptoms of calcium deficiency have a lower calcium content than the leaves of the plant that bears them.[9-11] On the other hand, in sugar beet plants, much of the calcium absorbed is retained in the fibrous root system and the swollen beet; rather little passes to the leaves which, therefore, become the first organs to show symptoms of deficiency.[12] Another indication of uneven distribution comes from experiments in which plants are transferred from a medium rich in calcium to one without calcium, for here it is the new leaves and growing points that exhibit deficiency symptoms, evidence that calcium has become immobilized in the older tissues.[13]

As the nature of the calcium-related disorders has become clearer, the literature has grown both on the practical aspect of preventing calcium deficiency, and on the more fundamental problems of calcium uptake, its mobility within plants, and its role in plant life. Reviews published in the last decade concern the disorders of apple and pear,[14] the central role of calcium in the disorders,[3] the calcium nutrition of plants,[10] and the development of symptoms of calcium deficiency.[15] Bangerth has contributed two general reviews.[16,17] The proceedings of a symposium on the calcium nutrition of economic crops held at Beltsville, Maryland in 1977 have been published[8] as 37 papers, some of them short reviews and others presenting original work.

The first section of the present review considers the factors that govern the movement of root-absorbed calcium through the plant and the resulting patterns of distribution. Next, we consider the etiology of calcium deficiency, trying to bridge the gap between the low calcium status of the tissue and the appearance of the symptoms that are observed. In considering what effects calcium deficiency has on events at the molecular and cell levels, attention will be directed especially to those aspects which seem to be significant in the etiology of the symptoms. Inevitably, other aspects have been ignored — such as the uptake of calcium from the soil, calcium in relation to plant enzymes, and effects on fruit respiration. The final section of this review returns to considerations at the whole plant level to discuss what can be done in practice to ward off potential deficiencies and produce fruit and vegetables free of blemish and fit for market.

II. TRANSPORT OF CALCIUM WITHIN PLANTS

Vascular plants have two long-distance transport systems, distinct in their anatomy, and in the mechanism by which they operate. The xylem contains what are essentially hollow tubes made up of the cell walls of vessel elements joined end to end. Their principal role in transport derives from transpiration, the flow of xylem sap up to the aerial parts of the plant where the water evaporates and passes out to the atmosphere through stomata or lenticels. Inorganic solutes may be secreted into the vessels at root level and swept up in the xylem stream, passing in greatest amount to those tissues of the plant which transpire the most rapidly, generally the leaves. Small and possibly inadequate quantities of solute will arrive in regions where transpiration is very slow, like the enclosed inner leaves of some vegetables or bulky fruits with a thick cuticle and few pores.

Xylem flow on its own would result in an accumulation of solutes in the distal regions. However, solutes may also pass through the sieve tubes of the phloem in response to the predominant concentration gradient, flowing for instance out of mature photosynthesizing leaves to the growing apices of root and shoot, and to developing seeds and fruits. The principal solute in the sieve tubes, sugar, is accompanied by some of the inorganic solutes that were delivered to the leaves by the xylem, but now pass back out of the leaf to other destinations.

Phloem transport is a symplastic process, occurring within the cytoplasm of the sieve cells, while movement along cell walls or within the xylem vessels is apoplastic. A solute that is freely able to move in both xylem and phloem, readily transferring from apoplast to symplast and vice versa, could thus be swept up from the roots to the leaves in the cytoplasm and then pass down again in the phloem, so circulating through the plant. Movement through each channel may be rapid enough to enable a particular atom of phosphorus or potassium, for instance, to make several complete cycles around a herbaceous plant in a day.[18]

An additional and much slower reversal of flow occurs in leaves at the time of senescence, for a substantial proportion of their solutes then move up to the apex supporting the growth of new leaves or fruits. One consequence of this apical drift of inorganic solutes is that symptoms of nutrient deficiency first present themselves in the older, basal leaves.

Not all mineral elements can circulate as freely as this. Potassium, magnesium, phosphorus, and sulfur are among those most mobile, iron has an intermediate status, while calcium (along with boron) stand out as the mineral elements which depart most markedly from the general pattern set out above.[19] The constraints upon the movement of calcium are discussed next; for a general review see Hanger.[20]

The ascent of calcium up the stem of young bean and apple shoots does not follow the pattern that would be expected from a mass flow of xylem sap, for calcium ions become bound in the xylem. Labeled calcium can be released from vessels preloaded with ^{45}Ca by perfusing them with a solution containing calcium, magnesium, or barium ions.[21-23] Such divalent ions enhance the mobility of calcium in the xylem to an extent dependent on their concentration.[20] The vessels evidently act like ion exchange columns endowed with negatively charged sites on their walls. Incoming calcium ions absorb onto these sites, moving upwards if displaced from these sites by fresh supplies of calcium (or other divalent cation).

Calcium ions move up the stem more slowly than the water in the xylem,[24] taking for instance 3 days to reach the top of a 30-cm apple plant.[22] The rate of calcium movement is, nevertheless, influenced to some extent by the rate of water flow in the xylem,[25] suggesting that calcium ions are swept along by the xylem flow in the intervening periods between displacement from one site and capture by a site further up the stem. On this model, the movement of calcium ions must be predominantly in the direction of the transpiration stream leading to the mature leaves. However, there seems to be a preferential migration of calcium towards the meristematic zones and younger tissues of apple[22,26] which suggests some form of regulation, such as hormone-directed transport.[22,27,28]

Some of the calcium moving up the xylem vessels leaks out laterally into neighboring tissues.[29] It seems likely that calcium remains in the apoplast as it migrates away from the xylem, for in oat leaves, this leakage of ^{45}Ca is restricted to the small transverse veins; the major longitudinal veins have a suberized lamella in the radial walls of the bundle sheath which would block such apoplastic movement.[20,30] Calcium lost from the xylem vessels of stems in this way may be accumulated in the epidermis and hypodermis of herbaceous plants,[31] or in the bark of trees (which, in fact, bear a substantial proportion of the calcium present,)[20,32] while some may be leached out by rain.[33]

The most important constraint on the movement of calcium in plants is undoubtedly that exercized by the phloem. When a pulse of ^{45}Ca is supplied to beans through the root system, it moves to the young developing leaves, and remains in them, apparently suffering little or no export to other parts of the plant in the next few days.[34] Likewise, when ^{45}Ca is applied to leaves, it is rapidly immobilized in them, relatively little being redistributed — less than 5% in a 3-week experiment with oat plants.[20] Experiments of this sort (and also the low concentration of calcium in sieve-tube exudates) have led many authors to conclude that calcium is immobile in the phloem.[19]

There is, nevertheless, clear evidence that under some circumstances calcium can be mobilized and transferred away from its initial site of deposition, as for instance, in trees. Labeled calcium injected into trees[35,36] or supplied to their roots[37] can be detected in the new growth made in the following year. In apple trees, about 25% of the calcium that appears in the new season's growth was already present in the dormant, overwintering tree.[32] According to the evidence available, this long-distance transport of calcium from the basal region of the tree up into the new growth occurs through the xylem rather than the phloem.[20,36,38]

Calcium may become more mobile as plants age.[39] It is possible that initial supplies of calcium are immobilized as crystalline deposits or held on exchange sites, supplies arriving later (when these locations have become saturated) being more free to move within the plant.[31] This suggestion is in accord with the finding that calcium can be made more mobile by adding chelators,[40] or additional calcium or other divalent ions.[41] It has been argued that these treatments are evidence in favor of phloem transport because they would ensure delivery of high concentrations of calcium to the loading sites, but Hanger[20] concludes from a critical examination of the literature that the xylem was responsible for transport in these experiments. The results of girdling experiments are also equivocal.[20] It will be sufficient for the present purposes to conclude that in the shortterm little calcium is transported through the phloem and to consider the consequences this has for the nutrition of fruit and vegetables.

At an early stage of development, while a fruit is still quite small, there may be sufficient transpiration to allow calcium import through the xylem but later, during the main period of growth, transpiration declines in amount as the surface/volume ratio falls and the cuticle becomes thicker and less permeable.[9] The phloem is then the main agency of supply to the fruit, transporting sugars and significant amounts of most nutrients (but not calcium), together with water. Most of the calcium present in fruits is, thus, accumulated at an early stage, thereafter suffering dilution as the fruit makes further growth; 90% of the calcium ultimately present in apples reaches them in the first 6 weeks after full bloom.[42] If young tomatoes are enclosed in plastic bags, transpiration is restricted, the supply of calcium correspondingly reduced, and the incidence of blossom-end rot more severe.[9]

The same principles apply to organs that develop underground. The peanut *Arachis* receives calcium from the xylem stream while it is still aerial, but once buried in the soil and exposed to a high humidity, the xylem supply stops and calcium can only be absorbed directly from the soil.[20] Likewise, labeled calcium supplied to the root system of a potato plant will only move to the developing tubers if they are maintained at a low humidity, effectively encouraging a flow of xylem sap.[43] At high humidity, tuber development is dependent on the exogenous supply of calcium to the tuber surface.[44]

The movement of calcium into apples has been investigated by autoradiography of fruit growing on plants to which ^{45}Ca had been administered. More radioactivity appeared in the pedicel than in the fruit itself, which is consistent with the very much higher calcium concentration in the pedicel.[32] Although some calcium might be diverted into the calcium-containing crystals in the pedicel, so diminishing yet further the supply to the fruit, the pedicel proved to be no barrier to the flow of ^{45}Ca in the first 6 weeks after anthesis.[20] Once within the fruit, ^{45}Ca supplied 4 to 5 weeks after full bloom accumulates most in the core towards the stem end, and least in the cortex;[45] this is the region with the lowest calcium concentration, and the region where bitter pits most commonly occur.[46]

This pattern of calcium supply stands in contrast to that for potassium and magnesium which are both mobile in the phloem and can, therefore, enter fruit and storage organs throughout the period of growth. Such organs are, therefore, characterized by a low calcium content and a high ratio of K + Mg/Ca.[20,47,48] In paprika, apple, and potato, the ratio K/Ca exceeds 10, but it remains around unity in the leaves of these plants, for the leaves receive a continuous supply of calcium through the xylem.[10]

A state of calcium deficiency may thus arise in one part of a plant, although other regions are well-supplied. Even within a single cell, apparently well-supplied with calcium, deficiency conditions may arise if so much of the calcium is sequestered that too little remains free for the really essential roles. Thus, treatment with citric acid induces symptoms of blossom-end rot in tomato[49] and of tipburn in lettuce,[50] presumably because a localized deficiency arose following chelation of the calcium. It seems possible that this could happen in the field under conditions such as high temperature which may cause a rise in the level of organic acids within the tissues.[50]

III. THE SYMPTOMS OF CALCIUM DEFICIENCY

Several general accounts of the symptoms of calcium deficiency have appeared.[51,52] In addition, there are color photographs of the symptoms exhibited in apple[11] and other crop plants[53] and anatomical descriptions of the lesions developed in tomato[54] and sunflower.[55] The precise nature of the deficiency symptoms depends on the organ in question (leaf, root, fruit, etc.) and on the species. The literature abounds with descriptions of the symptoms that develop in a particular species or in a particular fruit or vegetable, but a comparative approach will be made here, with emphasis on the way the symptoms develop.

A. Water-Soaked and Necrotic Lesions
1. Stems

A disorder known as *topple* is liable to occur in tulips forced to rapid growth under greenhouse conditions. Glassy, water-soaked regions a few centimeters long appear near the top of the flower stem in double tulips. Drops of sugary liquid may be forced out through tears in the epidermis and roll down the outside of the stem. Within days or even hours, the stem shrinks and becomes furrowed at the site of the disorder, as the soft tissues dry out. Once the parenchymatous cells have collapsed, the remaining vascular tissues can no longer support the weight of the flower, and the stem topples over so that the flower hangs upside down.[56,57] Similar symptoms develop in calcium-deficient sunflowers.[58]

Seeds set to germinate in media containing little or no calcium may also develop necrotic lesions in the stem. Some varieties of *Phaseolus* are liable to hypocotyl necrosis within a few days of the start of germination. First, a water-soaked translucent area is seen near the top of the hypocotyl, drops of liquid may then appear on the surface, and soon the hypocotyl shrinks as the cells of cortex and pith collapse. The disorder may be alleviated by the addition of calcium to the germination medium.[59,60] Potato sprouts may show comparable symptoms in the absence of exogenous calcium, elongating cells just below the apex becoming necrotic and black.[61]

2. Leaves

Although calcium-deficient Brussels sprouts may appear to be perfectly healthy from the outside, sectioning shows that the young leaves around the meristem have become necrotic and brown. Even a low incidence of such internal browning may render an entire lot of sprouts unsaleable, as the affected sprouts cannot be identified and discarded.[62] The healthy outer leaves obscure the lesion from view, and also prevent it from drying out. In more exposed leaves, not protected in this way, calcium-deficient lesions tend to develop further, the typical syndrome being a tipburn. In strawberry,[63] cabbage,[64] and lettuce,[65] the first visible symptom is the presence of water-soaked areas near the tips and margins of the young emerging leaves. These regions subsequently become necrotic and dry. As they can make no further growth, while the central part of the leaf blade continues to expand, the lamina becomes curled or cupped;[66] in maize, the leaf edge becomes serrated and torn.[6] Successively younger leaves show more pronounced symptoms, the lamina being reduced in size, scorched, and sometimes blackened.[67] The entire heart of celery plants may be destroyed, secondary infection with *Erwinia* yielding a slimy black rot.[68] In some species, there may be areas of chlorosis near the leaf margin and also chlorotic spots in the lamina.[52]

3. Roots

Under conditions of calcium deficiency roots fail to extend and become stunted. If established plants are transferred to media containing no calcium, the roots become translucent and waterlogged, eventually dying away.[52]

Little or no calcium is translocated to root tips from the rest of the plant. Even when radioactive calcium was supplied to a section of root a few centimeters behind the root tip of a maize or bean seedling, none was moved towards the tip; if the tip was in a calcium-free medium, root extension ceased within a day or two and the roots died soon after.[69]

To ensure the continued, healthy growth of roots, the tips must be allowed to extend into a calcium-containing medium. The roots of young maize seedlings cease growth in distilled water but grow normally in about 10^{-5} M calcium chloride,[70] while the roots of established plants only penetrate slowly through media with a low calcium content.[71] It is a matter of some interest that exogenous calcium must also be supplied for the normal growth of root hairs[72] and pollen tubes.[73]

Cavity spot is a disorder of the fleshy roots of carrot and parsnip. It appears first as a cavity in the cortex below the epidermis, but the covering tissues eventually collapse, leaving a laterally elongated, shallow scar lined with cork cells and often showing signs of infection.[74] The status of this disorder is now open to question, as recent evidence[75] suggests it is due to anaerobic, pectolytic bacteria rather than a deficiency of calcium.

4. Fruits

Several of the calcium deficiency disorders in fruits involve the appearance of very obvious water-soaked regions. Blossom-end rot of tomatoes and peppers has been studied in some detail. Symptoms first appear in immature fruits, beginning with a green water-soaked area under the fruit wall at the stylar end of the fruit. Within a few days the lesion attains its full size then drying out to leave a blackened, dry, and sunken spot occupied by the collapsed remains of necrotic cells.[76,77]

It is hardly surprising that the physiological disorders of apple have received particular attention because the apple crop is a big one and its market value enhanced if the fruit can be stored for some months after harvest. Watercore is a condition of apples (and pears) which develops while the fruit is still on the tree. Parts of the flesh, initially around the vascular bundles,[78] appear translucent or "glassy" because the intercellular spaces have become filled with sap instead of air. Analysis of this intercellular fluid shows that it contains sorbitol which is synthesized in the leaves and translocated to the fruit.[79] Such an apoplastic sorbitol solution might be delivered to the apple through the phloem if there were some breakdown in the phloem cells allowing the sieve tube sap to empty out into the intercellular spaces. Alternatively, sorbitol delivered from the leaves to the root system and transferred there to the xylem sap[80] might subsequently accumulate within apples as a result of root pressure flow. The suggestion that watercore develops more readily if there is rain at harvest[81] lends some support to this possibility.

The cells in a watercored apple swell up and eliminate the space between them during a period of storage.[78] Alternatively, watercore may lead to the development of internal breakdown,[82] a condition in which the pulp is spongy in texture and brown in color,[83] indicating a breakdown in compartmentation that allows polyphenoloxidase to gain access to its substrate.

Bitter pit is a disorder that may appear late in the season while the fruit is still on the tree or only after a period of storage. It is characterized by brown dry areas, a few millimeters in diameter, generally located just below the skin, but extending in severe cases throughout the cortex. Each pit region is intimately associated with a vascular bundle.[84] The cells in the pit are dead, their walls collapsed so that the pits contain cavities, "once occupied by four to six or more normal cells".[84]

The distribution of bitter pits within apple fruits corresponds with the distribution of calcium, the pits being more prevalent where the calcium content is low.[46] Nevertheless, analysis of sound and pitted tissue shows rather more calcium (and considerably more magnesium) in the pitted regions.[85,86] The accumulation of calcium in pitted areas has also been observed in autoradiographs of fruit from trees supplied with ^{45}Ca to the roots.[87] As radioactivity only appeared in the cortex in regions that were already pitted, its accumulation there seems to be a secondary response following the disorganization of cortical cells, rather than a stage in the development of the pit.[87,88]

B. Cracking of Fruit

Soft fruits such as apple,[11] tomato,[89] sweet cherry,[90] and prune[91] are liable to crack and split at the time of ripening if they are deficient in calcium. A number of observations suggest that the internal tissues enlarge in such fruits as the result of an osmotic uptake of water. Thus, fruit cracking is most prevalent after a period of rain or when the humidity is relatively high.[16,89,90] Presumably, water may enter a fruit as vapor through the lenticels, but it can also enter through the skin as evidenced by the way apples may crack if simply immersed in water for a few days.[92] Likewise, tomato fruits can be induced to crack within a matter of minutes by vacuum infiltrating water into them.[89] Once cracking has occurred, intake of external moisture by the exposed tissue often results in much more severe cracking.[93] Cherries and grapes behave like osmometers, swelling more in water than in sugar solutions.[94,95]

IV. CONSEQUENCES OF CALCIUM DEFICIENCY AT THE MOLECULAR LEVEL

The previous section described the presenting symptoms of calcium deficiency. The next step must be to consider the cellular events which underly the observed symptoms in the hope of identifying the molecular mechanisms involved in the development of the various deficiency lesions.

A recurring theme in many of the descriptions is the appearance of a water-soaked condition. Some lesions progress no further than this, as for instance watercore and internal breakdown of apple, and internal browning of sprouts. These lesions, as their name implies, are located within tissue masses and would therefore be very slow to dry out. The majority of the calcium deficiency lesions are not protected in this way and the water-soaked condition is, therefore, likely to be a penultimate stage, followed by desiccation. The evidence for such a sequence is clearly established for blossom-end rot and tulip topple, for instance, but enough is known to suggest that other dry, shrunken lesions also develop in this way. For cells within the lesion to dry out while those at a distance remain hydrated and alive, there must be a breakdown of membrane integrity. There seem to be two ways in which this situation could arise under conditions of calcium deficiency (Figure 1). One, set out on the left suggests that water enters the tissue from outside, is taken up osmotically, and causes the cells to expand. Such an expansion might crack open a soft fruit. Alternatively, the cell may eventually burst, the tissue then losing its turgor and becoming a sodden mass which may later dry out. According to the pathway on the right, cell membranes become so permeable in the absence of calcium that they no longer act as osmotic barriers allowing cell contents to flow out into the intercellular spaces from which the water may finally evaporate.

A. Cell Bursting

The basic observation underlying this route of lesion development is that if discs prepared from the flesh of soft, ripe fruit are placed in water, their electrolytes and sugars leak out rapidly. In a typical experiment with apple, 90% of the electrolytes were lost in 2.5 hr, whereas in M glycerol, only about 20% of that originally present was lost — no more than is likely to be present in the apoplast and in cells that were broken in preparing the discs of fruit.[96] It seems that massive solute release only occurs when tissue is placed in water or a solution of low molarity, that is to say, when the pressure gradient across the cell wall is at its greatest. The suggestion that the cells actually burst under this pressure is supported by the finding that apple disc cells readily become colored with Evan's blue, a dye which does not enter living cells with intact membranes.[96] The notion that some plant cells may burst when placed in water is not a new one.[97] Among fruits, the phenomenon is restricted to those that are soft and fleshy; it was not observed in unripe apples,[96] preclimacteric banana,[98] or in the relatively tough-walled fruit of cucumber and marrow.[96] Perhaps, only those fruits that have been cultivated and bred for their soft flesh and sweet taste have cell contents so sugary as to generate a high turgor pressure and walls so weak that they cannot stand the full extent of it. It would be interesting to compare for example the behavior of crab apple or rose hip with apple or grape.

The pathway of lesion development being discussed here (Figure 1) is in accord with the observation that fruits may crack open under conditions of calcium deficiency, for cracking is an indication that the tissues inside are swelling. The entry of water into fruit seems to be a prelude to cracking and the development of several other calcium deficiency lesions. Thus, tomatoes may crack open soon after a period of rain or irrigation[93] or following infiltration with water.[89] Exogenous water may be delivered to

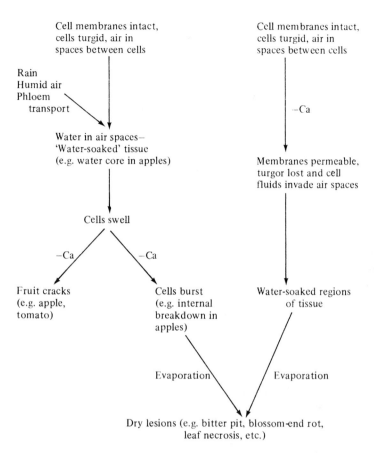

FIGURE 1. Pathways of development of calcium deficiency symptoms.[15] The left-hand pathway applies to soft-walled, succulent fruits. The cells of such fruits may swell as they absorb water taken up from the atmosphere or supplied by the phloem. If the tissue enlarges as a whole, the fruit will crack; if the union between cells weakens, individual cells may burst leading to the appearance of other symptoms. The right-hand pathway describes the situation in which membranes become so permeable that cell fluids invade the intercellular spaces. (From Simon, E. W., *New Phytol.*, 80, 1, 1978. With permission.)

apples through the vascular system, leading to watercore which, in turn, may be the precursor of internal breakdown. Spartan apples may develop internal breakdown if they are held in an atmosphere with relative humidity of 90% or more — but not in a drier atmosphere — suggesting that it is the condensation of water within the fruit that initiates their osmotic swelling and ultimately the bursting of cells. Holding apples for 2 to 3 days at 21°C prior to cold storage or subjecting them to 14 days hypobaric treatment reduced internal breakdown, possibly because these treatments would reduce the water content of the apples.[99]

If a lesion-like blossom-end rot is initiated by cells bursting within a tomato, then one might expect a low apoplast water potential to increase the incidence of lesions. There is no doubt that water stress maintained for long periods aggravates blossom-end rot,[100] but this may simply reflect the lowered calcium status of the fruit as xylem sap is diverted to the leaves. There is, however, one report[101] in the literature indicating that blossom-end rot may appear in a tomato plant supplied with complete fertilizer but kept under dry conditions with only just enough water to prevent wilting. In these experiments, blossom-end rot appeared in fruits containing 0.07% calcium, which is about the level in normal, healthy fruit from well-watered plants. (In calcium-deficient plants, blossom-end rot fruit had less than half this quantity of calcium.) Further work would be needed to test the possibility that water stress might so enhance the pressure gradient across cell walls as to cause the cells to burst, yielding symptoms of blossom-end rot.

It has long been known that calcium strengthens cell walls, making them more rigid and resistant to extension,[102] and it is perhaps relevant that calcium delays abscission[103] for this is a process, that like ripening involves a loosening of the bonds between adjacent cell walls. Even exogenous supplies of calcium can strengthen the cell wall in cherries.[16] After 4-hr immersion in water all the fruit were cracked, but after 4 hr in 0.05 M magnesium or barium chloride, cracking was only seen in about 75% of the fruit, and with calcium chloride it was down to 28%. The ultimate site of action of these cations must be readily accessible, 4 hr being sufficient for them to diffuse to the site and establish the requisite bonds or links. In another experiment, the degree of cracking was increased by treatment with ethylenediaminetetraacetic acid (EDTA) for 4 hr.

The cells in most plant tissues are held firmly in position by the many links established through the middle lamella with neighboring cells. No one cell is likely to become independent enough to swell and enlarge more than its neighbors under hypotonic conditions, as it would be restrained by the lattice of cells around it. However, as fruits ripen they become softer, less force then being necessary to separate the constituent cells of the flesh.[104] Both ultrastructural[105] and biochemical[106,107] studies show that softening is associated with loss of the insoluble pectic materials of the middle lamella. Calcium is associated with these substances, forming microcrystallites which are regions where the buckled polygalacturonate chains lie alongside one another in such a way that regular calcium-filled interstices appear, like eggs fitting into an egg-box. The calcium is coordinated to the polymer chains and so forms a relatively tight link between them.[108] Calcium deficiency might, therefore, loosen the association between the polygalacturonate chains and perhaps also render them more accessible to hydrolytic enzymes, so weakening the links between adjacent cells.

Loss of polygalacturonate would, thus, create a permissive situation in which a cell could push aside its neighbors and so find room to enlarge. However, such an enlargement would only take place if the wall around the cell became more plastic and able to yield under the turgor pressure developed within it. Calcium ions have a second regulating role here, for they are thought to prevent wall extension by inhibiting the H^+-induced wall-loosening process.[109]

It is possible on this view of a dual role for calcium in cell wall physiology to envisage what might cause fruits to crack. If calcium deficiency only loosened the cell walls allowing them to expand in response to the turgor pressure within, the tissue mass as a whole would enlarge and crack the skin; but if it also catalyzed breakdown of polygalacturonate, a different type of lesion would develop as individual cells enlarged to the point at which they burst. This proposition leads to the suggestion that there should be far less polygalacturonate breakdown in calcium-deficient apples that were cracking than in those exhibiting for instance, bitter pit.

B. Enhanced Membrane Permeability

There is abundant evidence that calcium plays some role in stabilizing the membranes of plant cells, reducing the passive leakage of substances through them. Thus, the presence of calcium ions in the bathing medium reduces the leakage of sugars from maize scutellum slices[110] and of betacyanin from beet slices.[111] In addition, calcium can reduce or prevent the leakage that would otherwise occur when the cells of various vegetative tissues are exposed to extremes of temperature[112,113] or to various chemical insults such as exposure to lysine,[114] short-chain alcohols,[115] or anaerobiosis.[112] It is particularly relevant in the present context that leakage is encouraged by EDTA which chelates calcium and so presumably removes it from membranes; and that this leakage can, in turn, be reversed by the subsequent addition of calcium.[110,116,117]

As well as rendering membranes less permeable, calcium also affects the process of ion uptake,[118] abolishing its selectivity. According to Epstein,[119] calcium is essential for maintenance of the conformation of the ion-binding site; in its absence the geometry of this site is so changed as to permit the binding of other ions, which, in the absence of calcium, would not be bound. Possible molecular mechanisms underlying these calcium-membrane interactions are set out by Bangerth.[17]

Electron microscopy of calcium-deficient tissues has also produced evidence of changes in membrane organization. In the shoot apex of barley,[120] the nuclear envelope, plasmalemma, and tonoplast appear to break up and later on, mitochondria and Golgi apparatus become disorganized. In a final stage there were no more than traces of fragmented membrane at the periphery of the cells. Comparable images have been seen in barley root cells[121] and potato sprouts.[122] Cells in which membrane organization has deteriorated to this point with visible discontinuities must certainly be leaky, but even at a much earlier stage, some minor change in molecular architecture might already enhance membrane permeability, although it had little effect on the appearance of the membrane.

Two reports on apples indicate some degree of cytoplasmic degeneration in calcium-deficient fruit developing bitter pit. The chief symptoms observed in a New Zealand study[123] were loss of Golgi bodies, vesiculation of endoplasmic reticulum, and degeneration of the internal structure of mitochondria and chromoplasts. Plasmalemma and tonoplast did not seem to undergo much change except for the presence of vacuolar inclusions indicating lysosomal activity. More severe symptoms were observed in an English study[124] of low calcium apples stored at 3°C — a separation of the plasma membrane from the cell wall and disintegration of the vacuolar membrane in some cells leading to disorganization of the cytoplasm. In addition, many of the tangential walls of cells underlying the epidermis became reticulated and disorganized in fruit stored for 99 days. This cell wall breakdown was particularly severe in fruit of low calcium content and it occurred "before appreciable signs of *membrane* breakdown".[124,125] This breakdown of wall structure seems to be a visible manifestation of the weakening of the union between neighboring cells.

C. Discussion

We do not yet have an assured understanding of the way in which calcium deficiency gives rise to the symptoms observed in plants. The mechanisms proposed here (Figure 1) are supported by circumstantial evidence only: first, the observed timing of events (watercore in apples preceding internal breakdown, lesions that are initially water-soaked, later drying out, etc.), and second, the evidence that calcium has functions in both membranes and walls. These mechanisms are mutually exclusive in that there will be little or no turgor pressure in a cell if its membranes have become relatively permeable, while the membranes are so fragile that they will certainly be disrupted if the whole cell expands to the point of bursting. It is, nevertheless, possible to envisage situations in which fruit cells might suffer from a *membrane* disorder if calcium deficiency arose at an early stage of development, while later, at the time of ripening, the *walls* might become much more sensitive to a deficiency. Likewise, fruits might swell and crack if they were not yet ripe and had little or no polygalacturonase, with a different type of lesion appearing in the fully ripe fruit.

As it is possible, in some cases at least, to prevent the appearance of deficiency symptoms by treating tissues with sprays or dips of calcium salts (see below), an interesting question can be put to the test: is magnesium as effective as calcium? This question owes its relevance to the different chemical and biochemical properties of calcium and magnesium. In particular, the coordination geometry of calcium is relatively flexible allowing it, for instance, to bind to the irregular geometries of the coordination sites in proteins and to form reversible crosslinks, whereas magnesium makes much stronger demands for a particular geometry and is only able to form weak crosslinks as in its association with the enzymes of phosphate transfer.[126] If calcium has a key role in crosslinking the possibility arises that magnesium would have the opposite effect, for it would not crosslink, although it might bind equally well to any single group or surface.[127] It follows that when calcium-deficient plants are treated with solutions of magnesium salts, magnesium might displace calcium from such crosslinking sites, to an extent dependent on the initial calcium status of the plant. Treatment with magnesium would, therefore, be tantamount to reducing yet further the level of calcium at crosslinking sites. This would have the effect of weakening the middle lamella and so might lead to a more severe display of the symptoms that arise when cells burst. There is, in fact, clear evidence that although calcium salts offer some degree of control over bitter pit in apples, treatment with magnesium salts makes the condition worse, both in orchard trials,[47,128] and laboratory experiments.[86,129] Similarly, incidence of watercore breakdown in apples is positively correlated with the magnesium status of the fruit and negatively related to the calcium content.[130]

Another role that magnesium plays less well than calcium is that of stabilizing membranes and reducing their permeability.[100-104,116] If treatment with magnesium displaces calcium from crosslinking sites in the membrane, it could, therefore, enhance their permeability and might in consequence exacerbate deficiency lesions attributed to enhanced permeability. This view is encouraged by reports that internal browning of sprouts[131] and blackheart of celery[132] are indeed worsened by sprays of magnesium salts. (Hypocotyl necrosis in beans is no worse when magnesium is supplied to the rooting medium,[59] but this might be due to effects of calcium and magnesium on transport within the plant, rather than events at the tip of the hypocotyl.)

The third role of calcium discussed above concerns cell wall extension. The view was put that fruits in which adjacent cell walls were still firmly linked through the middle lamella might crack open when calcium was deficient, as this allowed the adhering mass of cell walls to expand. Magnesium is a less efficient inhibitor of cell wall extension than calcium,[133] and it is unlikely that the amounts present in magnesium-treated fruits would suppress cell extension altogether. In other words, one could expect magnesium-treated fruits to show some degree of cracking, perhaps less than in the calcium-deficient controls, but certainly not more. This is borne out by experimental results; magnesium chloride does little[16] or nothing[90] to prevent cherries from cracking, while calcium chloride is very effective.

In summary, it appears that magnesium has little effect on fruit cracking, but it makes other types of calcium deficiency symptoms worse. This finding is in accord with the hypothesis that cracking has a quite different etiology to the other symptoms.

V. ENHANCING CALCIUM LEVELS IN PLANTS

As calcium reserves are not readily mobilized within plants, a continuous supply of calcium is necessary for normal growth. Many attempts have been made to improve the supply of calcium through the roots.[8,17] An alternative strategy is to apply calcium chloride or nitrate as a spray to celery,[132] tomato,[16] or apple;[14] or to give apples a postharvest dip in the solution.[129,134] Sprays are not effective against internal tipburn of Brussels sprouts presumably because too little of the calcium that falls on the outer leaves is translocated to those inside — but calcium-rich xylem sap can be caused to flow up to these leaves under the influence of root pressure.[135]

REFERENCES

1. DeLong, W. A., Variations in the chief ash constituents of apples affected with blotchy cork, *Plant Physiol.*, 11, 453, 1936.
2. Raleigh, S. M. and Chucka, J. A., Effect of nutrient ratio and concentration of growth and composition of tomato plants and on the occurrence of blossom-end rot of the fruit, *Plant Physiol.*, 19, 671, 1944.
3. Shear, C. B., Calcium-related disorders of fruits and vegetables, *HortScience*, 10, 361, 1975.
4. Pinkhof, M., Untersuchungen über die Umfallkrankheit der Tulpen, *Recl. Trav. Bot. Neerl.*, 26, 134, 1929.
5. Pinkhof, M., Untersuchungen über die Umfallkrankheit der Tulpen, *Proc. K. Ned. Akad. Wet.*, 32, 1248, 1929.
6. Kawaski, T. and Moritsugu, M., A characteristic symptom of calcium deficiency in maize and sorghum, *Commun. Soil Sci. Plant Anal.*, 10, 41, 1979.
7. Drake, M., Bramlage, W. J., and Baker, J. H., Correlations of calcium content of "Baldwin" apples with leaf calcium, tree yield, and occurrences of physiological disorders and decay, *J. Am. Soc. Hortic. Sci.*, 99, 379, 1974.
8. Shear, C. B., International Symposium on Calcium Nutrition of Economic Crops, *Commun. Soil Sci. Plant Anal.*, 10, 1, 1979.
9. Wiersum, L. K., Calcium content of fruits and storage tissues in relation to the mode of water supply, *Acta Bot. Neerl.*, 15, 406, 1966.
10. Marschner, H., Calcium nutrition of higher plants, *Neth. J. Agric. Sci.*, 22, 275, 1974.
11. Shear, C. B., Symptoms of calcium deficiency on leaves and fruit of "York Imperial" apple, *J. Am. Soc. Hortic. Sci.*, 96, 415, 1971.
12. Mostafa, M. A. E. and Ulrich, A., Absorption, distribution and form of Ca in relation to Ca deficiency (tip burn) of sugarbeets, *Crop Sci.*, 16, 27, 1976.
13. Loneragan, J. F. and Snowball, K., Calcium requirements of plants, *Aust. J. Agric. Res.*, 20, 466, 1969.
14. Wilkinson, B. G. and Fidler, J. C., Physiological disorders, in *The Biology of Apple and Pear Storage*, Part 2, Research Review No. 3, Fidler, J. C., Wilkinson, B. G., Edney, K. L., and Sharples, R. O., Eds., Commonwealth Agricultural Bureaux, 1973.
15. Simon, E. W., The symptoms of calcium deficiency in plants, *New Phytol.*, 80, 1, 1978.
16. Bangerth, F., Investigations upon Ca related physiological disorders, *Phytopathol. Z.*, 77, 20, 1973.
17. Bangerth, F., Calcium-related physiological disorders of plants, *Ann. Rev. Plant Pathol.*, 17, 97, 1979.
18. Pate, J. S., Exchange of solutes between phloem and xylem and circulation in the whole plant, in *Encyclopedia of Plant Physiology*, Vol. 1, Zimmerman, M. H. and Milburn, J. A., Eds., Springer-Verlag, Berlin, 1975, chap. 19.
19. Ziegler, H., Nature of transported substances, in *Encyclopedia of Plant Physiology*, Vol. 1, Zimmerman, M. H. and Milburn, J. A., Eds., Springer-Verlag, Berlin, 1975, chap. 3.
20. Hanger, B. C., The movement of calcium in plants, *Commun. Soil Sci. Plant Anal.*, 10, 171, 1979.
21. Bell, C. W. and Biddulph, O., Translocation of calcium. Exchange versus Mass flow, *Plant Physiol.*, 38, 610, 1963.
22. Shear, C. B. and Faust, M., Calcium transport in apple trees, *Plant Physiol.*, 45, 670, 1970.
23. Van de Geijn, S. C., Petit, C. M., and Roelofsen, H., Measurement of the cation exchange capacity of the transport system in intact plant stems. Methodology and preliminary results, *Commun. Soil Sci. Plant Anal.*, 10, 225, 1979.
24. Biddulph, O., Nakayama, F. S., and Cory, R., Transpiration stream and ascent of calcium, *Plant Physiol.*, 36, 429, 1961.
25. Emmert, F. H., Retention and passage of calcium and strontium in stems of *Phaseolus vulgaris* as mediated by xylem stream flow rate and dinitrophenol, *Physiologia Pl.*, 22, 246, 1969.
26. Stebbins, R. L. and Dewey, D. H., Role of transpiration and phloem transport in accumulation of ^{45}calcium in leaves of young apple trees, *J. Am. Soc. Hortic. Sci.*, 97, 478, 1972.
27. Bangerth, F., A role for auxin and auxin transport inhibitors on the Ca content of artificially induced parthenocargic fruits, *Physiologia Pl.*, 37, 191, 1976.
28. Marschner, H. and Ossenberg-Neuhaus, H., Effect of 2,3,5-triiodobenzoic acid (TIBA) on calcium translocation and cation exchange capacity in sunflower, *Z. Pflanzenphysiol.*, 85, 29, 1977.
29. Wieneke, J., Calcium transport and its microautoradiographic localization in the tissue, *Commun. Soil Sci. Plant Anal.*, 10, 237, 1979.
30. Ferguson, I. B., The movement of calcium in non-vascular tissue of plants, *Commun. Soil Sci. Plant Anal.*, 10, 217, 1979.

31. Läuchli, A., Translocation of inorganic solutes, *Ann. Rev. Plant Physiol.*, 23, 197, 1972.
32. Terblanche, J. H., Wooldridge, L. G., Hesebeck, I., and Joubert, M., The redistribution and immobilization of calcium in apple trees with special reference to bitter pit, *Commun. Soil Sci. Plant Anal.*, 10, 195, 1979.
33. Tukey, H. B., The leaching of substances from plants, *Ann. Rev. Plant Physiol.*, 21, 305, 1970.
34. Biddulph, O., Cory, R., and Biddulph, S., Translocation of calcium in the bean plant, *Plant Physiol.*, 34, 512, 1959.
35. Ferrell, W. K. and Johnson, F. D., Mobility of calcium-45 after injection into western white pine, *Science*, 124, 364, 1956.
36. Thomas, W. A., Retention of calcium-45 by dogwood trees, *Plant Physiol.*, 45, 510, 1970.
37. Führ, F. and Wieneke, J., Secondary translocation of ^{45}Ca to the fruit of apple trees the year after dormancy, in *Mechanisms of Regulation of Plant Growth*, Bieleski, R. L., Ferguson, A. R., and Cresswell, M. M., Eds., Royal Society of New Zealand, Wellington, 1974, 171.
38. Tromp, J., Seasonal variations in the composition of xylem sap of apple with respect to K, Ca, Mg and N, *Z. Pflanzenphysiol.*, 94, 189, 1979.
39. Greene, D. W. and Bukovac, M. J., Redistribution of calcium in *Phaseolus vulgaris* L., *Proc. Am. Soc. Hortic. Sci.*, 93, 368, 1968.
40. Millikan, C. R. and Hanger, B. C., Effects of chelation and of certain cations on the mobility of foliar-applied ^{45}Ca in stock, broad bean, peas and subterranean clover, *Aust. J. Biol. Sci.*, 18, 211, 1965.
41. Millikan, C. R. and Hanger, B. C., Movement of previously deposited ^{45}Ca in subterranean clover (*Trifolium subterraneum* L.) by foliar injections of certain cations, *Aust. J. Biol. Sci.*, 19, 1, 1966.
42. Shear, C. B., Calcium nutrition and quality in fruit crops, *Commun. Soil Sci. Plant Anal.*, 6, 233, 1975.
43. Kraus, A. and Marschner, H., Langstreckentransport von Calcium in Stolonen von Kartoffelpflanzen, *Z. Pflanzenernaehr. Bodenkd.*, 136, 228, 1973.
44. Krauss, A. and Marschner, H., Einfluss des Calcium-Angebotes auf Wachstumrate und Calcium-Gehalt von Kartoffelknollen, *Z. Pflanzenernaehr. Bodenkd.*, 138, 317, 1975.
45. Millikan, C. R., Mid-season movement of ^{45}Ca in apple trees, *Aust. J. Agric. Res.*, 22, 923, 1971.
46. Bramlage, W. J., Drake, M., and Baker, J. H., Changes in Ca level in apple cortex tissue shortly before harvest and during post harvest storage, *Commun. Soil Sci. Plant Anal.*, 10, 417, 1979.
47. Sharples, R. O., Orchard and climatic factors, in *The Biology of Apple and Pear Storage*, Part 4, Research Review No. 3, Fidler, J. C., Wilkinson, B. G., Edney, K. L., and Sharples, R. O., Eds., Commonwealth Agricultural Bureaux, 1973.
48. van Lune, P. and van Goor, B. J., Ripening disorders of tomatoes as affected by the K/Ca ratio in the culture solution, *J. Hortic. Sci.*, 52, 173, 1977.
49. Evans, H. J. and Troxler, R. V., Relation of calcium nutrition to the incidence of blossom-end rot in tomatoes, *Proc. Am. Soc. Hortic. Sci.*, 61, 346, 1953.
50. Misaghi, I. J. and Grogan, R. G., Physiological basis for tipburn development in head lettuce, *Phytopathology*, 68, 1744, 1978.
51. Baumeister, W., Hauptnährstoffe, in *Encyclopedia of Plant Physiology*, Ruhland, W., Ed., Springer-Verlag, Berlin, 1958, 482.
52. Hewitt, E. J., The essential nutrient elements: requirements and interactions in plants, in *Plant Physiology: A Treatise*, Steward, F. C., Ed., Academic Press, New York, 1963, chap. 2.
53. Wallace, T., *The Diagnosis of Mineral Deficiencies in Plants: A Colour Atlas and Guide*, Her Majesty's Stationery Office, London, 1951.
54. Kalra, G. S., Responses of the tomato plant to calcium deficiency, *Bot. Gaz. (Chicago)*, 118, 18, 1956.
55. Bussler, W., Gewebe- und Zellschädigungen bei Calciummangel-Sonnenblumen, *Z. Pflanzenernaehr. Bodenkd.*, 99, 215, 1962.
56. Algera, L., Topple disease of tulips, *Phytopathol. Z.*, 62, 251, 1968.
57. Doss, R. P., Christian, J. K., and Langager, J. M., Calcium deficiency and occurrence of topple disorder in bulbous iris, *Can. J. Plant Sci.*, 59, 185, 1979.
58. Bussler, W., Ca-Mangelsymptome bei Sonnenblumen, *Z. Pflanzenernaehr. Bodenkd.*, 99, 207, 1962.
59. Shannon, S., Natti, J. J., and Atkin, J. D., Relation of calcium nutrition to hypocotyl necrosis of snap bean (*Phaseolus vulgaris* L.), *J. Am. Soc. Hortic. Sci.*, 90, 180, 1962.
60. Helms, K., Calcium deficiency of dark-grown seedlings of *Phaseolus vulgaris* L., *Plant Physiol.*, 47, 799, 1971.
61. Dyson, P. W. and Digby, J., Effects of calcium on sprout growth and subapical necrosis in Majestic potatoes, *Potato Res.*, 19, 290, 1975.
62. Maynard, D. N. and Barker, A. V., Internal browning of Brussels sprouts: a calcium deficiency disorder, *J. Am. Soc. Hortic. Sci.*, 97, 789, 1972.

63. Mason, G. F. and Guttridge, C. G., The role of calcium, boron and some divalent ions in leaf tipburn of strawberry, *Sci. Hortic.*, 2, 299, 1974.
64. Hori, Y., Yamasaki, K., Kamihama, T., and Aoki, M., Calcium nutrition of vegetable crops. II. Calcium deficiency symptoms of chinese cabbage and the effect of the composition and salt concentration of culture solution on its occurrence, *J. Hortic. Assoc. Jpn.*, 29, 169, 1960.
65. Misaghi, I. J. and Grogan, R. G., Effect of temperature on tipburn development in head lettuce, *Phytopathology*, 68, 1738, 1978.
66. Millikan, C. R. and Hangar, B. C., Calcium nutrition in relation to the occurrence of internal browning in Brussels sprouts, *Aust. J. Agric. Res.*, 17, 863, 1966.
67. Berry, W. L. and Ulrich, A., Calcium nutrition of sugar beets as affected by potassium, *Soil Sci.*, 110, 389, 1970.
68. Takatori, F. H., Lorenz, O. A., and Connell, G. H., Strontium and calcium for the control of blackheart of celery, *Proc. Am. Soc. Hortic. Sci.*, 77, 406, 1961.
69. Marschner, H. and Richter, Ch., Calcium-transport in Wurzeln von Maisund Bohnenkeimpflanzen, *Plant Soil*, 40, 193, 1974.
70. Jones, R. G. W. and Lunt, O. R., The function of calcium in plants, *Bot. Rev.*, 33, 407, 1967.
71. Pohlman, G. G., Effect of liming different soil layers on yield of alfalfa and on root development and nodulation, *Soil Sci.*, 62, 255, 1946.
72. Cormack, R. G. H., Lemay, P., and Maclachlan, G. A., Calcium in the root-hair wall, *J. Exp. Bot.*, 14, 311, 1963.
73. Brewbaker, J. L. and Kwack, B. H., The essential role of calcium ion in pollen germination and pollen tube growth, *Am. J. Bot.*, 50, 747, 1963.
74. Guba, E. F., Young, R. E., and Ui, T., Cavity spot disease of carrot and parsnip roots, *Plant Dis. Rep.*, 45, 102, 1961.
75. Perry, D. A. and Harrison, J. G., Cavity spot of carrots. II. The effect of soil conditions and the role of pectolytic anaerobic bacteria, *Ann. Appl. Biol.*, 93, 109, 1979.
76. Spurr, A. R., Anatomical aspects of blossom-end rot in the tomato with special reference to calcium nutrition, *Hilgardia*, 28, 269, 1959.
77. Hamilton, L. C. and Ogle, W. L., The influence of nutrition on blossom-end rot of pimiento peppers, *Proc. Am. Soc. Hortic. Sci.*, 80, 457, 1962.
78. Birth, G. S. and Olsen, K. L., Nondestructive testing of water core in Delicious apples, *Proc. Am. Soc. Hortic. Sci.*, 85, 74, 1964.
79. Williams, M. W., Martin, G. C., and Stahly, E. A., The movement and fate of sorbitol-C^{14} in the apple tree and fruit, *Proc. Am. Soc. Hortic. Sci.*, 90, 20, 1967.
80. Hansen, P. and Grauslund, J., Levels of sorbitol in bleeding sap and in xylem sap in relation to leaf mass and assimilate demand in apple trees, *Physiologia Pl.*, 42, 129, 1978.
81. Faust, M., Shear, C. B., and Williams, M. W., Disorders of carbohydrate metabolism of apples (Watercore, internal breakdown, low temperature and carbon dioxide injuries), *Bot. Rev.*, 35, 168, 1969.
82. Bramlage, W. J. and Shipway, M. R., Loss of watercore and development of internal breakdown during storage of "Delicious" apples, as determined by repeated light transmittance measurements of intact apples, *Proc. Am. Soc. Hortic. Sci.*, 90, 475, 1967.
83. Lord, W. J. and Southwick, F. W., The susceptibility of two Delicious strains to some pre- and post-harvest physiological disorders, *Proc. Am. Soc. Hortic. Sci.*, 84, 65, 1964.
84. Smock, R. M. and van Doren, A., The histology of bitter pit in apples, *Proc. Am. Soc. Hortic. Sci.*, 35, 176, 1938.
85. Perring, M. A. and Plocharski, W., Differences in the mineral composition of sound and disordered apple fruits and of sound and pitted tissue, *J. Sci. Food Agric.*, 26, 1819, 1975.
86. Hopfinger, J. A. and Poovaiah, B. W., Calcium and magnesium gradients in apples with bitter pit, *Commun. Soil Sci. Plant Anal.*, 10, 57, 1979.
87. Ford, E. M., The distribution of calcium in mature apple fruits having bitter pit disorder, *J. Hortic. Sci.*, 54, 91, 1979.
88. Faust, M., Shear, C. B., and Smith, C. B., Investigation of corking disorders of apples. II. Chemical composition of affected tissues, *Proc. Am. Soc. Hortic. Sci.*, 92, 82, 1968.
89. Dickinson, D. B. and McCollum, J. P., The effect of calcium on cracking in tomato fruits, *Proc. Am. Soc. Hortic. Sci.*, 84, 485, 1964.
90. Bullock, R. M., A study of some inorganic compounds and growth promoting chemicals in relation to fruit cracking of bing cherries at maturity, *Proc. Am. Soc. Hortic. Sci.*, 59, 243, 1952.
91. Cline, R. A. and Tehrani, G., Effects of boron and calcium sprays and of mulch on cracking of Italian prune, *Can. J. Plant Sci.*, 53, 827, 1973.
92. Verner, L., A physiological study of cracking in stayman winesap apples, *J. Agric. Res.*, 51, 191, 1935.

93. Frazier, W. A. and Bowers, J. C., A final report on studies of tomato fruit cracking in Maryland, *Proc. Am. Soc. Hortic. Sci.*, 49, 241, 1942.
94. Kertesz, Z. I. and Nebel, B. R., Observations on the cracking of cherries, *Plant Physiol.*, 10, 763, 1936.
95. Considine, J. A. and Kriedemann, P. E., Fruit splitting in grapes: determination of the critical turgor pressure, *Aust. J. Agric. Res.*, 23, 17, 1972.
96. Simon, E. W., Leakage from fruit cells in water, *J. Exp. Bot.*, 28, 1147, 1977.
97. Simon, E. W., Membranes in ripening and senescence, *Ann. Appl. Biol.*, 85, 417, 1977.
98. Brady, C. J., O'Connell, P. B. H., Smydzuk, J., and Wade, N. L., Permeability, sugar accumulation, and respiration rate in ripening banana fruits, *Aust. J. Biol. Sci.*, 23, 1143, 1970.
99. Porritt, S. W., Lidster, P. D., and Meheriuk, M., Postharvest factors associated with the occurrence of breakdown in Spartan apple, *Can. J. Plant Sci.*, 55, 743, 1975.
100. Cerda, A., Bingham, F. T., and Lahanauskas, C. K., Blossom-end rot of tomato fruit as influenced by osmotic potential and phosphorous concentrations of nutrient solution media, *J. Am. Soc. Hortic. Sci.*, 104, 236, 1979.
101. Ward, G. M., Causes of blossom-end rot of tomatoes based on tissue analysis, *Can. J. Plant Sci.*, 53, 169, 1973.
102. Burstrom, H. G., Calcium and plant growth, *Biol. Rev.*, 43, 287, 1968.
103. Poovaiah, B. W. and Leopold, A. C., Inhibition of abscission by calcium, *Plant Physiol.*, 51, 848, 1973.
104. Nelmes, B. J. and Preston, R. D., Wall development in apple fruits: a study of the life history of a parenchyma cell, *J. Exp. Bot.*, 19, 496, 1968.
105. Ben-Arie, R., Kisley, N., and Frenkel, C., Ultrastructual changes in the cell walls of ripening apple and pear fruit, *Plant Physiol.*, 64, 197, 1979.
106. Gross, K. C. and Wallner, S. J., Degradation of cell wall polysaccharides during tomato fruit ripening, *Plant Physiol.*, 63, 117, 1979.
107. Poovaiah, B. W. and Nukaya, A., Polygalaturonase and cellulose enzymes in the normal Rutgers and mutant *rin* tomato fruits and their relationship to the respiratory climacteric, *Plant Physiol.*, 64, 534, 1979.
108. Rees, D. A., *Polysaccharide Shapes*, Chapman & Hall, London, 1977, 52.
109. Cleland, R. E. and Rayle, D. L., Reevaluation of the effect of calcium ions on auxin-induced elongation, *Plant Physiol.*, 60, 709, 1977.
110. Garrard, L. A. and Humphreys, T. E., The effect of divalent cations on the leakage of sucrose from corn scutellum slices, *Phytochemistry*, 6, 1085, 1967.
111. Poovaiah, B. W. and Leopold, A. C., Effects of inorganic salts on tissue permeability, *Plant Physiol.*, 58, 182, 1976.
112. Christiansen, M. N., Carns, H. R., and Slyter, D. J., Stimulation of solute loss from radicles of *Gossypium hirsutum* L. by chilling, anaerobiosis and low pH, *Plant Physiol.*, 46, 53, 1970.
113. Toprover, Y. and Glinka, Z., Calcium ions protect root cell membranes against thermally induced changes, *Physiologia Pl.*, 37, 131, 1976.
114. Siegel, S. M. and Daly, O., Regulation of betacyanin efflux from beet root by poly-L-lysine, Ca-ion and other substances, *Plant Physiol.*, 41, 1429, 1966.
115. Grunwald, C., Effect of starch on the permeability of alcohol-treated red beet tissue, *Plant Physiol.*, 43, 484, 1968.
116. van Steveninck, R. F. M., The significance of calcium on the apparent permeability of cell membranes and the effects of substitution with other divalent ions, *Physiologia Pl.*, 18, 54, 1965.
117. Rehfeld, D. W. and Jensen, R. G., Metabolism of separated leaf cells, *Plant Physiol.*, 52, 17, 1973.
118. Nissen, P., Uptake mechanisms: inorganic and organic, *Ann. Rev. Plant Physiol.*, 25, 53, 1974.
119. Epstein, E., Mechanisms of ion transport through plant cell membranes, *Int. Rev. Cytol.*, 34, 123, 1973.
120. Marinos, N. G., Studies on submicroscopic aspects of mineral deficiencies. I. Calcium deficiency in the shoot apex of barley, *Am. J. Bot.*, 49, 834, 1962.
121. von Marschner, H. and Günther, I., Ionenaufnahme und Zellstruktur bei Gerstenwurzeln in Abhängigkeit von der Calcium-Versorgung, *Z. Pflanzenernaehr. Bodenkd.*, 107, 118, 1964.
122. Hecht-Bucholz, Ch., Calcium deficiency and plant ultrastructure, *Commun. Soil Sci. Plant Anal.*, 10, 67, 1979.
123. Mahanty, H. K. and Fineran, B. A., The effects of calcium on the ultrastructure of Cox's Orange apples with reference to bitter pit disorder, *Aust. J. Bot.*, 23, 55, 1975.
124. Fuller, M. W., The ultrastructure of the outer tissues of cold-stored apple fruits of high and low calcium content in relation to cell breakdown, *Ann. Appl. Biol.*, 83, 299, 1976.
125. Buchloch, G., Baxter, P., and Neubeller, J., Zur Ätiologie der Stipigkeit von Apfelfruchten, *Angew. Bot.*, 35, 259, 1961.

126. **Williams, R. J. P.,** Calcium chemistry and its relation to biological function, *Symp. Soc. Exp. Biol.,* 30, 1, 1976.
127. **Williams, R. J. P.,** Calcium ions: their ligands and their functions, *Biochem. Soc. Symp.,* 39, 133, 1974.
128. **Martin, D., Lewis, T. L., and Cerny, J.,** Bitter pit in the apple variety Cleopatra in Tasmania in relation to calcium and magnesium, *Aust. J. Agric. Res.,* 11, 742, 1960.
129. **Sharples, R. O., Reid, M. S., and Turner, N. A.,** The effects of post-harvest mineral element and lecithin treatments on the storage disorders of apples, *J. Hortic. Sci.,* 54, 299, 1979.
130. **Sharples, R. O.,** A note on the occurrence of watercore breakdown in apples during 1966, *Plant. Pathol.,* 16, 119, 1967.
131. **Millikan, C. R. and Hanger, B. C.,** Calcium nutrition in relation to the occurrence of internal browning in Brussels sprouts, *Aust. J. Agric. Res.,* 17, 863, 1966.
132. **Geraldson, C. M.,** The control of blackheart of celery, *Proc. Am. Soc. Hortic. Sci.,* 63, 353, 1954.
133. **Cooil, B. J. and Bonner, J.,** The nature of growth inhibition by calcium in the *Avena* coleoptile, *Planta,* 48, 696, 1957.
134. **Betts, H. A. and Bramlage, W. J.,** Uptake of calcium by apples from postharvest dips in calcium chloride solutions, *J. Am. Soc. Hortic. Sci.,* 102, 781, 1977.
135. **Palzkill, D. A., Tibbitts, T. W., and Williams, P. H.,** Enhancement of calcium transport to inner leaves of cabbage for prevention of tipburn, *J. Am. Soc. Hortic. Sci.,* 101, 645, 1976.

Chapter 14

INTERACTIONS BETWEEN VIRUSES AND CALCIUM

Anthony C. H. Durham

TABLE OF CONTENTS

I.	Introduction	194
II.	Ion Binding to Viruses	194
III.	Virus Entry into Cells	197
IV.	Membrane Permeability Changes Due to Input Virions	200
V.	Takeover and Transformation	201
VI.	Late Permeability Changes in Infected Cell Membranes	202
VII.	Conclusion	202
Acknowledgment		203
References		203

I. INTRODUCTION

This brief review presents evidence for the importance of virus interactions with ions in general and with Ca^{2+} in particular. The text follows the progress of viruses through their life cycle: surviving outside a cell, penetrating into it, causing early and late metabolic changes, and then leaving the cell. Figure 1 recalls the shapes of some of the viruses discussed.

It has been clearly demonstrated in various individual cases that viruses bind cations, that virus infections alter membrane permeability to cations, and that production of progeny viruses requires adequate concentrations of divalent cations in the growth media. The logical jump from there to asserting that calcium is particularly important to viruses depends mainly upon the preeminence of Ca^{2+} as a signal in other systems and upon the large pCa gradient across membranes. Many published studies claiming an essential role for Ca^{2+} have not demonstrated adequate specificity vs. biologically relevant concentrations of other cations (H^+, Na^+, K^+, Mg^{2+}, Mn^{2+}, and polyamines); but, equally, most studies that have failed to find a role for Ca^{2+} did not rule out the possibility that a requirement for Ca^{2+} was satisfied by the traces of calcium unavoidably present in the experimental system.

II. ION BINDING TO VIRUSES

For the direct study of ion binding to virus particles (virions), plant viruses are the material of choice; they are generally simple structures made of RNA and one or a few types of coat protein, and they can often be obtained in gram quantities. Early experiments showed that tobacco mosaic virus binds several H^+ ions per protein subunit, with an anomalous pK_H of about seven, at a site where Pb^{2+} can also bind. Caspar,[1] therefore, deduced that the protein of this virus contains juxtaposed carboxylate groups, constituting a special binding site for H^+ (or metal) ions that control the assembly of the viral protein into complete virions.

Accumulated observations suggest that many plant viruses change in various ways in response to H^+ binding near pH 7 or to the binding of Ca^{2+} or Mg^{2+}. From this, the author induced that all plant viruses contain chelating sites capable of binding many different cations — H^+, Ca^{2+}, Mg^{2+}, Pb^{2+}, etc. — and hypothesized that these sites had evolved primarily to bind calcium.[2] This hypothesis proposes that in extracellular fluids virions are stabilized by bound calcium, whereas in cytoplasm they lose their calcium, become unstable, and release their nucleic acid. In other words, viruses interpret the surrounding Ca^{2+} concentration as a sign of whether they are inside or outside a cell, just as many secreted enzymes do.[3]

To test the hypothesis, I and several colleagues examined the competition between H^+, Ca^{2+}, Mg^{2+}, and some other cations to bind to 18 virus strains and 7 of their proteins.[2,4-11] Some of the results are shown in Figure 2. The H^+ and Ca^{2+}-binding constants derived[9] for the three binding sites per protein subunit of the vulgare strain of tobacco mosaic virus, in 50 mM KCl solution, are as follows:

pK_H	pK_{Ca}
5.7	2.3
7.6	1.5
8.2	5.5

The pK_H and pK_{Ca} values are equivalent to the pH and pCa values at which half the sites are saturated (in the absence of competition from other cations).

The possession of distinctive cation-binding sites does indeed seem to be a property of all plant viruses. Some sites are "weak" (with pK_H usually five to six and pK_{Ca} up

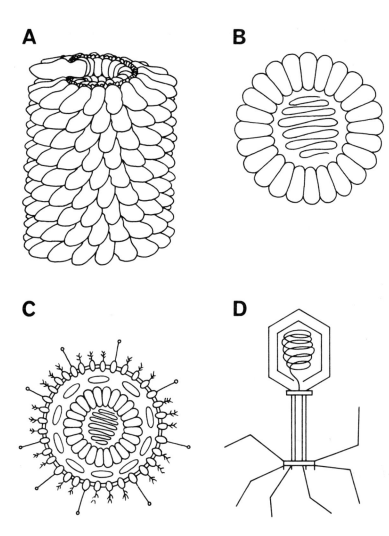

FIGURE 1. Schematic diagrams of some of the viruses discussed here, not drawn to a single scale. (A) Part of a rod-shaped virus, such as tobacco mosaic virus, papaya mosaic virus, or the internal nucleocapsid of certain animal viruses; (B) a simple spherical (actually icosadeltahedral) virus, such as bromegrass mosaic, southern bean mosaic, polio, or polyoma virus; (C) a generalized lipid-enveloped spherical virus with some features commonly found in many animal viruses (e.g., Sendai or Semliki Forest virus); the component parts, from the outside inward, are as follows: spikes with enzymic activity, glycoproteins, lipid bilayer membrane, M protein, inner capsid (which can be spherical, as drawn here, or coiled-up rods), and nucleic acid plus some tightly associated proteins; (D) a bacterial virus with a contractile tail like the T-even bacteriophages of *Escherichia coli;* most bacterial viruses have tails, but the tails are often much less mechanically sophisticated than this.

to three) and probably consist of juxtaposed protein carboxylate groups just as in many extracellular enzymes. More interesting are the "strong" sites, also found in all plant viruses so far examined in detail. These sites have pK_H between seven and nine and pK_{Ca} of about five; they are present in the whole virion, which contains RNA, but are absent from the structurally similar reassembled protein capsid, which lacks RNA.

Protein carboxylates and RNA phosphates are both implicated in the construction of the strong sites, but there is no direct evidence about the detailed configurations of

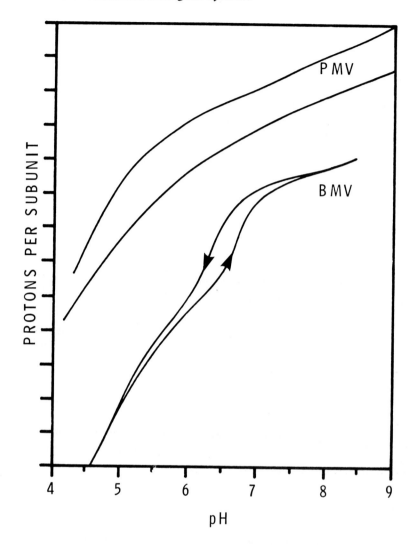

FIGURE 2. H⁺ titration curves for the helical papaya mosaic virus (top pair of curves, marked PMV) and for the spherical bromegrass mosaic virus (bottom pair of curves, marked BMV). The lower PMV curve is for the virus alone and the upper curve is for the virus plus 25 mM CaCl$_2$; the vertical distance between the two curves at any given pH represents the number of hydrogen ions that can be displaced by Ca^{2+} and is the primary datum from which affinity constants can be calculated. Both the BMV curves were measured in the absence of divalent cations; the lower one was measured in the direction with increasing pH and the upper one with decreasing pH. The step in the curves corresponds to the reversible swelling/recompaction of the virion, while the hysteresis loop reflects the different pathways for the two processes. Ca^{2+} removal and readdition would control the size of BMV, and of many other simple spherical viruses, just as H⁺ does in this figure.

the sites, since none of the high-resolution structures recently determined for viruses[12] have yet clearly shown the RNA-protein interfaces. The arrangement of charged groups near the axis of tobacco mosaic virus hints that the RNA phosphate groups make themselves felt from a distance by imposing order on a number of carboxylates rather than by direct proximity to a bound metal ion.

The number of cations bound to a virion determines its stability. A necessary condition for nucleic acid release is that all multivalent cations, including any nonbiological trace metals, be removed (e.g., with a chelating agent) and that the key hydrogen ions be removed (pH above seven suffices for many viruses). Many spherical viruses then swell and change configuration; different expansion and contraction pathways generate hysteresis loops in the H^+ titration curves. Contrary to my original hypothesis, cation removal is not a sufficient condition for a plant virus to release its nucleic acid in an aqueous solution. The required extra impetus to disassembly is probably provided in vivo by a membrane and can be provided in vitro by heat, urea, detergents, etc.

Knowledge of cation binding can be useful in virus purification; however, it may be difficult to decide which of various competing ionic influences are at work.[10] Ca^{2+} ions, although they tend to stabilize viruses, are generally best excluded from purification buffers, lest they activate degradative enzymes, reduce the virus charge enough to permit isoelectric precipitation, promote absorption of host-cell material, or tear out the RNA and precipitate it as the calcium salt. The virus can behave in unexpected ways when it is isolated in an unrecognized, swollen form or when the capsid is impermeable to small ions and the binding sites are therefore hidden.

The capsids of simple spherical viruses have gaps between the protein subunits, through which ions can pass. The gaps are smaller when the virion is compact than when it is swollen, and Ca^{2+} ions make the virion compact.[10] In this behavior lies the essence of an ion channel that closes by negative feedback, as found for some animal-virus proteins in membranes.[13]

Many animal viruses seem to have cation-binding sites analogous with those of plant viruses, although the evidence is limited because such small amounts of their virions can be obtained for direct study. A clear example is polyoma virus, which dissociates if calcium is removed by chelating agents.[14] Mg^{2+} stabilizes poliovirus, although the impermeable shell of this virus complicates matters.[15]

There is no rigid dividing line between animal and plant viruses. The family of rhabdoviruses includes some that multiply in plants and some that multiply in animals, both with suggestions of typical responses to cations.[16,17] (For more references to animal-virus responses to cations, see References 2 and 18.)

The calf rotavirus studied by Cohen et al.[19] is particularly interesting. The removal of Ca^{2+} (but not of Mg^{2+}) with a chelating agent releases two polypeptides from the surface of the virion, thus unmasking an RNA polymerase activity in the inner particle of the virion. As Figure 3 shows, this animal virus responds to pH and divalent ions very much as a plant virus does. In the case of the rotavirus, however, the calcium-binding site cannot lie at the RNA-protein interface; evidently the Ca^{2+} response is functionally important to the life of viruses, rather than a trivial chemical consequence of protein-nucleic acid binding.

Bacterial viruses have frequently been observed to be stabilized by divalent cations;[20] some are less stable at pH 7 than at pH 6 when divalent cations are absent.[21-23] Filamentous bacterial viruses have the weak but not the strong cation-binding sites of the types found in plant viruses.[11]

III. VIRUS ENTRY INTO CELLS

It has long been customary to add calcium salts to viral inocula to favor the initial steps of infection, i.e., adsorption to and penetration into the host cell. Experimentally, it is difficult to distinguish all the initial steps, and viruses can differ widely in the extent to which a specific ionic requirement can be detected for any of these multiple steps.

Both viruses and cells generally carry a net negative surface charge at neutral pH. Ions act both as a mobile shield and as bound neutralizers to reduce the consequent electrical repulsion, though not necessarily on average to zero,[24] so that the initial

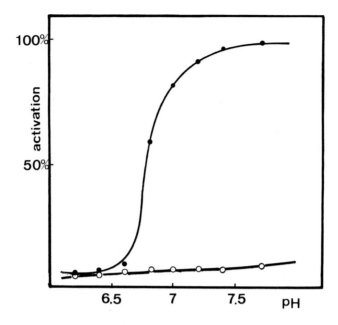

FIGURE 3. The unmasking of an RNA polymerase activity in a calf rotavirus as a function of pH in the presence (closed circles) and absence (open circles) of EDTA. This figure follows the same pattern as with many plant viruses, except that then the vertical axis would be marked with some structural parameter such as hydrodynamic radius. (From Cohen, J., Laporte, A., Charpillienne, A., and Scherrer, R., Arch. Virol., 60, 177, 1979. With permission.)

attachment between virus and cell often takes place via a protuberance on one or the other. Ca^{2+} contributes greatly to this extracellular neutralization, along with Na^+ and Mg^{2+}. From the binding constants measured for plant viruses one can actually calculate the amount by which Ca^{2+} reduces their charge (Figure 4).

The short-range binding of a virion to a cell usually involves the recognition of a carbohydrate receptor on the cell surface. Despite the involvement of Ca^{2+} in the analogous binding of many lectins and the geometrical adaptability of Ca^{2+} for cross-bridging, relatively few publications claim a role for Ca^{2+} in virus-to-cell binding,[25-27] and most find no requirement for a particular ion. The binding to receptors is usually reversible; next there occurs an irreversible "eclipse" that commits a virus to injecting its nucleic acid into the cell.

Animal viruses vary in the extent to which they penetrate into a cell at the external surface or after being taken inside the cell ("viropexis") in a lipid vesicle and delivered to lysosomes,[28] nuclear membrane, etc. Ca^{2+} is implicated with actomyosin action in endocytosis, and with transaminase action[29] in the formation of coated pits.

Many viruses contain enzymes that they use to cut through the external carbohydrate layers of their host cells (both on entry and on exit). Like most extracellular enzymes, these viral enzymes are frequently Ca^{2+}-activated.[30-33]

The penetration of viral nucleic acid (or of an inner nucleocapsid) into a cell generally requires that the cell membrane be energized.[34,35] In animal cells, the acidic pH inside lysosomes has been implicated in the uncoating of viral nucleic acid;[28] the antiviral agent amantadine inhibits uncoating;[36] it and other weakly antiviral amines drive up the pH inside lysosomes.[37] In bacteria, the electrical potential across lysosomal membranes rather than the pH gradient seems to be involved.[38]

Evidence that Ca^{2+} drives the penetration of nucleic acid into cells has been presented

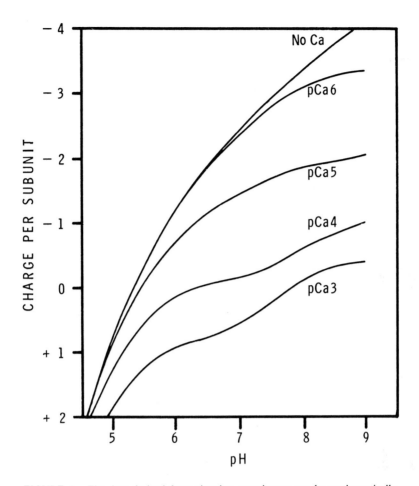

FIGURE 4. The data derived from titration experiments can be mathematically transformed to show how the electrical charge on a virion varies with pH and pCa. In this figure, for papaya mosaic virus one can see that in solutions containing extracellular levels of Ca^{2+} the charge on the virion becomes about zero at pH seven and ceases to be a sensitive function of pH near that value. This diagram suggests a potential new experimental technique: isoelectric focusing of viruses in a pCa gradient. (From Durham, A. C. H. and Bancroft, J. B., Virology, 93, 246, 1979. With permission.)

in connection with several types of bacterial virus.[39-42] A calcium-stimulated ATPase may be involved in one DNA injection process.[42] The case of bacteriophage T5 is particularly interesting: Ca^{2+} is not needed for attachment, nor for injection of the first 8% of the DNA and expression of the viral "pre-early" genes on this DNA. However, adequate extracellular Ca^{2+} is needed for the pre-early proteins to repair some single-strand nicks in the DNA, before the rest of the DNA is pulled into the cell and the "early" viral proteins can be expressed.[43]

The measured Ca^{2+} affinities of plant viruses imply that in crossing a membrane a virion could cotransport several hundreds or thousands of calcium ions down their concentration gradient. This would make very large amounts of chemiosmotic energy available to drive any process coupled to the penetration. I therefore, suggested that plant viruses (and by extension many animal viruses) disassemble in the act of crossing a membrane, while using this energy to drive the reaction.[44]

IV. MEMBRANE PERMEABILITY CHANGES DUE TO INPUT VIRIONS

Kohn[45] has thoroughly reviewed the early effect of viruses upon animal cells. In general, at the moment of virus entry there are rapid changes in the cell membrane that can be detected by various spectroscopic means, followed shortly by changes in the transmembrane flux of many ions and small molecules, notably K^+ leaking out. The underlying cause seems to be relatively nonspecific damage to the permeability barrier normally posed by the cell membrane. There is, of course, great variation, depending on virus and cell type, in the intensity (and observability) of early permeability changes; in part this variation reflects the relative amounts of virus that penetrate cell membranes before internalization.

Sendai virus produces easily observable permeability changes that allow Ca^{2+} to flow into cells but that are partially prevented by extracellular Ca^{2+} concentrations such as those in serum. Pasternak's group have extensively studied the action of Sendai virus (Reference 13, and references therein) and have discussed at length the significance of viral permeability lesions, with particular reference to the negative-feedback action of Ca^{2+}.

Animal viruses with lipid envelopes, such as Sendai virus, enter cells by fusing their membrane with a cell membrane. The subsequent permeability lesions are probably due to actions of the proteins from the viral envelope diffusing within the cell membrane. There has been much published discussion of the role of Ca^{2+} in membrane fusion, but there is evidence that the virus-induced fusion process itself (as distinct from subsequent changes in cell shape) is independent of calcium[46] and arises essentially when one virion tries to enter two cells at once.

Bacterial viruses deenergize the host-cell membrane at the moment of nucleic acid entry and cause nonspecific leakiness to ions and small molecules. The effect is due to proteins of the entering virions, as is evidenced by the increased leakiness with increased multiplicity of infection, by the particularly pronounced leakage with virus "ghosts", empty of nucleic acid,[47,48] and by detailed genetic studies.[49] Some publications have described a specific requirement for extracellular Ca^{2+} persisting beyond the attachment/penetration phase,[25] notably in the case of T5 mentioned above.[43]

Some colicins, which are plasmid-coded and resemble defective lysogenic viruses, induce "voltage-dependent relatively nonselective ion-permeable channels" in membranes.[50] Plant pathogens in general induce major fluxes of ions, among which calcium is prominent;[51] one plant virus so far has been linked to this pattern, in that it depolarizes membranes.[52]

The hypothesis of Ca^{2+}-driven disassembly mentioned above suggests that there should exist plant viruses whose proteins (which can be available in huge quantities, but have not generally been thought of as interacting with membranes) produce observable permeability changes in model membranes. Alfalfa mosaic virus is a promising candidate, because its coat protein (and that of other plant viruses, such as tobacco streak virus) is essential for the infectivity of the viral RNA in some capacity other than just wrapping up the nucleic acid.[53] Furthermore, its coat-protein gene has extensive sequence homology[54] with the gene for the M ("matrix" or "membrane") protein that forms the interface between the nucleocapsid and the lipid envelope of vesicular stomatitis virus.

The author has screened many viruses for prima facie evidence of membrane permeabilizing activity, using as assay a suspension of vesicles made from soy bean lecithin, with a pH 7 solution of the calcium indicator Arsenazo III in 100 mM sodium EDTA inside and an isotonic NaCl/imidazole chloride pH 7 solution with added $CaCl_2$ outside (like the technique of Reference 55). Six viruses allowed Ca^{2+} to reach the dye

— alfalfa mosaic virus, tobacco streak virus, frog virus 3, vaccinia virus, cytoplasmic irridescent virus, and southern bean mosaic virus. Most interestingly, the coat protein of southern bean mosaic virus precipitated the vesicles very much as many plant lectins do.

These unfinished experiments strengthen the frequently made analogy between viruses and other proteins that first bind to cell surface receptors and are then taken into the cell.[56] In particular, many toxins have two subunits: one binds to a receptor and makes a transmembrane channel through which the other enters the cell and poisons some sensitive metabolic process. Indeed, poliovirus infection allows the poisonous subunit of some toxins to enter cells,[57] and some viruses act as mitogens.[58]

Infected cells often appear to recover after the immediate damage associated with virion entry (except at very high multiplicity of infection). With animal cells, this may reflect the removal of virus proteins from the surface by endocytosis, but in bacteria, synthesis of new virus-coded proteins is needed for recovery (as demonstrated by the effect of virus ghosts).

V. TAKEOVER AND TRANSFORMATION

Many viruses, though not all, shut off the synthesis of host-cell macromolecules during virus multiplication. The idea has been independently proposed several times that this "takeover" process works partly through changes in ionic concentration. According to this hypothesis, viruses prolong the changes in membrane permeability beyond the unavoidable damage associated with penetration and/or synthesize new membrane proteins, specifically in order to produce intracellular ion concentrations optimal for virus-coded enzymes but not for host enzymes.

The significance of virus-induced ion fluxes in bacteria has been discussed often. For example, Guttmann and Begley[59] studied the effects of Mg^{2+} on *Escherichia coli* infected with the virus T4 and proposed that " . . . the ionic environment has to be adjusted to conditions that are optimal for phage-induced polymerases". The clearest exposition of an ionic takeover mechanism in bacteria was by Schweiger's group,[60] who noticed that virus-coded enzymes had different salt requirements from host-coded enzymes.

Carrasco[61] explicitly stated essentially the same idea for animal viruses, proceeding from the often-repeated observation that the synthesis of animal-virus proteins both in vivo and in vitro is less inhibited by environments rich in Na^+ than is the synthesis of most cell proteins. His paper summarizes much relevant evidence; various publications have demonstrated the reality of the permeability lesions induced by several families of animal viruses, notably the small nonenveloped viruses like poliovirus.[57,62,63]

Both Carrasco and Schweiger emphasize monovalent cations as the agents of takeover, though it is not yet clear to what extent the concentration of Na^+ or of Na^+ plus K^+, or the Na^+/K^+ ratio, is the relevant factor in changes in enzymic activity. While independently proposing an analogous ionic hypothesis for takeover,[44] I have pointed out that Ca^{2+} would be the most potent ion passing through a nonspecific "hole" and could well induce K^+ fluxes as a secondary effect.

There is still only scarce direct evidence of prolonged changes in Ca^{2+} flux during lytic virus infection. Mitochondrial Ca^{2+} transport alters after virus infection.[64,65] Calcium enhances the production of interferon,[66] whose messenger RNA is one of the few host-coded ones to be translated efficiently in the altered ionic environment after infection.[67] The production of virus in plant protoplasts is inhibited by gentamicin or chelating agents, unless Ca^{2+} or Mn^{2+} ions are supplied;[68] aminoglycoside antibiotics such as gentamicin or streptomycin are frequently used in cell-growth media, but are known to interact with Ca^{2+} fluxes.[69]

The clearest instance of a prolonged Ca^{2+} permeability change occurs in virus-induced cancer. As cells become malignant, they become able to grow in media with less Ca^{2+} (see the accompanying article by Boynton et al.). A few remarks are relevant here in this connection.

In some oncogenic viruses (papova- and adeno-viruses), the transforming genes are early genes, corresponding to the takeover genes of lytic viruses. The reported sizes and other characteristics of the proteins encoded by oncogenic viruses and implicated in transformation are at present (August 1980) extraordinarily diverse; there are much greater similarities in the altered physiologies that viruses impose upon cells.

The progression to malignancy is a multi-step process, and viruses and viral proteins differ in their abilities to perform the various steps. Cells, too, differ in their abilities to complement the transforming potentials of viruses. It is not yet clear how many of the steps involve Ca^{2+}.

The establishment of a virus-transformed clone of cells is quite a rare event, whose probability can be influenced by the Ca^{2+} concentration in the growth medium, notably with adenoviruses[70] and poxviruses.[71]

The bacterial analogue of the cancerous state — lysogeny — generally involves the insertion of a virus-coded pore protein, continuously synthesized like the well-known cytoplasmic repressor, into the external bacterial membrane.[72] The resulting state of "lysogenic conversion" involves subtle changes of cell physiology[73] and availability for infection by other viruses. Sometimes, membrane proteins from two viruses or plasmids can interact to produce an ion-flux catastrophe that lyses the cell.[74]

VI. LATE PERMEABILITY CHANGES IN INFECTED CELL MEMBRANE

It becomes relatively easy to demonstrate permeability lesions and ion fluxes[75,76] in many cell/virus systems once viral proteins have built up in a cell membrane late in infection. Substantial membrane leakiness can occur long before a cell is moribund and depleted of ATP.[77]

The outcome of a lytic virus replication cycle is by definition that the infected cell lyses. The ultimate moment of death for any cell occurs when its membrane can no longer prevent an overwhelming entry of Ca^{2+} ions.[78] In infected bacteria, the timing of lysis is actively determined by virus-coded proteins.[79]

Many animal viruses do not kill the host-cell, but bud off through the cell membrane. If budding occurs into an internal vesicle, which must fuse with the external membrane to release the progeny virions, the discharge, just like any other form of exocytosis, requires Ca^{2+}. It is reasonable to speculate that during virus budding a localized entry of Ca^{2+} provoked by a high local concentration of viral-envelope proteins would cause the condensation of the complete virion by crosslinking the interior nucleocapsid to the envelope.

Extracellular Ca^{2+} concentration has been shown to influence the yield of complete progeny virions in the Dengue virus of animals[80] and in the PM2 virus of bacteria.[81]

VII. CONCLUSION

This survey of virus interactions with Ca^{2+} poses three main questions:

1. What is the structure of the strong cation-binding sites in virions?
2. What happens, in molecular terms, when a virus thrusts its nucleic acid across a membrane?
3. What is the configuration of the permeability lesions induced in membranes by many virus proteins?

Once one accepts that the actual capsid proteins of certain viruses can permeabilize membranes, it becomes clear that the answer to any one of these questions may clarify the other two.

Virus-induced lesions in membranes sometimes behave like narrow channels with defined cutoff sizes for the molecules they let through, but at other times they allow large molecules, including the virus's own nucleic acids, to pass. For more than 10 years, writers about viruses have been puzzled by this paradox, usually without explicitly discussing it in these terms. As a working hypothesis, I think of the virus proteins as being able to disrupt membranes locally and form a "hole" of indeterminate size in the otherwise closed surface of a membrane; calcium ions help determine how big this hole can grow.

ACKNOWLEDGMENT

I thank the Cancer Research Campaign for financial support.

REFERENCES

1. Caspar, D. L. D., Assembly and stability of the tobacco mosaic virus particle, *Adv. Protein Chem.*, 18, 37, 1963.
2. Durham, A. C. H., Hendry, D. A., and Von Wechmar, M. B., Does calcium ion binding control plant virus disassembly?, *Virology*, 77, 524, 1977.
3. Clemente, F. and Meldolesi, J., Calcium and pancreatic secretion: subcellular distribution of calcium and magnesium in the exocrine pancreas of the guinea pig, *J. Cell. Biol.*, 65, 88, 1975.
4. Durham, A. C. H. and Hendry, D. A., Cation binding by tobacco mosaic virus, *Virology*, 77, 510, 1977.
5. Durham, A. C. H. and Abou Haidar, M., Cation binding by tobacco rattle virus, *Virology*, 77, 520, 1977.
6. Pfeiffer, P. and Durham, A. C. H., The cation binding associated with structural transitions in bromegrass mosaic virus, *Virology*, 81, 419, 1977.
7. Durham, A. C. H., Vogel, D., and DeMarcillac, G. D., Hydrogen-ion binding by tobacco mosaic virus polymers, *Eur. J. Biochem.*, 79, 151, 1977.
8. Durham, A. C. H. and Bancroft, J. B., Cation binding by papaya mosaic virus and its protein, *Virology*, 93, 246, 1979.
9. Hendry, D. A. and Durham, A. C. H., Titration behavior of three strains of tobacco mosaic virus, *Virology*, 100, 65, 1980.
10. Durham, A. C. H., Kruse, K. M., and DeMarcillac, G. D., Purification and some sedimentation properties of southern bean mosaic virus, 1982, submitted.
11. Durham, A. C. H., Bancroft, J. B., and Bourque, D. P., The generality of cation-binding sites in rod-shaped viruses, *Bioscience Rep.*, 1, 547, 1981.
12. Harrison, S. C., Virus crystallography comes of age, *Nature (London)*, 286, 558, 1980.
13. Impraim, C. C., Foster, K. A., Micklem, K. J., and Pasternak, C. A., Nature of virally mediated changes in membrane permeability to small molecules, *Biochem. J.*, 186, 847, 1980.
14. Brady, J. N., Winston, V. D., and Consigli, R. A., Dissociation of polyoma virus by the chelation of calcium ions found associated with purified virions, *J. Virol.*, 23, 717, 1977.
15. Fujioka, R. S. and Ackermann, W. W., The inhibitory effects of $MgCl_2$ on the inactivation kinetics of poliovirus by urea, *Proc. Soc. Exp. Biol. Med.*, 148, 1063, 1975.
16. Ahmed, M. E., Sinha, R. C., and Hochster, R. M., Purification and some morphological characters of wheat striate mosaic virus, *Virology*, 41, 768, 1970.
17. Heggeness, M. H., Scheid, A., and Choppin, P. W., Conformation of the helical nucleocapsids of paramyxoviruses and vesicular stomatitis virus: reversible coiling and uncoiling induced by changes in salt concentration, *Proc. Natl. Acad. Sci. USA*, 77, 2631, 1980.
18. Floyd, R. and Sharp, D. G., Viral aggregation: effects of salts on the aggregation of poliovirus and reovirus at low pH, *Appl. Environ. Microbiol.*, 35, 1084, 1978.

19. Cohen, J., Laporte, A., Charpillienne, A., and Scherrer, R., Activation of rotavirus RNA polymerase by calcium chelation, *Arch. Virol.*, 60, 177, 1979.
20. Adams, M. H., The stability of bacterial viruses in solutions of salts, *J. Gen. Physiol.*, 32, 579, 1949.
21. Thorne, C. B. and Holt, S. C., Cold lability of *Bacillus cereus* bacteriophage CP-51, *J. Virol.*, 14, 1008, 1974.
22. Bryner, J. H., Ritchie, A. E., Foley, J. W., and Berman, D. T., Isolation and characterization of a bacteriophage for *Vibrio* fetus, *J. Virol.*, 6, 94, 1970.
23. Incardona, N. L., Blonski, R., and Feeney, W., Mechanism of adsorption and eclipse of bacteriophage phiX174. I. In vitro conformational change under conditions of eclipse, *J. Virol.*, 9, 96, 1972.
24. Weiss, L. and Horoszewicz, J. S., Some biophysical aspects of E-B virus adsorption to the surfaces of three types of mammalian cells, *Int. J. Cancer*, 7, 149, 1971.
25. Neubauer, R. H. and Young, B. G., ST-1 bacteriophage: some properties of a host range mutant, *J. Gen. Virol.*, 22, 411, 1974.
26. Schmidt, L. S., Ven, H.-C., and Gest, H., Bioenergetic aspects of bacteriophage replication in the photosynthetic bacterium *Rhodopseudomonas* capsulata, *Arch. Biochem. Biophys.*, 165, 229, 1974.
27. Kalyanaraman, V. S., Sarngadharan, M. G., and Gallo, R. C., Characterization of Rauscher murine leukemia virus envelope glycoprotein receptor in membranes from murine fibroblasts, *J. Virol.*, 28, 686, 1978.
28. Helenius, A., Kartenbeck, J., Simons, K., and Fries, E., On the entry of Semliki Forest virus into BHK-21 cells, *J. Cell. Biol.*, 84, 404, 1980.
29. Davies, P. J. A., Davies, D. R., Levitzki, A., Maxfield, F. R., Milhaud, P., Willingham, M. C., and Pastan, I. H., Transglutaminase is essential in receptor-mediated endocytosis of α_2-macroglobulin and polypeptide hormones, *Nature (London)*, 283, 162, 1980.
30. Baker, N. J. and Gandhi, S. S., Effect of Ca^{++} on the stability of influenza virus neuraminidase, *Arch. Virol.*, 52, 7, 1976.
31. Barnet, Y. M. and Humphrey, B., Exopolysaccharide depolymerases induced by *Rhizobium* bacteriophages, *Can. J. Microbiol.*, 21, 1647, 1975.
32. Benchetrit, L. W., Gray, E. D., Edstrom, R. D., and Wannamaker, L. W., Purification and characterization of a hyaluronidase associated with a temperate bacteriophage of group A, type 49 streptococci, *J. Bacteriol.*, 134, 221, 1978.
33. Tsukagoshi, N., Schäfer, R., and Franklin, R. M., Structure and synthesis of a lipid-containing bacteriophage: an endolysin activity associated with bacteriophage PM2, *Eur. J. Biochem.*, 77, 585, 1977.
34. Lonberg-Holm, K. and Whiteley, N. M., Physical and metabolic requirements for early interaction of poliovirus and human rhinovirus with HeLa cells, *J. Virol.*, 19, 857, 1976.
35. Hancock, R. E. W. and Braun, V., Nature of the energy requirement for the irreversible adsorption of bacteriophages T1 and $\phi 80$ to *Escherichia coli*, *J. Bacteriol.*, 125, 409, 1976.
36. Hoffman, C. E., Neumayer, E. M., Haff, R. F., and Goldsby, R. A., Mode of action of the antiviral activity of amantadine in tissue culture, *J. Bacteriol.*, 90, 623, 1965.
37. Ohkuma, S. and Poole, B., Fluorescence probe measurement of the intralysosomal pH in living cells and the perturbation of pH by various agents, *Proc. Natl. Acad. Sci. USA*, 75, 3327, 1978.
38. Labedan, B. and Goldberg, E. B., Requirements for membrane potential in injection of phage T4 DNA, *Proc. Natl. Acad. Sci. USA*, 76, 4669, 1979.
39. Watanabe, K. and Takesue, S., The requirement for calcium in infection with *Lactobacillus* phage, *J. Gen. Virol.*, 17, 19, 1972.
40. Paranchych, W., Stages in phage R17 infection: the role of divalent cations, *Virology*, 28, 90, 1966.
41. Steensma, H. Y. and Blok, J., Effect of calcium ions on the infection of *Bacillus subtilis* by bacteriophage SF6, *J. Gen. Virol.*, 42, 305, 1979.
42. Karnik, S. S. and Gopinathan, K. P., Possible involvement of a calcium-stimulated ATP-hydrolysing activity associated with mycobacteriophage I3 in the DNA injection process, *J. Virol.*, 33, 969, 1980.
43. Herman, R. C. and Moyer, R. W., In vivo repair of bacteriophage T5 DNA: an assay for viral growth control, *Virology*, 66, 393, 1975.
44. Durham, A. C. H., The role of small ions, especially calcium, in virus disassembly, takeover, and transformation, *Biomedicine*, 28, 307, 1978.
45. Kohn, A., Early interactions of viruses with cellular membranes, *Adv. Virus Res.*, 24, 223, 1979.
46. Knutton, S. and Pasternak, C. A., The mechanisms of cell-cell fusion, *Trends Biochem. Sci.*, 4, 220, 1979.
47. Duckworth, D. H., Biological activity of bacteriophage ghosts and "take-over" of host functions by bacteriophage, *Bacteriol. Rev.*, 34, 344, 1970.
48. Takeishi, K. and Kaji, A., Protein synthesis in bacteriophage ghost-infected cells, *J. Virol.*, 18, 103, 1976.
49. Ponta, H., Gratzel, M., Pfennig-Yeh, M., Hirsch-Kauffmann, M., and Schweiger, M., Membrane alteration induced by T7 virus infection, *FEBS Lett.*, 73, 207, 1977.

50. Schein, S. J., Kagan, B. L., and Finkelstein, A., Colicin K acts by forming voltage-dependent channels in phospholipid bilayer membranes, *Nature (London)*, 276, 159, 1978.
51. Wheeler, H., *Plant Pathogenesis*, Springer-Verlag, Berlin, 1975.
52. Stack, J. P. and Tattar, T. A., Measurement of transmembrane electropotentials of Vigna sinensis leaf cells infected with tobacco ringspot virus, *Physiol. Plant Pathol.*, 12, 173, 1978.
53. Bol, J. F., Van Vloten-Doting, L., and Jaspars, E. M. J., A functional equivalence of top component a RNA and coat protein in the initiation of infection by alfalfa mosaic virus, *Virology*, 46, 73, 1971.
54. Rose, J. K., Complete intergenic and flanking gene sequences from the genome of vesicular stomatitis virus, *Cell*, 19, 415, 1980.
55. Weissmann, G., Anderson, P., Serhan, C., Samuelsson, E., and Goodman, E., A general method, employing Arsenazo III in liposomes, for study of calcium ionophores: results with A23187 and prostaglandins, *Proc. Natl. Acad. Sci. USA*, 77, 1506, 1980.
56. Neville, D. M., Jr. and Chang, T.-M., Receptor-mediated protein transport into cells: entry mechanisms for toxins, hormones, antibodies, viruses, lysosomal hydrolases, asialoglycoproteins and carrier proteins, *Curr. Top. Memb. Transp.*, 10, 65, 1978.
57. Fernandez-Puentes, C. and Carrasco, L., Viral infection permeabilizes mammalian cells to protein toxins, *Cell*, 20, 769, 1980.
58. Butchko, G. M., Armstrong, R. B., Martin, W. J., and Ennis, F. A., Influenza A viruses of the H2N2 subtype are lymphocyte mitogens, *Nature (London)*, 271, 66, 1978.
59. Guttman, B. S. and Begley, L., Evidence for a magnesium pump induced by bacteriophage T4, *Virology*, 36, 687, 1968.
60. Herrlich, P., Rahmsdorf, H.-J., and Schweiger, M., Regulation of macromolecular synthesis by membrane changes, *Adv. Biosci.*, 12, 523, 1973.
61. Carrasco, L., The inhibition of cell functions after viral infection; a proposed general mechanism, *FEBS Lett.*, 76, 11, 1977.
62. Nair, C. N., Stowers, J. W., and Singfield, B., Guanidine-sensitive Na^+ accumulation by poliovirus-infected cells, *J. Virol.*, 31, 184, 1979.
63. Garry, R. F., Bishop, J. M., Parker, S., Westbrook, K., Lewis, G., and Waite, M. R. F., Na^+ and K^+ concentrations and the regulation of protein synthesis in Sindbis virus-infected chick cells, *Virology*, 96, 108, 1979.
64. Balcavage, W. X., Baxter-Gabbard, K. L., Ko, M., Rea, M., Padgett, F., and Levine, A. S., Mitochondrial alterations associated with avian reticuloendotheliosis virus (strain T) pathogenicity, *Biochem. Biophys. Res. Comm.*, 48, 605, 1972.
65. Peterhans, E., Haenggeli, E., Wild, P., and Wyler, R., Mitochondrial calcium uptake during infection of chicken embryo cells with Semliki Forest virus, *J. Virol.*, 29, 143, 1979.
66. Meager, A., Graves, H. E., and Bradshaw, T. K., Stimulation of interferon yields from cultured human cells by calcium salts, *FEBS Lett.*, 87, 303, 1978.
67. Garry, R. F. and Waite, M. R. F., Na^+ and K^+ concentrations and the regulation of the interferon system in chick cells, *Virology*, 96, 121, 1979.
68. Kassanis, B., White, R. F., and Woods, R. D., Inhibition of multiplication of tobacco mosaic virus in protoplasts by antibiotics and its prevention by divalent metals, *J. Gen. Virol.*, 28, 185, 1975.
69. Schacht, J., Biochemistry of neomycin ototoxicity, *J. Acoust. Soc. Am.*, 59, 940, 1976.
70. Panigrahy, B., McCormick, K. J., Stenbach, W. A., and Trentin, J. J., In vitro transformation of hamster embryo cells by bovine adenovirus type 3 (strain WBR-1), *J. Gen. Virol.*, 26, 141, 1975.
71. Milo, G. E., Jr. and Yohn, D. S., Alterations of enzymes associated with plasma membranes and cellular organelles during infection of CV-1 cells with Yaba tumor poxvirus, *Cancer Res.*, 35, 199, 1975.
72. Pugsley, A. P. and Schnaitman, C. A., Outer membrane proteins of *Escherichia coli*. VII. Evidence that bacteriophage-directed protein 2 functions as a pore, *J. Bacteriol.*, 133, 1181, 1978.
73. Lin, L., Bitner, R., and Edlin, G., Increased reproductive fitness of *Escherichia coli* lambda lysogens, *J. Virol.*, 21, 554, 1977.
74. Cheung, A. K.-M. and Duckworth, D. H., Membrane damage in abortive infections of colicin Ib-containing *Escherichia coli* by bacteriophage T5, *J. Virol.*, 23, 98, 1977.
75. Foster, K. A., Gill, K., Micklem, K., and Pasternak, C. A., Survey of virally mediated permeability changes, *Biochem. J.*, 190, 639, 1980.
76. Thompson, S. and Wiberg, J. S., Late effect of bacteriophage T4D on the permeability barrier of *Escherichia coli*, *J. Virol.*, 25, 491, 1978.
77. Norkin, L. C., Cell killing by simian virus 40: impairment of membrane formation and function, *J. Virol.*, 21, 872, 1977.
78. Schanne, F. A. X., Kane, A. B., Young, E. E., and Farber, J. L., Calcium dependence of toxic cell death: a final common pathway, *Science*, 206, 700, 1979.
79. Rolfe, B. G. and Campbell, J. H., Genetic and physiological control of host cell lysis by bacteriophage lambda, *J. Virol.*, 23, 626, 1977.

80. **Matsumura, T. and Yamashita, H.**, Effects of Ca^{++} ion on the liberation of Dengue virus from BHK-21 cells in culture, *Microbiol. Immunol.*, 22, 803, 1978.
81. **Snipes, W., Cupp, J., Sands, J. A., Keith, A., and Davis, A.**, Calcium requirement for assembly of the lipid-containing bacteriophage PM2, *Biochim. Biophys. Acta*, 339, 311, 1974.

Calcium in Pharmacology

Chapter 15

COMPETITIVE ANTAGONISM BETWEEN CALCIUM AND ANTIBIOTICS

A. P. Corrado, W. A. Prado, and I. Pimenta de Morais

TABLE OF CONTENTS

I.	Biological Roles of Calcium Ions	210
II.	Generalities of the Groups of Antibiotics	210
III.	Competitive Antagonism Between Calcium and Aminoglycoside Antibiotics	210
	A. Release of Neurotransmitters	210
	1. Ganglionic Transmission	210
	2. Adrenergic Transmission	211
	3. Neuromuscular Transmission	212
	B. ECC	213
	1. Skeletal Muscle	213
	2. Cardiac Muscle	215
	3. Smooth Muscle	216
	C. Secretory Cells	218
	D. The Interaction of Aminoglycoside Antibiotics With Membrane Structures	218
References		220

I. BIOLOGICAL ROLES OF CALCIUM IONS

The importance of calcium in several biological processes such as neurotransmitter release in autonomic and motor nervous systems, excitation-contraction coupling (ECC) in skeletal, cardiac, and smooth muscles and excitation-secretion coupling (ESC) in endocrine and exocrine glands is well-established.[1] However, the exact role played by calcium in each one of these processes is not yet very well-understood. The development of calcium antagonists has helped in part to elucidate the mechanism by which calcium participates in these biological activities and explains the increased interest on new classes of calcium antagonists. In this chapter, a series of antibiotics known to inhibit calcium-dependent biological processes are discussed and the possibility that some of them might be used as pharmacological tools in order to unravel the participation of calcium in the aforementioned biological events is raised.

II. GENERALITIES OF THE GROUPS OF ANTIBIOTICS

Ionophores, tetracyclines, polymyxines, and aminoglycosides are groups of antibiotics known to interfere with calcium ions either by increasing or by decreasing calcium-dependent effects. The facilitatory action may be obtained with ionophores which are antibiotics able to increase the permeability of membranes to divalent cations.[2] On the other hand, the blockade of the effects may be the result of either a chelation or a competition of the drug with calcium-specific sites at the cellular membrane. Some tetracyclines (chlortetracycline, oxytetracycline, and rolitetracycline) are known to have a calcium and other divalent-cations complexant property,[3] which is accepted as the explanation of some of their therapeutic and other pharmacological effects. The well-known neuromuscular blocking property of the tetracyclines is difficult to be reverted by calcium salts;[4] the antagonistic effect of calcium is inconsistent and sometimes inexistent.[5] Among the antibiotics, the polymyxines such as colistin, colistimethate, and polymyxin B are the most potent neuromuscular blockers. These antibiotics depress the motor-end plate sensitivity to Ach and decrease the response of muscle to direct stimulation.[6] These effects, however, are also inconsistently reverted by calcium salts.[4,7-8] The aminoglycosides seem to be the only group of antibiotics which really compete with calcium in several biological activities. Among them, streptomycin, dihydrostreptomycin, kanamycin, neomycin, gentamicin, tobramycin, and sisomicin were the most studied on this subject. They form the scope of this chapter and the competitive nature of their calcium-blocking effects will be considered afterwards.

III. COMPETITIVE ANTAGONISM BETWEEN CALCIUM AND AMINOGLYCOSIDE ANTIBIOTICS

A. Release of Neurotransmitters
1. Ganglionic Transmission

The ganglioplegic action of aminoglycoside antibiotics was first demonstrated in cats and dogs by Corrado[9] who observed that streptomycin reduced drastically the salivary flow induced by stimulation of the chorda tympani and the cardiovascular effects obtained by stimulation of the peripheral end of the vagus nerve. On the other hand, the hypotension that follows the i.v. injection of Ach was not inhibited by the antibiotic. In atropinized dogs, the nicotinic-pressor effect induced by Ach injection was also blocked by streptomycin. Both the sympathetic and parasympathetic ganglionic blocking actions of the antibiotics were promptly reversed by i.v. infusion of calcium salts. These results prompted Corrado[9] to suggest a competitive interaction of streptomycin and Ach at the receptors of the autonomic ganglia. Similar results were obtained with neomycin[10] and also kanamycin.[11]

In 1977, Wright and Collier[12] studied the actions induced by neomycin when administered into the perfused superior cervical ganglion of the cat. They observed that concentrations of the antibiotic sufficient to block by 50 to 100% the response of the nictitating membrane to preganglionic nerve stimulation, had no effect on the same response induced by submaximal concentration of nicotine, injected close to the ganglion, a result which indicates a predominant preganglionic action of the antibiotic. In fact, Wright and Collier[12] also demonstrated a reduction of about 60% of the evoked release of Ach by the antibiotic, an action that was significantly increased when calcium concentration in the bathing fluid was reduced from 2.5 to 0.5 mM. Under the latter experimental condition, an abolition of the increased accumulation of ^{45}Ca induced by preganglionic stimulation in the presence of the antibiotic was also demonstrated. This action was never demonstrated with conventional ganglioplegic agents such as d-tubocurarine.[12-13]

More recently, Alkhadi and McIsaac[14] studied the electrophysiological changes induced by streptomycin on the isolated superior cervical ganglia of the rabbit. The authors demonstrated that streptomycin 1.0 mg/mℓ selectively blocked S_1 major presynaptic nerve terminal spike, induced by supramaximal stimulation of the preganglionic nerve. Increasing the dose of streptomycin up to 3.0 mg/mℓ, the S_2 spike was also blocked. Selective depression of S_1 spikes is a characteristic of agents that either increase the stability of the postsynaptic ganglionic membrane or that decrease the amount of transmitter released.[15] Since the selective block of the S_1 spike induced by streptomycin was antagonized by calcium, Alkhadi and McIsaac[14] interpreted the action of the antibiotic as being due, mainly, to a presynaptic effect resulting in a decreased release of Ach.

2. Adrenergic Transmission

Up to now, only one experimental study about the action of aminoglycoside antibiotics on adrenergic transmission has been done. Working with the rat isolated anococcygeus muscle, Wright and Collier[6] showed that neomycin inhibited the responses of the muscle to either field stimulation or noradrenaline administration, both effects being reversed by increasing the calcium concentration in the nutrient fluid. The increased release of ^{3}H-noradrenaline induced by field stimulation on the same preparation was inhibited in a reversible way by neomycin, showing again that the antibiotic may act at both pre- and postjunctional levels.

At the adrenal medulla, this double effect of the antibiotic could once more be demonstrated. In continuation of the first report by Corrado,[9] Guimarães et al.[16] observed that small doses of neomycin which blocked the hypertension induced by the stimulation of the distal end of splanchnic nerve, did not inhibit the hypertensive response to close-retrograde injection of nicotine into the adrenolombar vein. When higher doses of neomycin were employed, the hypertensive responses to both electrical and pharmacological stimuli were blocked. The injection of calcium chloride by the same route antagonized the presynaptic action of low doses of neomycin but failed to reverse the blocking action obtained with higher doses of the antibiotic. It should be mentioned that in this last situation, the injection of calcium chloride was able to restore the hypertensive response to nicotine. The authors conclude that the action of neomycin at the adrenal medulla is primarily presynaptic; when the dose is increased, the cromaffin cell membrane is also involved with a consequent interference in the calcium-dependent releasing processes of both Ach and catecholamines.

As described below for the neuromuscular transmission, the ganglionic, adrenergic, and medullar adrenal actions of aminoglycoside antibiotics are quite similar to those described for Mg^{++}, in that the antibiotics compete with Ca^{++} in the process of neurotransmitter release. However, no specific experiments have yet been done to test this hypothesis.

3. Neuromuscular Transmission

The mechanism of the well-known neuromuscular blocking properties of the aminoglycoside antibiotics[17-19] is not yet clear. It was initially demonstrated that the myographic characteristics of the neuromuscular blockade produced by streptomycin in dogs were quite similar to those induced by Mg^{++}.[20] It should be recalled that Mg^{++} besides inhibiting the action of Ca^{++} in transmitter release, have also a depressing effect on the excitability of muscle fibers.[21] These two properties were also demonstrated to be induced by streptomycin[20,22] and other related aminoglycosides.[18] A group of authors believe that the most important effect of these antibiotics occurs at the presynaptic level.[22-24] On the contrary, other investigators claim that the postsynaptic membrane is the main site of action of the drugs, since they observed that neomycin, streptomycin, and kanamycin did not affect Ach release from the frog neuromuscular junction.[25-26] A more detailed study of the neuromuscular effects of streptomycin in the same preparation has led to a confirmation of the predominant postsynaptic inhibitory action by low doses of the antibiotics, which is accompanied by a facilitatory prejunctional effect; when high doses were employed, the release of Ach was also inhibited by the drugs.[27]

Bearing these controversial observations in mind, the neuromuscular actions of streptomycin and neomycin were analyzed on the isolated rat diaphragm preparation.[12] It became evident that at 1.6 mM calcium concentration, the neuromuscular blockade induced by neomycin had a greater presynaptic component than the neuromuscular blockade produced by streptomycin; neomycin was somehow more effective than streptomycin as an inhibitor of Ach release. According to Wright and Collier,[12] these differences might explain why some authors were more impressed with the presynaptic blocking effect of neomycin,[23-24] while the group which studied streptomycin[25-26] was more impressed with its postsynaptic blocking action. The discrepancies in the literature may also result from differences in the types of muscles employed; for instance, it has been demonstrated that the antibiotics induce a predominant presynaptic effect in fast-twitch muscles, while acting predominantly at postjunctional sites when used slow-twitch muscles.[28] However, these data do not explain the presynaptic facilitatory effect of streptomycin,[27] an action that indicates a possible intracellular effect on nerve terminals. Even though this explanation has already been suggested to explain a similar action of Mg^{++},[29-30] it was not accepted by some authors in relation to antibiotics.[12]

Based on the results of Hava et al.[31] who proposed a reduction on the ionized calcium by neomycin, Corrado[17] and also Sobek et al.[32] attributed the mechanism of action of streptomycin, neomycin, and kanamycin to their ability to reduce calcium blood level. However, neomycin has not been proven to reduce ionized calcium when blocking concentrations were used.[6,33-34] Tetracyclines, which are typical calcium chelating antibiotics,[3] induce a neuromuscular block on the isolated diaphragm of the rat which is followed by contracture when high doses of the antibiotic are employed. These effects, although partially antagonized by calcium salts, reveal dose-response curves with significant deviation from parallelism, an indication of a noncompetitive type of interaction.[35] As will be discussed later, these results have never been demonstrated with the aminoglycoside antibiotics. Oppositely, the analysis of the neuromuscular block induced by streptomycin and kanamycin on the isolated rat diaphragm has shown that the increase in calcium concentration induces a progressive shift to the right of the curves relating the dose of antibiotic to the percentage block.[36-37] These data, which are similar to those obtained with neomycin,[38] do not show any significant deviation from parallelism in the dose-response curves which is highly suggestive of competitive antagonism between calcium and antibiotics.[39]

Del Castillo and Katz[40] have suggested a competition between Mg^{++} and Ca^{++} for specific presynaptic receptive sites called X. They form respectively, irreversible MgX

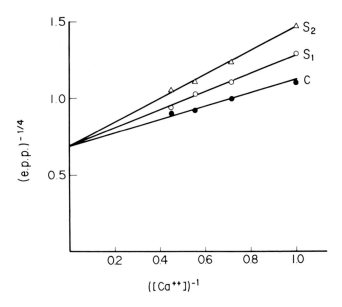

FIGURE 1. Double reciprocal plot of the fourth root of the amplitude of epp obtained on the isolated sartorius muscle of the toad, as a function of the logarithm of the external calcium concentration (mM) obtained in absence (C) and presence of streptomycin sulfate in the bath fluid, 0.03 and 0.06 mM at S_1 and S_2, respectively. Each point represents the average of 16 epp measured in mV. (From Prado, W. A., Corrado, A. P., and Marseillan, R. F., *Arch. Int. Pharmacodyn. Ther.*, 231, 297, 1978. With permission.)

and reversible CaX complexes, the latter being the only one effective for transmitter release.[41] Dodge and Rahamimoff[42] postulated the necessity of a cooperative action of four CaX complexes in order to guarantee the release of Ach, and that Mg^{++} competitively interfere with this cooperative interaction. Working with the isolated sartorius muscle of the toad, Prado[36] and Prado et al.[37] demonstrated that in the presence of streptomycin or kanamycin there is a change in the amplitude of the intracellularly recorded epp that leads to plots quite similar to those reported by Dodge and Rahamimoff.[42] When the reciprocals of the fourth root of the amplitude of epp were plotted against the reciprocals of the calcium concentrations used, the straight lines obtained in the presence and in the absence of the antibiotic had a common intercept in the ordinate axis (Figure 1). This result is highly suggestive of an interaction of these drugs with the proposed site X in a manner quite similar to that of Mg^{++}, with the formation of antibiotic-X reversible complexes which are ineffective for Ach release.

B. ECC
1. Skeletal Muscle

In addition to the afore-mentioned competitive inhibition of the inward movement of calcium at the motor nerve terminal exerted by the aminoglycoside antibiotics, Fairhurst and Macri,[38] analyzing the uptake of ^{45}Ca by isolated sarcoplasmic reticulum fragments (SRF) of rabbit's skeletal muscle, described the same type of competition at the postsynaptic level for neomycin and streptomycin. It was observed that in media containing initial free calcium concentrations of 10^{-7} and 10^{-6} M, the level of the ionized calcium markedly influences the inhibitory effect of neomycin, the greater inhibition being seen at calcium 10^{-7} M. The double reciprocal plot of the dose-response curves indicates that neomycin competitively inhibits calcium-uptake by SRF (Figure 2).

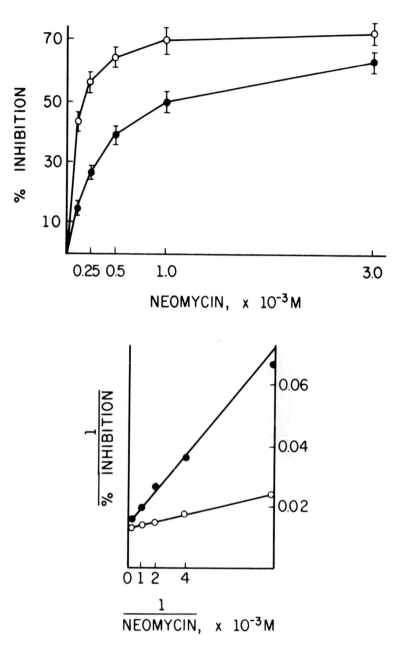

FIGURE 2. Inhibition of initial rate of calcium uptake in FSR by neomycin at initial concentrations of free calcium 10^{-7} M (o) and 10^{-6} M (•). Each point represents the average of four preparations. Vertical bars denote SE. (From Fairhurst, A. S. and Macri, J., *Life Sci.*, 16, 1321, 1975. With permission.)

Early evidences for this type of antagonism were described by Corrado et al.[43] who showed that concentrations of neomycin higher than those necessary to cause neuromuscular blockade of a presynaptic nature, were able to induce an increase in the amplitude of the muscular contractions of the directly stimulated denervated rat diaphragm. This facilitatory muscular effect became more evident in chronically denervated diaphragms with tobramycin causing a stronger and more prolonged increase in the amplitude of muscular contractions.[44]

Although paradoxical for a neuromuscular blocking agent, this effect can be easily understood if we assume the following possible ways of action for these antibiotics:

1. To cross the plasma membrane and reach intracellular structures
2. To displace calcium from sites or stores in the cell membrane leading to an increase of available intracellular Ca^{++}
3. To act as ionophores facilitating the translocation of calcium from the extra- to the intracellular medium

The two latter possibilities are the least probable ones if we take into account the *"depolarization-induced release of calcium"* hypothesis,[45] according to which the activation of the transverse tubules triggers Ca^{++} release from the sarcoplasmic reticulum (SR) by an extent sufficient to saturate the contractile system by a mechanism that does not involve any flux of Ca^{++} across the sarcolemma. An experimental confirmation of the first possibility could easily explain the facilitatory muscular effect via a competitive antagonism between the antibiotics and the uptake of calcium by the SR with a consequent increase in available myoplasmic calcium.

Therefore, the two opposite effects resulting from the postjunctional muscular actions of the antibiotics (inhibition[27] and facilitation[43]) resemble those reported elsewhere for the action of the antibiotics at the motor nerve terminal, i.e., reduction and increase of the neurotransmitter release; both probably depend on the same property of the drug to interfere with calcium flow at either the cellular membrane or membranes of intracellular structures.

2. Cardiac Muscle

Earlier in vivo and in vitro studies of the cardiovascular effects induced by aminoglycoside antibiotics showed that streptomycin, dihydrostreptomycin, and kanamycin produced a depressant effect[46-49] that was promptly antagonized by calcium.[46-47] These studies did not provide any conclusive evidence about the side of action of the antibiotics. This has been demonstrated by Adams[50] on isolated, electrically driven rat left atria, in which gentamicin produced a maintained and concentration-dependent depression of myocardial contractile tension which was easily reversed by simple washing of the preparation with antibiotic-free solution. The negative inotropic response elicited by gentamicin was competitively antagonized by calcium, since the maximum or near maximum calcium-induced contractile response could still be elicited even in the presence of severe gentamicin-produced negative inotropic effect, by simply increasing the calcium concentration. The positive inotropic response to calcium in isolated preparations is believed to be due to an increase in the inward movement of the ion, with the consequent increase of its intracellular concentrations, which according *to the hypothesis of "calcium induced-release of Ca^{++}"*[45] triggers a regenerative release of additional calcium from cellular storage sites. More recently, Adams and Durret[51] worked out an experimental model that accentuates the contractile dependence of myocardial fibers on the influx of extracellular Ca^{++} through specific "slow channels" of the sarcolemma. The method is based on the inactivation of fast sodium channels by excess potassium and on the restoration of the lost contractility by isoproterenol. In this situation, the depressant effect of gentamicin was much greater, indicating a direct cardiac depressant action of the antibiotic and suggesting either a blockade of the transport system for calcium ions through slow channels of the sarcolemma, an interference with the availability of Ca^{++} for translocation to these sites, or both effects.

The suggestion of a competitive antagonism[50] was stressed by Annovazzi et al.[52] working with isolated and electrically driven guinea pig left atria. These authors demonstrated that the dose-response curves to the cardiodepressor effect induced by gentamicin were shifted to the right by increasing calcium concentration and did not observe any significant deviation from parallelism facts which are highly suggestive of competitive antagonism.[39] In the same isolated preparation, it was demonstrated that gentamicin and, in a greater degree tobramycin, can induce an increase of the maximal response to calcium.[53] This apparently discrepant result is probably due to an inhibition of the uptake of calcium by intracellular organelles, in analogy to what has been previously discussed in relation to the facilitation produced by these antibiotics on skeletal muscle.

3. Smooth Muscle

Early reports on alterations of contraction induced on smooth muscle by aminoglycoside antibiotics, have shown that streptomycin caused an inhibitory effect on spontaneous and agonist-induced contractions, as well as on peristaltic reflex or electrically evoked contractions of rat and rabbit's uterus both in vitro and *in situ,* on the guinea pig ileum and on the rabbit duodenum.[47,54-57]

Although it has been demonstrated that calcium ion antagonizes streptomycin inhibition of spontaneous contractions of the rat uterus[56] and also contractions of the guinea pig ileum elicited by coaxial stimulation,[57] the nature of the calcium-antibiotic interaction has not yet been elucidated. The quantitative analysis of the antagonism has recently been performed[19,58-59] using the guinea pig ileum depolarized by high potassium in which contractility is a hyperbolic function of the external Ca^{++} concentration;[60] the peak amplitude of the calcium-evoked contractions can be represented by the equation $Ca = K R/R_T - R$; where Ca is the concentration of the ion in the medium, R is the peak amplitude of the contractile response, R_T is the maximal response, and K is a proportionality constant. The reciprocal form of this equation gives a straight line represented by the equation $1/R_T = K/Ca + 1$, where $R_T = 1$ is given by the intercept of the straight line on the ordinate axis. In the presence of aminoglycoside antibiotics or Mg^{++}, a new straight line is obtained with no change in ordinate intercept but with a bigger slope. This indicates a competitive antagonism between calcium and the antibiotics or magnesium (Figure 3, I to III) which has been confirmed by the method reconized by Arunlakshana and Schild[61] (Figure 3, IV). The pA_2 index calculated from the plot of Figure 3 has been used to compare the relative potencies between Mg^{++} and aminoglycoside antibiotics. On a molar basis, the inhibitory potency decreases in the following way: ($pA_2 \pm$ SEM; n = number of experiments); gentamicin (3.90 ± 0.05; n = 6) > magnesium (3.73 ± 0.08; n = 26) > neomycin (3.40 ± 0.07; n = 6) > streptomycin (3.19 ± 0.04; n = 21) > kanamycin (2.2 ± 0.01; n = 24).[19,59] It was also observed[19,59] that the action of the calcium chelating agent EDTA and tetracycline did not display the characteristic of competitive antagonism such as mentioned above in relation to Mg^{++} and aminoglycoside antibiotics. This is a confirmation of the conclusion already described that a mechanism involving calcium chelation by the aminoglycoside antibiotics is very unlikely.

The blockade of the calcium-elicited contractions of the intestinal smooth muscle induced by the antibiotics could be attributed to a competition between the antibiotics and calcium for membrane binding sites involved in the transport of the ion.[19,59] This would result in a decreased influx and consequently in a reduction of the myoplasmic calcium concentration necessary for the activation of the contractile proteins. Radiocalcium studies performed in the same isolated preparation[19,59] give further support to this interpretation. As a matter of fact, streptomycin produced a dose-dependent decrease in the uptake of ^{45}Ca with reduction of both the fast and slow cellular components. Therefore, the sustained increase of ^{45}Ca efflux observed after the addition of streptomycin on the washout experiments seems to reflect a displacement of bound

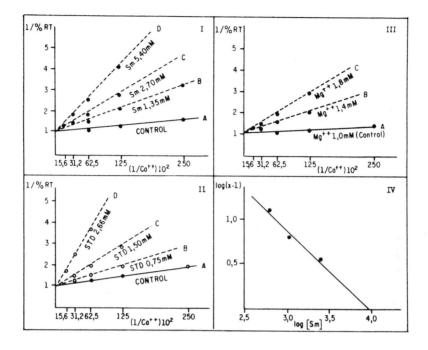

FIGURE 3. Double reciprocals plot of the contractile responses of guinea pig ileum (plotted as % of maximal effects) vs. calcium concentration in the extracellular fluid (mM) in the absence (solid lines) or presence (dashed lines) of competitive (I, II, and III) antagonists. Note the common intercept on the ordinates of the straight lines with streptomycin (Sm), streptidine (STD), and magnesium (Mg^{++}). At IV, Arunlakshana and Schild's plot of the antagonism between Sm and Ca^{++}. At the abcissa, Sm represents the molar concentration of the antibiotic and at the ordinate, x is the dose ratio of calcium to elicit 50% of the maximal effect in the presence of three different concentrations of streptomycin. Note that the straight line makes an approximate 45° angle with the abscissa axis. (From Pimenta de Morais, I., Corrado, A. P., and Suarez-Kurtz, G., Arch. Int. Pharmacodyn. Ther., 231, 317, 1978. With permission.)

^{45}Ca or a decrease of its uptake, rather than an increase in the membrane permeability to calcium or an activation of the postulated energy-dependent mechanism for calcium efflux in the intestinal smooth muscle.[62]

The inhibitory effect of aminoglycoside-antibiotics has also been demonstrated on the contractile response elicited by norepinephrine and potassium on isolated preparations of vascular smooth muscles of rabbits, dogs, and nonhuman primates.[63-67] The ^{45}Ca movements in these vascular preparations did not differ qualitatively from that of the above-mentioned experiments with the guinea pig ileum. It was also concluded that neomycin, streptomycin, gentamicin, and kanamycin selectively inhibited the uptake, binding, and/or availability of the superficial calcium fraction in vascular smooth muscles.[63-65] On the other hand, the slight but sustained increase of tension produced by neomycin in isolated rabbit aortic strips found by Adams et al.,[63] was explained by the possible displacement of "some membrane-bound calcium" by the antibiotic which, in turn, induces an inward shift of Ca^{++} with a consequent increase in muscle tension. However, these authors did not take into account the possible role played by the richly developed SR of rabbit aortic smooth muscle.[68] Accordingly, as we have already suggested for skeletal and cardiac muscles, if the possibility of the antibiotic crossing the plasma membrane and interfering with the uptake of calcium by SR exists, it could reasonably explain the unexpected increase of muscular tension caused by neomycin. The absence of such phenomena on the guinea pig ileum is probably due to its poorly developed SR which seems to play no role as a regulator of the contractile activity; this type of muscle mobilizes primarily the extracellular or membrane superficially bound calcium.

The uniform pattern of action of the aminoglycoside antibiotics on skeletal, cardiac, and smooth muscles is related to a basic interference with the transmembrane movement of calcium and confer to the group a broader spectrum of action when compared to substances such as verapamil which selectively blocks the calcium inflow in smooth and cardiac muscles. Therefore, it seems to us that the so-called "specific" calcium antagonists as referred to the group of verapamil,[69] fits better with the aminoglycoside antibiotics. In fact, the recent reviews by Putney and Ashkari[70] and Putney[71] on calcium antagonists show that besides lanthanum, only with the aminoglycoside antibiotics is it possible to obtain an easily reversible antagonism of the competitive type; this makes them useful tools for pharmacological analysis. The interference with this transmembrane calcium movement is a side effect of the antibiotics with no therapeutic application and seen only when high doses are employed.

C. Secretory Cells

It was reasonable to expect, following the demonstration that aminoglycoside antibiotics interfere with the influx of calcium necessary to trigger neurotransmitter release, that it could also interfere with the calcium-dependent release of other endogenous substances. In fact, secretory cells other than the adrenal cromaffin cells which have been analyzed elsewhere are affected by the aminoglycoside antibiotics. It has been shown on the isolated neurohypophysis of the rat that neomycin inhibits both the spontaneous and the potassium-evoked release of vasopressin, an effect which is accompanied by a decrease in the rate of ^{45}Ca efflux.[72] Although the character of this interaction has not yet been analyzed, the reduction in plasma membrane permeability to calcium ions could be a reasonable explanation for the reported results.

More recently, Boschero[73] demonstrated that sisomicin can inhibit glucose-induced insulin release in rat pancreatic islets. By determining the amount of insulin released at each calcium chloride concentration used, this author also demonstrated a parallel shift to the right of the control log dose-response curves when the antibiotic was added to the nutrient fluid. It is possible, therefore, that the calcium-antibiotic interaction at the endocrine pancreas might occur in a competitive manner. In our laboratory, we are analyzing the EDTA-pretreated isolated rat peritoneal mast cells as a model of a secretory cell for the study of drugs that might interfere with calcium-induced secretion. The preliminary results indicate that streptomycin competitively inhibits the A23187 ionophore-induced release of histamine, since the control log dose-response curve was parallely shifted to the right when streptomycin was added to the nutrient fluid.[74]

D. The Interaction of Aminoglycoside Antibiotics With Membrane Structures

Prado[36] and Prado et al.[37] extended their studies to the neuromuscular effects induced by streptidine, the inositolic moiety of streptomycin which has been reported to be responsible for the blocking action exhibited by the antibiotic.[75] The authors observed that streptidine induced a neuromuscular block with characteristics quite similar to those reported with streptomycin, including the competitive interaction with calcium. At the intestinal smooth muscle, the same similarities of effects of streptidine and streptomycin have been demonstrated[59] (Figure 3, II).

Bearing in mind the structural similarities between streptidine and the phosphatidylinositide components of the cellular membrane, Corrado et al.[19] suggested that the inositolic fraction acting as a cation, could either interact with calcium-binding sites or, alternatively, compete with calcium binding sites of phosphatidylinositides. In fact,

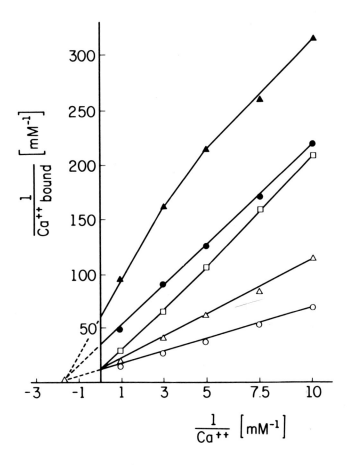

FIGURE 4. Inhibition by neomycin of calcium binding to synaptosomal membranes. Each point is the mean of duplicate determinations obtained with no neomycin o—o; $3 \cdot 10^{-5}$ M △—△; 10^{-5} M □—□; $3 \cdot 10^{-4}$ M •—•; and 10^{-3} M neomycin ▲—▲. (From Lodhi, S., Weiner, N. D., and Schacht, J., *Biochim. Biophys. Acta*, 426, 781, 1976. With permission.)

phosphatidylinositides have high affinity and binding capacity toward calcium[76-77] and probably play an important role in cholinergic transmission,[78] as well as in the regulation of membrane permeability and ion transport (for a review see Sanders and Sanders[79]). In agreement with the first suggestion of Corrado et al.[19] are reports showing that neomycin inhibits the turnover of phosphatidylinositides[80] and blocks the binding of ^{45}Ca in a competitive way.[81] These results have been obtained in homogenates of guinea pig inner ear tissues which are known to contain phosphatidylinositides. According to Schacht,[82] the strong cationic charge of the aminoglycoside could compete with calcium for the negatively charged phosphate groups of the phosphatidylinositides. An extension of these studies has demonstrated that neomycin, tobramycin, gentamicin, kanamycin, streptomycin, and dihydrostreptomycin are potent inhibitors of calcium binding to synaptosomal membranes.[83] Neomycin, the most potent antibiotic in this regard, competes with calcium binding when used in concentrations as 3×10^{-5} to 10^{-4} M. Higher concentrations of the antibiotics show a noncompetitive type of inhibition (Figure 4).

In summary, it is possible that at adequate concentrations, the aminoglycoside antibiotic might interfere competitively with the calcium-phosphatidylinositide interaction at cellular membranes, thus, explaining the results present elsewhere.

REFERENCES

1. Rubin, R. P., The role of calcium in the release of neurotransmitter substances and hormones, *Pharmacol. Rev.*, 22, 389, 1970.
2. Reed, P. W. and Lardy, H. A., A23187: a divalent cation ionophore, *J. Biol. Chem.*, 247, 6970, 1972.
3. Kelly, R. G. and Buyske, D. A., Metabolism of tetracycline in the rat and the dog, *J. Pharmacol. Exp. Ther.*, 130, 144, 1960.
4. Singh, Y. N., Harvey, A. L., and Marshall, I. G., Antibiotic-induced paralysis of the mouse phrenic nerve-hemidiaphragm preparation, and reversibility by calcium and by neotigmine, *Anesthesiology*, 48, 418, 1978.
5. Bowen, J. M., Influence of induced hypermagnesemia and hypocalcemia on neuromuscular blocking property of oxytetracycline in the horse, *Am. J. Vet. Res.*, 36, 1025, 1975.
6. Wright, J. M. and Collier, B., The site of the neuromuscular block produced by polymyxin B and rolitetracycline, *Can. J.Physiol. Pharmacol.*, 54, 926, 1976.
7. Van Nyhuis, L. S., Miller, R. D., and Fogdall, R. P., The interaction between d-tubocuranine, pancuronium, polymyxin B, and neotigmine on neuromuscular function, *Anesth. Analg. Cleveland*, 55, 224, 1976.
8. Lee, C., Chen, D., and Nagel, E. L., Neuromuscular block by antibiotics: Polymyxin B., *Anesth. Analg. Cleveland*, 56, 373, 1977.
9. Corrado, A. P., Ganglioplegic action of streptomycin, *Arch. Int. Pharmacodyn. Ther.*, 114, 166, 1958.
10. Corrado, A. P. and Ramos, A. O., Neomycin — its curariform and ganglioplegic actions, *Rev. Bras. Biol.*, 18, 81, 1958.
11. Corrado, A. P. and Ramos, A. O., Some pharmacological aspects of a new antibiotic-kanamycin, *Rev. Bras. Biol.*, 20, 43, 1960.
12. Wright, J. M. and Collier, B., The effects of neomycin upon transmitter release and action, *J. Pharmacol. Exp. Ther.*, 200, 576, 1977.
13. Blaustein, M. P., Preganglionic stimulation increases calcium uptake by sympathetic ganglia, *Science*, 173, 391, 1971.
14. Alkhadi, K. A. and McIsaac, R. J., Ganglion blocking effects of streptomycin, *Arch. Int. Pharmacodyn. Ther.*, 232, 58, 1978.
15. Kayaalp, S. O. and McIsaac, R. J., Differential blockade and potentiation of transmission in a sympathetic ganglion, *J. Pharmacol. Exp. Ther.*, 173, 193, 1970.
16. Guimarães, A. F., Prado, W. A., and Corrado, A. P., Interação cálcio x neomicina nas sinapses colinérgicas da adrenal de cão, *Cienc. Cult. (Sao Paulo)*, 29, 607, 1977.
17. Corrado, A. P., Respiratory depression due to antibiotics: calcium in treatment, *Anesth. Analg.*, 42, 1, 1963.
18. Pittinger, C. and Adamson, R., Antibiotic blockade of neuromuscular function, *Ann. Rev. Pharmacol.*, 12, 169, 1972.
19. Corrado, A. P., Prado, W. A., and Pimenta de Morais, I., Competitive antagonism between calcium and aminoglycoside antibiotics in skeletal and smooth muscles, in *Concepts of Membranes in Regulation and Excitation*, Rocha e Silva, M. and Suarez-Kurtz, G., Eds., Raven Press, New York, 1975, 201.
20. Vital Brazil, O. and Corrado, A. P., The curariform action of streptomycin, *J. Pharmacol. Exp. Ther.*, 120, 452, 1957.
21. Del Castillo, J. and Engbaek, L., The nature of the neuromuscular block produced by magnesium, *J. Physiol. (London)*, 124, 370, 1954.
22. Vital Brazil, O., Streptomycin effect on the skeletal muscle stimulation produced by acetylcholine, *Arch. Int. Pharmacodyn. Ther.*, 130, 136, 1961.
23. Elmqvist, D. and Josefsson, J. O., The nature of neuromuscular block produced by neomycin, *Acta Physiol. Scand.*, 54, 105, 1962.
24. Vital Brazil, O. and Prado-Franceschi, J., The nature of neuromuscular block produced by neomycin and gentamicin, *Arch. Int. Pharmacodyn. Ther.*, 179, 78, 1969.
25. Dretchen, K. L., Gergis, S. D., Sokoll, M. D., and Long, J. P., Effect of various antibiotics on neuromuscular transmission, *Eur. J. Pharmacol.*, 18, 201, 1972.
26. Dunkley, B., Sanghvi, I., and Goldstein, G., Characterization of neuromuscular block produced by streptomycin, *Arch. Int. Pharmacodyn. Ther.*, 201, 213, 1973.
27. Dretchen, K. L., Sokoll, M. D., Gergis, S. D., and Long, J. P., Relative effects of streptomycin on motor nerve terminal and endplate, *Eur. J. Pharmacol.*, 22, 10, 1973.
28. Adams, H. R., Mathew, B. P., Teske, R. H., and Mercer, H. D., Neuromuscular blocking effects of aminoglycoside antibiotics on fast- and slow-contracting muscles of the cat, *Anesth. Analg. Cleveland*, 55, 500, 1976.

29. Cooke, J. D., Okamoto, K., and Quastel, D. M. J., The role of calcium in depolarization secretion coupling at the motor nerve terminal, *J. Physiol. (London)*, 228, 459, 1973.
30. Miledi, R., Transmitter release induced by injection of calcium ions into nerve terminals, *Proc. R. Soc. London Ser. B*, 183, 421, 1973.
31. Hava, M., Sobek, V., and Mikuláškova, J., On the role of calcium ions in the toxic neomycin action, *Biochem. Pharmacol.*, 8, 76, 1961.
32. Sobek, V., Háva, M., Mikuláškova, J., and Waitzova, D., Uber den Mechanismus der toxischen Wirkungen von Neomycin und die Möglichkeiten ihrer Beeinflussung, *Arzneim. Forsch.*, 13, 391, 1963.
33. Pittinger, C. B., pH and streptomycin influences upon ionic calcium in serum, *Anesth. Analg.*, 49, 540, 1970.
34. Suarez-Kurtz, G., Inhibition of membrane calcium activation by neomycin and streptomycin in crab muscle fibres, *Pfluegers Arch. Gesamte Physiol. Menschen Tiere*, 349, 337, 1974.
35. Prado, W. A. and Corrado, A. P., 1978, unpublished.
36. Prado, W. A., Antagonismo cálcio-antibióticos na juncão neuromuscular, Ph.D. thesis, Faculty of Medicine of Ribeirão Preto, Sao Paulo, Brazil, 1975.
37. Prado, W. A., Corrado, A. P., and Marseillan, R. F., Competitive antagonism between calcium and antibiotics at the neuromuscular junction, *Arch. Int. Pharmacodyn. Ther.*, 231, 297, 1978.
38. Fairhurst, A. S. and Macri, J., Aminoglycoside — Ca^{++} interaction in skeletal muscle preparations, *Life Sci.*, 16, 1321, 1975.
39. Schild, H. O., Parallelism of log dose-response curves in competitive antagonism, *Pharmacol. Res. Commun.*, 1, 1, 1969.
40. Del Castillo, J. and Katz, B., The effect of magnesium on the activity of motor nerve endings, *J. Physiol. (London)*, 124, 370, 1954.
41. Jenkinson, D. H., The nature of antagonism between calcium and magnesium at the neuromuscular junction, *J. Physiol. (London)*, 138, 434, 1957.
42. Dodge, F. A., Jr. and Rahamimoff, R., Cooperative action of calcium ions in transmitter release at the neuromuscular junction, *J. Physiol. (London)*, 193, 419, 1967.
43. Corrado, A. P., Ramos, A. O., and Escobar, C. T., Neuromuscular blockade by neomycin, potentiation by ether anesthesia and d-tubocurarine and antagonism by calcium and prostigmine, *Arch. Int. Pharmacodyn. Ther.*, 121, 380, 1959.
44. Prado, W. A. and Corrado, A. P., Efeitos neuromusculares da tobramicina, *Cienc. Cult. (Sao Paulo)*, 32, 760, 1980.
45. Fabiato, A. and Fabiato, F., Calcium release from the sarcoplasmic reticulum, *Circ. Res.*, 40, 119, 1977.
46. Swain, H. H., Kiplinger, G. E., and Brody, R. M., Actions of certain antibiotics on the isolated dog heart, *J. Pharmacol. Exp. Ther.*, 117, 151, 1956.
47. Corrado, A. P., Considerações sobre os efeitos da estreptomicina na pressão arterial, Ph.D. thesis, Faculty of Medicine, University of São Paulo, Brazil, 1957.
48. Leaders, F., Pittinger, C. B., and Long, J. P., Some pharmacological properties of selected antibiotics, *Antibiot. Chemother.*, 11, 503, 1960.
49. Cohen, L. S., Wechsler, A. S., Mitchell, J. H., and Glick, G., Depression of cardiac function by streptomycin and other antimicrobial agents, *Am. J. Cardiol.*, 26, 505, 1970.
50. Adams, H. R., Direct myocardial depressant effects of gentamicin, *Eur. J. Pharmacol.*, 30, 272, 1975.
51. Adams, H. R. and Durrett, L. R., Gentamicin blockade of slow Ca^{++} channels in atrial myocardium of guinea pigs, *J. Clin. Invest.*, 62, 241, 1978.
52. Annovazzi, R., Prado, W. A., and Corrado, A. P., Antagonismo cálcio X amicacina no coração isolado de cobaia, *Cienc. Cult. (Sao Paulo)*, 29, 607, 1977.
53. Frick, W., Antonio, A., Prado, W. A., and Corrado, A. P., Alterações da contratilidade do miocárdio induzidas pela tobramicina, *Cienc. Cult. (Sao Paulo)*, 32, 760, 1980.
54. Molitor, H. and Graessle, O. E., Pharmacology and toxicology of antibiotics, *Pharmacol. Rev.*, 2, 1, 1950.
55. Džoljić, M. and Atanacković, D., Action of streptomycin on guinea pig ileum peristalsis, *Arch. Int. Pharmacodyn. (Ther.)*, 156, 279, 1965.
56. Džoljić, M. and Babic, M., Action of streptomycin on the uterus, *Eur. J. Pharmacol.*, 2, 123, 1967.
57. Rodriguez, M. G., Acción de la estreptomicina sobre las contracciones del ileon aislado de cobayo estimulado coaxialmente, *Arch. Pharmacol. Toxicol.*, I, 229, 1975.
58. Corrado, A. P., Pimenta de Morais, I., Prado, W. A., and Marseillan, R. F., Antagonismo competitivo y no competitivo calcio-antibióticos, in *Recientes Adelantos en Biologia*, Mejia, R. H. and Moguilevsky, J. A., Eds., Bona S.C.A., Buenos Aires, 385, 1971.

59. Pimenta de Morais, I., Corrado, A. P., and Suarez-Kurtz, G., Competitive antagonism between calcium and aminoglycoside antibiotics on guinea-pig intestinal smooth muscle, *Arch. Int. Pharmacodyn. (Ther.)*, 231, 317, 1978.
60. Hurwitz, L. and Suria, A., The link between agonist action and response in smooth muscle, *Ann. Rev. Pharmacol.*, 11, 303, 1971.
61. Arunlakshana, O. and Schild, H. O., Some quantitative uses of drugs antagonists, *Br. J. Pharmacol.*, 14, 48, 1959.
62. Hurwitz, L., Fitzpatrick, D. F., Derbas, G., and Landon, E. J., Localization of calcium pump activity in smooth muscle, *Science*, 79, 384, 1973.
63. Adams, H. R., Goodman, F. R., Lupean, V. A., and Weiss, G. G., Effects of neomycin on tension and ^{45}Ca movements in rabbit aortic smooth muscle, *Life Sci.*, 12, 279, 1973.
64. Adams, H. R., Goodman, F. R., and Weiss, G. G., Alteration of contractile function and calcium ion movements in vascular smooth muscle by gentamicin and other aminoglycoside antibiotics, *Antimicrob. Agents & Chemother.*, 5, 640, 1974.
65. Adams, H. R. and Goodman, F. R., Differential inhibitory effect of neomycin on contractile responses of various canine arteries, *J. Pharmacol. Exp. Ther.*, 193, 393, 1975.
66. Goodman, F. R., Weiss, G. B., and Adams, H. R., Alterations by neomycin of ^{45}Ca movements and contractile responses in vascular smooth muscle, *J. Pharmacol. Exp. Ther.*, 188, 472, 1974.
67. Goodman, F. R. and Adams, H. R., Contractile function and ^{45}Ca movements in vascular smooth muscle of nonhuman primates: effects of aminoglycoside antibiotics, *Gen. Pharmacol.*, 7, 227, 1976.
68. Hurwitz, L. and Joiner, P. D., Mobilization of cellular calcium for contraction in intestinal smooth muscle, *Am. J. Physiol.*, 218, 12, 1970.
69. Fleckenstein, A., Specific pharmacology of calcium in myocardium, cardiac pacemakers and vascular smooth muscle, *Ann. Rev. Pharmacol. Toxicol.*, 17, 149, 1977.
70. Putney, J. W., Jr. and Ashkari, A., Modification of membrane function by drugs, in *Physiology of Membrane Disorders*, Andreoli, T. E., Hoffman, J. F., and Fanestil, D. D., Eds., Plenum Press, New York, 1978, 417.
71. Putney, J. W., Jr., Stimulus-permeability coupling: role of calcium in the receptor regulation of membrane permeability, *Pharmacol. Rev.*, 30, 209, 1979.
72. Batalla-Sotello, L., Efectos de la neomicina sobre movimientos de calcio y liberación de vasopressina en neurohipofisis de rata, Ph.D. thesis, Faculty of Medicine, University of Rio de Janeiro, Brazil, 1974.
73. Boschero, A. C., 1980, unpublished.
74. Wenzel, A. M., Corrado, A. P., and Prado, W. A., Interação entre antibióticos aminoglicosidicos e Ca^{++} em mastócitos isolados de ratos, *Ciênc. Cult. (Sao Paulo)*, 32, 760, 1980.
75. Vital Brazil, O., Corrado, A. P., and Berti, F. A., Neuromuscular block produced by streptomycin and some of its degradation products, in *Curare and Curare-Like Agents*, Bovet, D., Bovet-Nitti, F., and Marini-Bettolo, G. B., Eds., Elsevier, Amsterdam, 1957, 415.
76. Piccinini, F., Galatulas, I., and Galli, C., The uptake of ^{45}Ca by individual phospholipids isolated from guinea pig heart mitochondria and the effects of various drugs, *Eur. J. Pharmacol.*, 10, 328, 1970.
77. Dawson, R. M. C. and Hauser, H., Binding of calcium to phospholipids, in *Calcium and Cellular Function*, Cuthbert, A. W., Ed., Macmillan, New York, 1970, 17.
78. Lunt, G. G. and Lapetina, E. G., Phospholipid metabolism of isolated nerve endings, *Brain Res.*, 17, 164, 1970.
79. Sanders, W. E. and Sanders, C. C., Toxicity of antibacterial agents: mechanism of action on mammalian cells, *Ann. Rev. Pharmacol. Toxicol.*, 19, 53, 1979.
80. Schacht, J., Interaction of neomycin with phosphoinositide metabolism in guinea pig inner ear and brain tissues, *Ann. Otol.*, 83, 613, 1974.
81. Orsulakova, A., Stockhorst, E., and Schacht, J., Effect of neomycin on phosphoinositide labelling and calcium binding in guinea pig inner ear tissues *in vivo* and *in vitro*, *J. Neurochem.*, 26, 285, 1976.
82. Schacht, J., Inhibition by neomycin of polyphosphoinositide turnover in subcellular fractions of guinea pig cerebral cortex *in vitro*, *J. Neurochem.*, 27, 1119, 1976.
83. Lodhi, S., Weiner, N. D., and Schacht, J., Interactions of neomycin and calcium in synaptosomal membranes and polyphosphoinositide monolayers, *Biochim. Biophys. Acta*, 426, 781, 1976.

Chapter 16

THE EFFECT OF CALCIUM ON THE TOXICITY OF HEAVY METALS

Sachiko Moriuchi

TABLE OF CONTENTS

I.	Introduction	224
II.	The Effect of Dietary Calcium on the Intestinal Absorption of Cadmium	224
III.	The Effect of Cadmium on the Calcium Transport in Intestine	225
IV.	Interaction of Cadmium With Vitamin D-Dependent Calcium Binding Protein and Alkaline Phosphatase in Intestine	226
V.	The Effect of Calcium on the Cadmium-Induced Changes in Intestinal Villi and Microvilli	227
VI.	Conclusions	230
	Acknowledgments	230
	References	230

I. INTRODUCTION

Many reports have indicated that the toxicity of heavy metals, such as cadmium and lead, is modulated by nutritional factors. Symptoms of chronic poisoning by cadmium include various forms of skeletal deformities, suggesting that cadmium interferes with calcium metabolism and that calcium modulates cadmium toxicity.[1-3] It has been reported that cadmium-exposed rats fed a low calcium diet show a greater toxicity than those fed a sufficient amount of calcium.[1-5] In the Itai-Itai disease, there is a depletion of calcium as a result of repeated pregnancy and low calcium intake under conditions of malnutrition.[6]

On the other hand, lead ingestion induces the disruption of heme biosynthesis, renal function, nervous system development, and reproductive function.[7-10] Experimental studies have shown that a low calcium diet leads to an increase in the intestinal absorption of lead.[11-12] The effects of various concentrations of dietary calcium on the manifestation of lead toxicity have been demonstrated in a number of species.[13-14] Moreover, high dietary calcium has a protective effect against the adverse effects of dietary lead in pigs.[14] Thus, calcium is thought to be one of the most important nutritional factors which modulate toxicity of heavy metals and there are numerous reports dealing with the effect of calcium on the toxicity of cadmium and lead.

This review will deal mainly with the recent literature concerning the role of calcium as a modulator of heavy metal toxicity. The discussion will specifically focus on the interaction of calcium and cadmium at the site of intestinal absorption.

II. THE EFFECT OF DIETARY CALCIUM ON THE INTESTINAL ABSORPTION OF CADMIUM

Larsson and Piscator found that cadmium-exposed rats fed a low calcium diet exhibited a greater reduction in body weight and bone mineral content, and accumulated approximately 60% more cadmium in the liver and kidney than cadmium-exposed rats fed a diet sufficient in calcium.[1] Itokawa et al.[2,3] reported similar effects, as well as anemia and degenerative changes in osteoid tissue in rats fed a low calcium diet. Other observations have supported many of these findings.[4-5] However, these experiments were mainly conducted in the presence of vitamin D. Animals raised on a low calcium diet in the presence of vitamin D showed the enhanced calcium absorption in intestine adapted with low calcium diet.[15,16] Thus, the greater toxicity of cadmium in animals raised on low calcium diet could be explained by two mechanisms: (1) cadmium could be co-transported with calcium in the vitamin D-stimulated calcium transport system in the animals raised on low calcium diet; or (2) cadmium could inhibit vitamin D-stimulated calcium transport system and cause a reduction in the availability of calcium in the animals raised on low calcium diet.

In our study, the effect of calcium on the intestinal absorption of cadmium was assessed by measuring intestinal cadmium transport and by the accumulation of cadmium in tissues in the presence or absence of vitamin D_3. We found that duodenal cadmium transport in vitro was significantly increased in vitamin D-deficient rats, but it was not influenced by dietary calcium.[17] Renal cadmium accumulation was unchanged by either dietary calcium or vitamin D_3. However, hepatic cadmium accumulation was significantly increased in vitamin D-deficient rats, although it was not influenced by dietary calcium.[18] These results suggest that cadmium-induced toxicity in the animals raised on the low-calcium diet was not due to the enhanced cadmium absorption as a result of co-transport with calcium, but was rather due to the inhibition of calcium transport by cadmium. This inhibition then produced a functional calcium deficiency, which in turn increased the accumulation and toxicity of the dietary cadmium.

III. THE EFFECT OF CADMIUM ON THE CALCIUM TRANSPORT IN INTESTINE

It is well-established that vitamin D_3 is one of the essential factors in intestinal calcium transport. However, vitamin D_3 must be first metabolized to a hydroxylated product, 1,25-dihydroxyvitamin D_3 (1,25-$(OH)_2$-D_3), before it can act on the process of intestinal calcium transport. 1,25-$(OH)_2$-D_3 is now regarded as a hormone, which is secreted from the kidney under the strict control of blood calcium, phosphorus, and parathyroid hormone levels; it is accumulated specifically in the nuclei of the target tissues, such as the intestine, by a specific cytoplasmic receptor for 1,25-$(OH)_2$-D_3.[19]

Recent evidence clearly demonstrates that cadmium interferes with vitamin D-stimulated calcium transport,[18,20] especially in rats raised on low calcium diets (Figure 1).[18] Cadmium may inhibit vitamin D-stimulated calcium transport by either of two mechanisms: it may directly affect the absorption of calcium in intestine, or it may act indirectly by influencing the metabolism of vitamin D.

There is evidence that cadmium inhibits the 1α-hydroxylation of 25-hydroxyvitamin D_3 (25-OH-D_3) in vitro,[21] but the concentration of cadmium necessary to inhibit the enzyme is higher than that obtained by cadmium feeding.[20] This finding agrees with the observation that serum 1,25-$(OH)_2$-D_3 levels in animals fed a diet low in calcium are not influenced by dietary cadmium, despite reduced intestinal calcium transport,[18] thus, it is very likely that cadmium interferes directly with calcium absorption in the intestine.

In vitro studies have shown that as little as 0.025 mM cadmium added to the mucosal medium resulted in a reduction in the vitamin D-stimulated calcium transport ratio. In contrast, magnesium, at a concentration of 0.25 mM or less, did not show any significant effect on calcium transport (Figure 2A).[22] This suggests that cadmium has a specific and direct effect on calcium transport. Furthermore, kinetic studies in vitro showed that the inhibition of calcium transport by cadmium at concentration of 0.025 to 0.05 mM was competitive (Figure 2B).[22]

In contrast, Hamilton and Smith reported that calcium uptake by intestinal slices was inhibited noncompetitively by cadmium at 0.5 and 1.0 mM.[23] In our studies, these levels of cadmium caused a complete inhibition of vitamin D-stimulted active transport of calcium. Therefore, we suggest that cadmium at low concentrations inhibits calcium transport by competing for the site of calcium transport, and this inhibition can be reversed by increasing the concentration of calcium. At higher cadmium concentrations, calcium transport may be impaired by an irreversible effect of cadmium on intestinal brush border membranes. Thus, the overall effect of dietary cadmium is to reduce the absorption of calcium, the consequences being more severe with increasing cadmium and decreasing calcium.

In other studies, Yuhas et al.[24] examined the effect of cadmium on intestinal calcium absorption in rats using an *in situ* single-pass intestinal perfusion technique. The addition of cadmium (10^{-5} to 10^{-2} M) to the perfusion medium caused a decrease in net calcium absorption. At cadmium concentrations greater than 10^{-5} M, there was a net negative calcium balance. This negative calcium balance resulted from both a decrease in the lumen-to-plasma flux of calcium (10^{-5} to 10^{-2} M cadmium) and an increase in the plasma-to-lumen flux of calcium (10^{-4} to 10^{-2} M). However, intestinal absorption of calcium was unaltered 72 hr following a single acute injection of cadmium acetate (2 mg/kg i.p.). Furthermore, in rats given cadmium (0, 1, 10, or 100 ppm) in the drinking water for 13 weeks, there was no effect of 1 to 10 ppm perfusate cadmium on the absorption, lumen-to-plasma flux, or plasma-to-lumen flux of calcium, even at 100 ppm of dietary cadmium. Thus, while cadmium added to the perfusate may exert a direct inhibitory effect on intestinal calcium absorption, the treatment of rats with cadmium in vivo, either acutely or chronically, did not decrease intestinal calcium

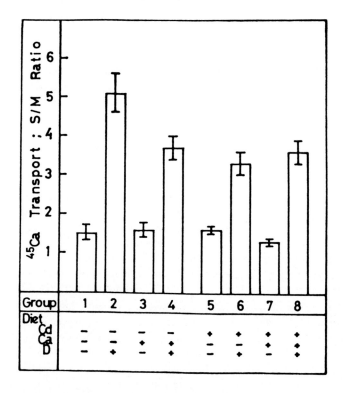

FIGURE 1. Effect of dietary cadmium and calcium on vitamin D-stimulated calcium transport in rat duodenum. Ca transport was expressed as ^{45}Ca radioactivity ratio of serosal to mucosal medium after incubation (S/M ratio). Values are means ± SE of 5 rats. Significantly different at $p < 0.001$ when comparison was made between group 1 and 2, group 3 and 4, group 5 and 6, group 7 and 8, respectively. Significantly different at $p < 0.05$ when comparison was made between group 2 and 4, and group 2 and 6, respectively. (Cd + :200 ppm, Ca−:0.002%, Ca + :0.47%, D + :100 IU × 5/2 weeks.) (From Tsuruki, F., Wung, H.-L., Tamura, M., Shimura, F., Moriucha, S., and Hosoya, N., Vitamins, 52, 161, 1978. With permission.)

transport. These findings are in contrast to ours.[22] The discrepancy must be due to the different experimental conditions used, i.e., the rats were fed a vitamin D- and calcium-sufficient diet.[24]

IV. INTERACTION OF CADMIUM WITH VITAMIN D-DEPENDENT CALCIUM BINDING PROTEIN AND ALKALINE PHOSPHATASE IN INTESTINE

The mechanism of vitamin D action in the intestinal calcium transport is not clear; however, it can be studied by stimulating the synthesis of specific proteins for calcium transport. Two kinds of specific proteins induced by 1,25-(OH)$_2$-D$_3$ have been demonstrated: the soluble calcium binding protein (CaBP) and proteins associated with microvilli.[25]

The intracellular localization of CaBP is controversial. Some investigators suggest that CaBP is localized inside the epithelial cells, but others contend that it is localized on the surface of those cells.[25] The proteins associated with microvilli, such as alkaline phosphatase, are found on the epithelial surface. One might, therefore, expect an interaction of cadmium with those proteins.

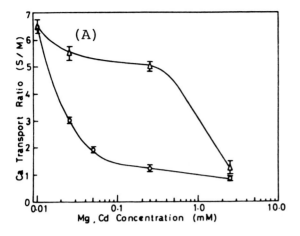
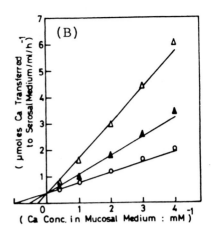

FIGURE 2. Effect of cadmium and magnesium on the vitamin D-stimulated calcium transport (A) in vitro and (B) its kinetic analysis. Calcium transport was studied in the rats that were raised on vitamin D-deficient diet (Ca:0.002%, P:0.3%) for 3 weeks followed by a dose of 100 IU of vitamin D_3 24 hr before sacrifice. (A) The mucosal medium consisted of 0.25 mM $CaCl_2$ containing ^{45}Ca (10,000 cpm/ml) and various concentrations of $CdCl_2$ or $MgCl_2$. The serosal medium was without $CdCl_2$ or $MgCl_2$. Incubation was carried out at 37°C for 2 hr. Each point represents mean ± SE of 5 rats. −o−: In the presence of $CdCl_2$; −△−: in the presence of $MgCl_2$. (B) Mucosal medium consisted of various concentrations of Ca in the absence (−o−) or in the presence of Cd (0.025 mM $CdCl_2$ (−△−), and 0.05 mM $CdCl_2$ (−▲−)). Initial serosal medium was free from Ca and Cd. Incubation was carried out at 37°C for 1 hr. The plots are according to the method of Lineweaver-Burk. The abscissa represents 1/S where S is the concentration of Ca in the initial mucosal medium. The ordinate represents 1/V where V is the amount of Ca transferred/ml/hr. Each point represents mean of 5 rats. (From Tsuruki, F., Otawara, Y., Wung, H.-L., Moriuchi, S., and Hosoya, N., *J. Nutr. Sci. Vitaminol.*, 24, 237, 1978. With permission.)

The binding of calcium to CaBP in vitro is inhibited by various metal ions, especially cadmium,[26,27] and the mechanism is thought to be competitive.[27] Alkaline phosphatase is also inhibited by cadmium in vivo and in vitro.[28,29] These observations provide strong evidence that the inhibition of vitamin D-stimulated calcium transport might be the result of an interaction of cadmium with CaBP or brush border membrane proteins, such as alkaline phosphatase.

V. THE EFFECT OF CALCIUM ON THE CADMIUM-INDUCED CHANGES IN INTESTINAL VILLI AND MICROVILLI

The inhibition of vitamin D-stimulated calcium transport in animals raised on a low calcium diet could be ascribed to an interaction of cadmium with the intestinal cellular sites for calcium transport via CaBP and/or alkaline phosphatase. However, after incubation of everted gut sacs in the presence of cadmium, significant amounts of cadmium bind to the intestinal surface, and the intestinal villi show remarkable changes.[17] Therefore, prolonged feeding of cadmium may be expected to induce both morphological and biochemical changes.

Observations with scanning electron microscopy revealed that vitamin D_3 and dietary calcium were required for the formation of normal intestinal villi and microvilli. Duodenal villi of rats raised on vitamin D-deficient or low calcium diets did not show a typical villous shape. Under those conditions, villi were connected to each other in a zigzag way or in linear bands along the long axis of the intestine which could probably restrict their movement.[28] This change in the villi appears as a kind of retrograde change in villous development. In the course of normal development, the longitudinal folds (previllous ridges) appear initially along the long axis of intestine, differentiate

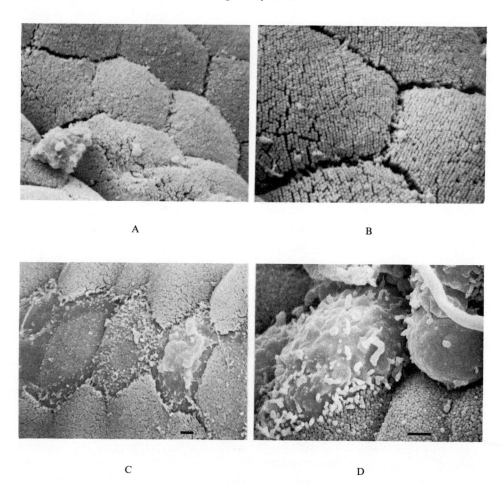

FIGURE 3. Surface of duodenal epithelial cells. The duodenal epithelial cells are covered by knob-like microvilli arranged in orders in the absence of cadmium (A and B). The border of cell is observed as the fissure of the microvilli (A and B). In the cadmium exposed rats, duodenal epithelial cells which have lost their microvilli are observed mainly in the region of middle to upper intestinal villous surface, especially in the vitamin D-deficient and calcium deficient group (C and D). Horizontal bars indicate 1 μm. (From Noda, S., Kubota, K., Yamada, K., Yoshizawa, S., Moriuchi, S., and Hosoya, N., *J. Nutr. Sci. Vitaminol.*, 24, 405, 1978. With permission.)

later into a zigzag pattern, and finally develop into distinctive tongue-shaped villi.[30] In the cadmium-exposed rats, there was a retrograde change in the villous development and the intestinal absorptive cells lost most of their microvilli. These changes were not seen in vitamin D_3-repleted and calcium sufficient rats (Figure 3).[28] In the low-calcium dietary group exposed to cadmium, damaged cells were frequently seen even in the presence of vitamin D_3.[28] This suggests that prolonged feeding of cadmium in a low-calcium diet, even in the presence of vitamin D_3, is more likely to produce cadmium toxicity than when cadmium is fed in a normal calcium diet.[1-5] The mechanism by which cadmium causes membrane deterioration is not known. However, its effects are enhanced under conditions producing fragile membranes, such as those in vitamin D or calcium deficiency.

Other studies revealed that the microvilli in cadmium-exposed rats were shorter than those of control rats.[28] It has also been reported that the microvilli in chick and rat intestine are elongated by vitamin D_3.[28,30] However, the effect of vitamin D_3 on the length of microvilli was suppressed by cadmium, especially in the absence of dietary calcium.

Table 1
THE EFFECT OF VITAMIN D₃, DIETARY CALCIUM LEVEL AND CADMIUM ON THE HYDROLYTIC ENZYMES IN RAT DUODENAL MICROVILLI

Group	Diet			Enzyme activities		
	Cd	Ca	D	Sucrase[a]	Maltase[a]	Alkaline phosphatase[b]
1	−	−	−	56.8 ± 6.5	213.5 ± 18.0	96.8 ± 4.2
2	−	−	+	45.6 ± 3.8	219.0 ± 20.1	206.0 ± 15.2[c]
3	−	+	−	50.3 ± 4.7	229.5 ± 18.5	114.7 ± 21.7
4	−	+	+	49.6 ± 3.4	192.0 ± 22.0	197.3 ± 27.0[c]
5	+	−	−	32.3 ± 6.1[d]	144.5 ± 14.5[d]	25.1 ± 2.9[d]
6	+	−	+	28.5 ± 3.2[d]	112.2 ± 18.9[d]	64.8 ± 7.3[a,c,d]
7	+	+	−	29.0 ± 3.2[d]	141.0 ± 14.0[d]	35.6 ± 6.2[d]
8	+	+	+	27.0 ± 2.2[d]	110.9 ± 14.1[d]	62.1 ± 5.8[a,c,d]

Note: Values are the means ± SE of three rats; Cd −, no addition; Cd +, 200 ppm added; Ca −, 0.002%; Ca +, 0.47%; D −, vitamin D-deficient; D +, 100 IU of vitamin D₃ was dosed five times in 2 weeks orally.

[a] Enzyme activities expressed as μmol substrate hydrolyzed/mg protein/hr.
[b] Enzyme activities expressed as μmol p-nitrophenol produced/mg protein/min.
[c] Significantly different from respective vitamin D deficient group at $p < 0.05$.
[d] Significantly different from respective Cd nonexposed group at $p < 0.05$.

From Noda, S., Kubota, K., Yamada, K., Yoshizawa, S., Moriuchi, S., and Hosoya, N., *J. Nutr. Sci. Vitaminol.*, 24, 405, 1978.

The effects of cadmium poisoning have also been examined using transmission electron microscopy.[32] These studies revealed swelling and destruction of mitochondria, sparseness and exfoliation of microvilli, and the disappearance of the terminal web. In the goblet cells, thought to be the target cells for cadmium, the boundaries of the granules became less distinct and there was a difference in electron density. The glycocalyx, their free suface side, became detached. These changes affected the adjacent cells, which lost microvilli and thus could absorb cadmium nonselectively.

All the above abnormalities were prevented when the rat was raised on a diet supplemented with normal amounts of calcium and vitamin D₃.[32]

The morphological changes in intestinal villi and microvilli were reflected in the biochemical properties of the cells. The activities of sucrase, maltase, and alkaline phosphatase were decreased in cadmium-exposed rats compared to controls (Table 1).[28] Among these enzymes, alkaline phosphatase is thought to be essential in calcium transport. Studies using polyacrylamide disc gel electrophoresis indicated that the alkaline phosphatase from the vitamin D-deficient and cadmium-exposed rats behaves differently and that the lowered activity of the enzyme could be ascribed to changes in the enzyme protein induced by cadmium.

The cadmium-induced change in the lipid composition of microvilli was also modulated by dietary calcium and vitamin D₃.[33] Vitamin D₃ caused an increase in the phospholipid content of duodenum, while cadmium ingestion resulted in decreased cholesterol and glycolipid. Furthermore, vitamin D₃ treatment increased the proportion of long chain polyunsaturated phospholipid esters and short chain cholesterol esters, but in the presence of cadmium, this caused an increase in the proportion of long chain saturated phospholipid esters and short chain saturated cholesterol esters.

VI. CONCLUSIONS

The modulation of cadmium toxicity by dietary calcium was discussed with special reference to vitamin D status. Vitamin D is an essential factor in the prevention of cadmium poisoning, especially in the initial exposure to cadmium. Since cadmium interferes with vitamin D-stimulated calcium transport, dietary calcium must be maintained at a high level to prevent further decreases of the availability of calcium and subsequent interference in calcium metabolism.

ACKNOWLEDGMENTS

The author would like to thank Professor Norimasa Hosoya (University of Tokyo) and Professor Kura Kubota (Tokyo Women's Medical College) for their valuable suggestions for preparing the manuscript and Dr. Setsuko Noda (Tokyo Women's Medical College) for the preparation of the electron microscopic photograph. The author also would like to thank Dr. Maija Zile for critical reading of this manuscript.

REFERENCES

1. **Larson, S. E. and Piscarter, M.**, Effect of cadmium on skeletal tissues in normal and calcium deficient rat, *Isr. J. Med. Sci.*, 7, 495, 1971.
2. **Itokawa, Y., Abe, T., and Tanaka, S.**, Bone changes in experimental chronic cadmium poisoning — radiological and biological approaches, *Arch. Environ. Health*, 26, 241, 1973.
3. **Itokawa, Y., Abe, T., Tabei, R., and Tanaka, S.**, Renal and skeletal lesions in experimental cadmium poisoning — histological and biological approaches, *Arch. Environ. Health*, 28, 149, 1974.
4. **Washko, P. W. and Cousins, R. J.**, Effect of low dietary calcium on chronic cadmium toxicity in rats, *Nutr. Rep. Int.*, 11, 113, 1975.
5. **Omori, M. and Muto, Y.**, Effect of dietary protein, calcium, phosphorus and fiber on renal accumulation of exogenous cadmium in young rats, *J. Nutr. Sci. Vitaminol.*, 23, 361, 1977.
6. **Emmerson, B. T.**, "Ouch-Ouch" disease: the osteomalacia of cadmium nephropathy, *Ann. Int. Med.*, 73, 854, 1970.
7. **Eisinger, J.**, Lead and man, *Trends Biochem. Sci.*, 2, N147, 1977.
8. **Hammond, P. B.**, Exposure of humans to lead, *Ann. Rev. Pharmacol. Toxicol.*, 17, 197, 1977.
9. **Grandjean, P.**, Widening perspectives of lead toxicity. A review of health effects of lead exposure in adults, *Environ. Res.*, 17, 303, 1978.
10. **Repko, J. D. and Corum, C. R.**, Critical review and evaluation of the neurological and behavioral sequelae of inorganic lead absorption, *Crit. Rev. Toxicol.*, 6(2), 135, 1979.
11. **Mahaffey, D. R.**, Nutritional factors and susceptibility to lead toxicity, *Environ. Health Perspect.*, 7, 107, 1974.
12. **Six, K. M. and Goyer, R. A.**, Experimental enhancement of lead toxicity by low dietary calcium, *J. Lab. Clin. Med.*, 76, 933, 1970.
13. **Mahaffey, K. R. and Goyer, R.**, Dose-response to lead ingestion in rats fed low dietary calcium, *J. Lab. Clin. Med.*, 82, 92, 1973.
14. **Hsu, F. S., Krook, L., Pond, W. G., and Duncan, J. R.**, Interaction of dietary calcium with toxic levels of lead and zinc in pigs, *J. Nutr.*, 105, 117, 1975.
15. **Nicolaysen, R., Eeg-Larsen, N., and Malm, O. J.**, Physiology of calcium metabolism, *Physiol. Rev.*, 33, 424, 1953.
16. **Ribovich, M. L. and DeLuca, H. F.**, The influence of dietary calcium and phosphorus on intestinal calcium transport in rats given vitamin D metabolites, *Arch. Biochem. Biophys.*, 170, 529, 1975.
17. **Moriuchi, S., Otawara, Y., Hosoya, N., and Noda, S.**, The effect of dietary calcium and vitamin D_3 on the duodenal cadmium transport in the rat, *Vitamins*, 52, 547, 1978.
18. **Tsuruki, F., Wung, H.-L., Tamura, M., Shimura, F., Moriuchi, S., and Hosoya, N.**, The effect of dietary calcium and vitamin D_3 on the cadmium accumulation in the tissues, *Vitamins*, 52, 161, 1978.

19. DeLuça, H. F., Vitamin D, in *The Fat-Soluble Vitamins, Handbook of Lipid Research*, Vol. 2, DeLuca, H. F., Ed., Plenum Press, New York, 1978, chap. 2.
20. Suda, T., Suzuki, M., Fujii, K., Sasaki, A., and Yoshiki, S., Abnormal calcium metabolism induced by oral administration of cadmium, *Bone Metab.*, 7, 65, 1975.
21. Suda, T., Horiuchi, N., Ogata, E., Ezawa, I., Otaki, N., and Kimura, M., Prevention by metallothionein of cadmium-induced inhibition of vitamin D activation reaction in kidney, *FEBS Lett.*, 42, 23, 1974.
22. Tsuruki, F., Otawara, Y., Wung, H.-L., Moriuchi, S., and Hosoya, N., Inhibitory effect of cadmium on vitamin D-stimulated calcium transport in rat duodenum in vitro, *J. Nutr. Sci. Vitaminol.*, 24, 237, 1978.
23. Hamilton, D. L. and Smith, M. W., Cadmium inhibits calcium absorption by rat intestine, *J. Physiol.*, 265, 54p, 1976.
24. Yuhas, E. M., Mita, T. S., and Schnell, R. C., Influence of cadmium on calcium absorption from the rat intestine, *Toxicol. Appl. Pharmacol.*, 43, 23, 1978.
25. Moriuchi, S. and Hosoya, N., Possible role of vitamin D in calcium absorption in intestine, in *Biochemical Aspects of Nutrition*, Yagi, K., Ed., Japan Scientific Societies Press, Tokyo, 1979, 321.
26. Ingersoll, R J. and Wasserman, R. H., Vitamin D_3-induced calcium-binding protein. Binding characteristics, conformational effects, and other properties, *J. Biol. Chem.*, 246, 2808, 1971.
27. Moriuchi, S., Vitamin D-stimulated intestinal calcium transport and cadmium, in *Biochemical Approaches in Toxicology — With Special Reference to Heavy Metals*, Imura, N., Nakao, M., and Suzuki, T., Eds., Shinohara-Shuppan, Tokyo, 1980, 129.
28. Noda, S., Kubota, K., Yamada, K., Yoshizawa, S., Moriuchi, S., and Hosoya, N., The effect of vitamin D_3 and dietary calcium level on the cadmium-induced morphological and biochemical changes in rat intestinal mucosa, *J. Nutr. Sci. Vitaminol.*, 24, 405, 1978.
29. Sugawara, N. and Sugawara, C., Effect of cadmium, in vivo and in vitro, on intestinal brush border ALPase and ATPase, *Bull. Environ. Contamin. Toxicol.*, 14, 653, 1975.
30. Grey, R. D., Morphogenesis of intestinal villi. I. Scanning electron microscopy of the duodenal epithelium of the developing chick embryo, *J. Morphol.*, 137, 193, 1972.
31. Jande, S. S. and Brewer, L. M., Effect of vitamin D_3 on the duodenal absorptive cells of chicks. An electron microscopic study, *Z. Anat. Entwicklunggesch.*, 144, 249, 1974.
32. Noda, S., Kubota, K., 1980, personal communication.
33. Tsuruki, F., Moriuchi, S., and Hosoya, N., 1980, unpublished.

Chapter 17

URINARY INHIBITORS OF CRYSTALLIZATION

L. Brandes, D. G. Oreopoulos, P. Crassweller, and A. G. Toguri

TABLE OF CONTENTS

I.	Introduction	234
II.	Modifiers of Crystallization	235
	A. Promoters of Crystallization and Epitaxial Growth	235
	B. Inhibitors of Crystallization	237
III.	Methods of Measurements of the Inhibitors	237
IV.	Nature of Inhibitors and Mechanism of Their Activity	239
V.	Therapeutic Use of the Inhibitors	240
References		242

I. INTRODUCTION

Formation of renal calculi is not a manifestation of a single disease but the result of various metabolic disorders. The majority of stones that form within the urinary tract contain calcium, in the form of either oxalate or phosphate salts. Three mechanisms are thought to be important in the formation of stones: an increase in urinary concentration (activity product) of the calcium salts; a decrease in the various urinary inhibitors; and a combination of the two. Some definitions used are given prior to reviewing the subject of inhibitors.

The process of formation of an inorganic solid from a solution containing constituents of the solid phase (ions) is called *phase transformation*. The initial event of phase transformation is called *nucleation*, i.e., when infinitesimal fragments of the solid are forming. If the fragments are formed in the absence of impurities the process is called *homogenous nucleation*. If on the other hand, the solid transformation is initiated by a foreign substance added to the system, the nucleation is called *heterogenous nucleation*.[1] An essential requirement of the mechanism in mineral phase transformation is the supersaturated state of the solution with respect to the salts involved. Varying degrees of supersaturation can be defined; thus, *metastable range* is the range of saturation in which precipitation may occur when induced by homogenous or heterogenous nucleation, while the *unstable region* is the saturation range in which rapid spontaneous precipitation can occur in the absence of nuclei. The critical level between the two regions is called the *spontaneous or formation product*. The dividing line between supersaturation and undersaturation, below which phase transformation will not occur, is called *solubility product*.

In an aqueous solution, such as urine, when the calcium salt saturation is above the formation product, crystals of these salts will be formed and can be easily detected.[5] If on the other hand, the ionic product is below the saturation product, formation of crystals will not occur. The mechanism of crystal formation can be divided into three stages:[6]

1. The initial formation of isolated small crystals as the result of homogenous or heterogenous nucleation from a supersaturated solution
2. The subsequent growth of the initial crystals and formation of larger crystals
3. The aggregation of crystals and formation of small or large crystal aggregates

Recently, crystal aggregation has attracted much attention as an important factor in kidney stone formation. It is known that for small particles, about 1 μm in diameter, adhesion forces are much greater than gravitational ones and, thus, it appears that adherence between particles themselves and particles to membranes is an important factor in stone formation. Aggregation is due to six basic mechanisms: electrostatic attraction, Van der Waal forces, liquid bridge, capillary, viscous immersion, and solid bridge. Only a few of these mechanisms seem to have relevance in the formation of urinary aggregates: these are the Van der Waal forces and the viscous binding, probably due to a monomolecular layer of proteins which coats 75% of each calcium oxalate particle. The electrostatic forces, if significant, would be repulsive as indicated in the zeta potential[1] in particles immersed in urine.

Robertson[2] was the first to establish a method for calculating the saturation level of urine. According to this method, the level of saturation of urine with calcium salts (octocalcium phosphate and calcium oxalate) is expressed as their activity product. The product depends on the concentration of the individual ions of calcium, oxalate, phosphate, and urine pH as the concentration of other monovalent and divalent ions present in urine such as sodium, potassium, sulfate, magnesium, etc. Using this

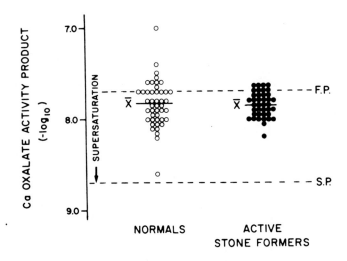

FIGURE 1. Calcium oxalate activity product in 24-hr urine samples of normal individuals and stone patients. There is no significant differences between groups.

method,[3,4] the state of saturation of urine in normal individuals and in patients with recurrent kidney stones was measured. It was found (Figure 1) that urine from both groups are supersaturated with respect to calcium oxalate; although most individuals are in the metastable zone; however, some individuals in both groups had urine saturation above the formation product. In contrast to Robertson's studies[71] in which he observed that patients with recurrent calcium oxalate stones had significantly more supersaturated urine than normal controls, we were unable to observe a significant difference between the mean Ca Ox activity products between the two groups. With respect to octocalcium phosphate, the mean value of stone formers was significantly higher than the controls, although there was overlap of values (Figure 2). The separation of the two groups was more distinct when the brushite instead of octocalcium phosphate was meaured in the two groups.[72,73]

Since urine is supersaturated with calcium oxalate in most individuals (normals and patients), it is surprising that formation of kidney stones is not a universal disease. This can probably be explained by the presence in the urine of certain substances known as inhibitors. These prevent the calcium salts from precipitating and keep them in solution. Each of the three phases of crystal formation (i.e., nucleation, growth, and aggregation) may be inhibited, or promoted, by compounds which one may call *modifiers of crystallization*.

II. MODIFIERS OF CRYSTALLIZATION

A. Promoters of Crystallization and Epitaxial Growth

The presence of organic compounds in urine and organic matrix in urinary calculi[7] raises the question whether these play a role in the formation of kidney stones.

Urine from stone formers have higher amounts of certain urinary proteins than normal subjects.[9,10] These proteins can bind calcium[8] and induce calcification in vitro.[11] Experiments suggest that these substances may be actively involved in the formation of kidney stones and that their presence in the stones may not be the result of passive precipitation.[10,12] There is, however, no concrete evidence that the urinary proteins stimulate the initial event of stone formation [4] (Figure 3).

Another mechanism which may play a role in the formation of crystals is the phe-

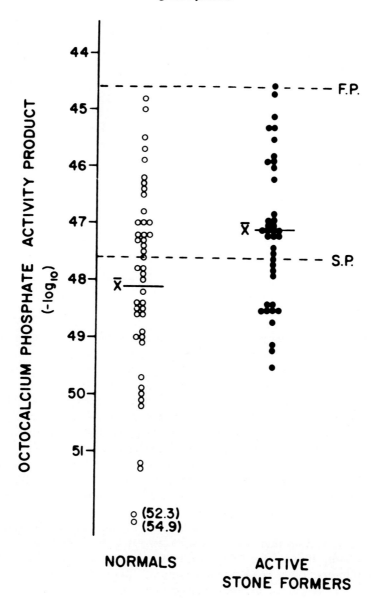

FIGURE 2. Octocalcium phosphate activity products in 24-hr urine samples of normal individuals and stone patients.

nomenon of epitaxy in which crystals of one salt induce crystallization of another salt. The phenomenon of *epitaxial induction* taking place between crystals having similar lattice dimensions is well-known in crystallography. Such similarities are found among crystals present in urine such as uric acid, calcium oxalate, and calcium phosphate[13] and epitaxial induction has been observed in crystals of these salts. Thus, the precipitation of sodium urate is favored by both hydroxyapatite and calcium oxalate[14] and the precipitation of calcium oxalate can be induced from metastable solutions by hydroxyapatite, brushite,[14,15] and urate.[14,16] Although an attractive theory, there is no direct evidence that epitaxis is the major mechanism for formation of renal calculi.[17] A recent study of epitaxis at the atomic level[17] identified the atomic arrays which can interact epitaxially, and reduced the number of possible matches to one or two interactions. For example, studies of Weddelite and Whewellite/uric acid showed that the

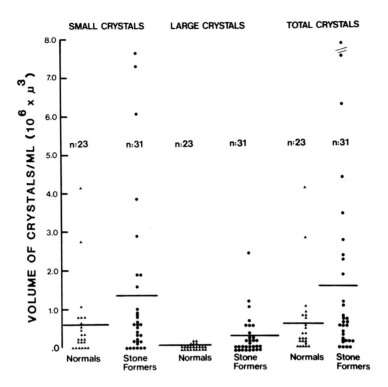

FIGURE 3. Comparison of volumes of small crystals, large crystals, and total crystals between healthy persons and stone-forming patients.

two interacting lattices are mutually supportive of each other. However, in the case of Whewellite/apatite, the atomic arrays did not match the predicted crystalline lattice section indicating that this system is not epitaxially mediated but probably serves only as a nucleating site for Whewellite crystal growth. These findings can explain the crystal growth inhibition effect of different phosphate ions. Closer examination of this type of atomic interactions taking place in epitaxial crystal growth in renal calculi might be helpful in devising new methods for altering crystal morphology and its structure or its atomic surface, thereby blocking the growth. It is already well documented that patients with calcium stones are often hyperuricosuric and that the treatment with Allopurinol, which decreases the excretion of urate, may reduce the formation of calcium stones.[18]

B. Inhibitors of Crystallization

The difficulties in identifying the part played by inhibitors in the process of crystal nucleation, growth, and aggregation are the defects in the methods of measurements. The techniques depend on measurements in diluted urine or in systems with unphysiological conditions. Extrapolation to fresh undiluted urine may not be justified, because the significance of the interaction between the concentration of inhibitors and other ions in urine is unknown.

III. METHODS OF MEASUREMENT OF THE INHIBITORS

The early technique for measurement of the urine inhibitory ability made use of the rachitic rat cartilage model[19] and, subsequently, a collagen matrix[20] test system was developed. Epiphyseal cartilage from rachitic rats were incubated in vitro in a supersaturated salt solution, in which urine or the substance to be tested was added and precipitation of calcium phosphate[21-23] on the cartilage was studied. However, since these test models are biologically active, the possibility exists that its intrinsic biochemical activity may interfere with the effect of inhibitors. Hence, the argument is that the results (collagen calcifications) are difficult to interpret and actually may be totally unrelated to stone formation (crystallization).[6]

According to a more recent technique,[24] the total inhibitory activity on crystal growth and aggregation is measured from the degree of disappearance of calcium oxalate or calcium phosphate from standard supersaturated solutions following induction of crystallization; this method has been used to compare urine and other additives under standard conditions. A variation of this method involves adjustment of calcium and phosphate concentrations in urine to a fixed concentration in the metastable range and to measure the amount of salt precipitated after the addition of various inhibitors or nucleators.[24] A number of different techniques using similar approaches have been described. Thus, Robertson[74] tested the inhibitory capacity by varying the product of calcium oxalate and calcium phosphate to determine the amount required to reach the formation product.

Similar methods can be used to determine the inhibitors of calcium phosphate and calcium oxalate precipitation.[22,23] With calcium oxalate, the kinetic analysis of the crystal growth is straightforward. According to one of these methods, the rate of precipitation after addition of a seed is determined by the rate of disappearance of each ion analyzed quantitatively.[26,27] Another method to study seeded calcium oxalate crystal growth and aggregation in metastable solutions was developed by Robertson.[28] In this method, the degree of retardation of growth and aggregation is used as an indicator of the inhibitor activity. It is a relatively simple method in which changes in particle size are measured with the Coulter Counter®. Different urine additives and diluted urines can be used with this system, and the degree of retardation of growth and aggregation is used as an indicator of the inhibiting activity. It is very attractive as a method for testing individual inhibitors, but requires tedious preparation of samples and glassware in order to obtain acceptable reproducibility. Using this technique, Robertson and his colleagues[8] were able to show a significant difference in the inhibitory activity in urine between normals and stone formers, but others were unable to reproduce their findings. The problem with most of these methods is that the concentration of the ions and pH of the solution change during the experiment. Thus, it must be kept in mind, especially in the calcium phosphate systems, that the change in the formation product or with the depletion of calcium or phosphate, which reflect inhibitory activity, refer only to the final phase of the reaction. However, very often in the solutions saturated with calcium phosphate, the first salt formed will transform during the time of the experiment into another type.[75] Thus, the inhibitor acting on the growth of the second salt might be missed, since the first phase will proceed normally.

A solution to this problem is the use of a system which maintains a constant composition of the salts and pH.[30] This method provides a means for measuring the rate of calcium oxalate crystal growth after a stable supersaturated solution has been seeded with calcium oxalate monohydrate. The concentrations of calcium and oxalate ions are maintained constant by simultaneous additions of reagents containing these ions. The rate of addition of the salts is controlled by a specific calcium electrode. This method is highly reproducible and can be used for the comparison of the inhibitory activity of multiple urine samples. Another promising method was developed by Miller and Randolph.[80] Based on the Finlayson[31] theory of continuous crystallization, this method is used to measure quantitatively the linear crystal growth rate and nucleation rate in an environment similar to that found in urine. The technique requires use of a mixed suspension and a mixed product removal crystallizer which generates supersaturation by a chemical reaction. Steady-state operation of the mixed suspension maintains constant flow rates, composition, temperature, crystal volume, and pH during the entire experimental time. This method provides an opportunity for a rapid, quantitative, and reproducible testing for the inhibitor characteristics of drugs and chemicals needed in renal stone research.

IV. NATURE OF INHIBITORS AND MECHANISM OF THEIR ACTIVITY

One of the first inhibitors to be isolated and identified from the urine was inorganic pyrophosphate;[39] this compound has been shown to be inhibitory in all tests systems. It reduces calcium phosphate formation product in seeded crystal growth studies,[39] diminishes the rate of calcium phosphate precipitation,[40,41] increases the amount of crystals needed to induce growth, and inhibits calcification of cartilage.[42] In addition, inorganic pyrophosphate prevents the transformation between different calcium phosphate phases, for example, change from amorphous calcium phosphate into its crystalline form.[43,44]

Other substances such as magnesium,[41,42,45] zinc,[46] fluoride,[45] and citrate[41,47] were also shown to be inhibitory but to a much lesser degree. With regard to magnesium, it seems that it competes with calcium to form soluble magnesium oxalate and our observations have shown that the ratio of magnesium to calcium in urine may be of greater importance than the absolute amounts of magnesium in urine.[76] This Mg/Ca ratio is significantly lower in stone formers than in urine of normal controls. Citric acid, which is a normal constituent of urine, has been shown to inhibit calcification in in vitro systems involving cartilage and tendon calcification experiments. This finding was supported by X-ray diffraction and electron microscopy studies which showed that only tiny amounts of citrate were needed to maintain Ca^{++} and inorganic phosphate in a noncrystalline state, which normally would transform to identifiable hydroxyapatite crystals.[69]

The known inhibitors of crystallization present in urine account only for part of the total inhibition, suggesting that there may be some additional unidentified inhibitors. One of the least documented of the inhibitor compounds is phosphocitrate. First reported by Howard,[69] it recently has gained more attention as a potential inhibitor of citrated hydroxyapatite formation. Howard and co-workers[69] came upon an interesting finding when testing the effect of artificial urine on tendon calcification. By elimination, it became evident that concentration of organic phosphate and citric acid played a significant role in determining the quantity of inhibitor material formed in the synthetic urines. Further efforts in identifying this hypothetical inhibitor included such methods as anion exchange chromatography with radioactively labeled components, gradient elution, and analysis by nuclear magnetic resonance (NMR) spectroscopy. What was even more interesting was that this compound, derived from a mixture of phosphate and citrate, was synergistic with citric acid. Phosphocitrate was shown to be a potent inhibitor. Synthetically prepared phosphocitrate is now available. Other research groups[70] have supported Howard's findings and have accumulated data on the relative potency of this compound in calcification systems. Thus, phosphocitrate has been shown to interfere, even at low concentrations, with the conversion of amorphous calcium phosphate to crystalline forms (hydroxyapatite) and reduces calcium deposition with tissues such as kidney. It is stronger than all the inhibitors examined to date which include pyrophosphate, imidodiphosphate, and ethene 1-hydroxy-1,1-diphosphone. In relation to calcium oxalate crystallization, phosphotocitrate action is not so definitive[70] and seems to be weaker than that of glycosaminoglycans.

Recently, glycosaminoglycans (GAGS) have attracted attention as an important inhibitor of the aggregation of calcium oxalate crystals. Chondroitin sulfate which is the major GAGS in normal urine[56] was used by Leal and Finlayson[57] in their adsorption of this substance by calcium oxalate crystals which could explain the inhibition of growth and aggregation of calcium oxalate crystals such as seen with the in vitro precipitation systems.

Other substances as hyaluroniadase polyacrylic acid ethylene, maleic acid, and RNA-like material[58] were also found to have an inhibitory activity.

The need for an oral administration of inhibitor, which would be excreted in urine, led to the investigation of polyphosphates, ATP[50] phosphorylated inositols,[51] and imidodiphosphate.[52] They were found to be effective inhibitors, and this raised the possibility that other structurally related compounds, which were more resistant to hydrolysis, could be found and used therapeutically. For awhile, diphosphonates seemed to have all the required properties,[23,25,53] but taken orally, they caused some bone demineralization which limited their usefulness.

Methylene blue was shown[54,55] that by altering the electrical potential of crystals, it may act as an inhibitor of calcium oxalate crystal aggregation, and may even lead to their disaggregation.

Although the majority of the urinary inhibitors can inhibit both calcium oxalate and calcium phosphate crystallization, it seems that some of them may inhibit preferentially one salt or the other. This can be exemplified by the observation that phosphocitrate inhibits predominantly calcium phosphate, whereas glycosaminoglycans inhibits calcium oxalate precipitation.

The distinction as to what stage of crystallization (nucleation, growth, and aggregation) that each inhibitor is affecting, is difficult to establish with the exception of chondroitin sulfate. This principal constituent of human urinary glycosaminoglycans, was found to be a far more potent inhibitor of crystal aggregation than of crystal growth.[28,37]

Teleologically, one would expect that the role of inhibitors in urine is to prevent renal stone formation[69] and that, at least in a certain percentage of patients, formation of stones is the result of a decrease in urinary inhibitors. The results of investigations in this area are conflicting, with some authors reporting a decrease,[71] others showing no difference,[73] and even others showing an increase in the inhibitory activity of stone patients compared with normal controls[29,79] (Figure 4). The reasons for these discrepancies are not known. One possibility may be that the high urine volume that stone patients have[73] results in a lower concentration of the inhibitor(s) at a time when the 24-hr excretion of these substances remains the same. Robertson and his colleagues suggested that the inhibitors should be looked at in relationship to the salt saturation level (saturation/inhibition index).[78] With this index they were able to separate their stone formers from normal controls, whereas there was a significant overlap when the two individual components (saturation and inhibition) were studied separately.

Finally, observations[38] showed that the levels of urine glycosaminoglycans showed a significant $(p < 0.001)$ positive correlation with the volume of small $(< 12\ \mu)$ crystals, but an inverse one with the large $(> 12\ \mu)$ ones (Figure 5). This fact and the observation[29,79] that the level of inhibitors is higher among stone patients indicates that this increase may be a protective response of the body to inhibit large crystal formation.

V. THERAPEUTIC USE OF THE INHIBITORS

Even though the role of the inhibitors has not been elucidated, it is accepted by most investigators that measures leading to an increase in endogenous or exogenous inhibitors will prevent kidney stone disease. Until now, only a few ways to increase urinary inhibitors were known. Oral administration of orthophosphates[63] results in a marked

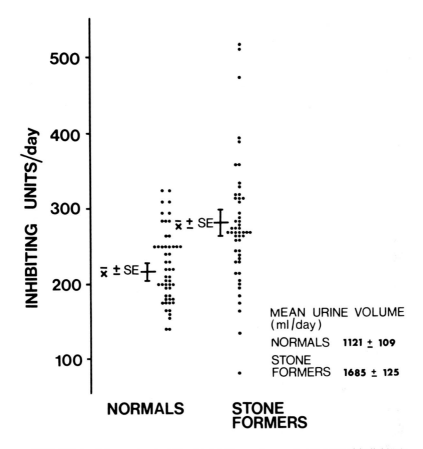

FIGURE 4. Mean values(± SE) of inhibiting units per day in normal individuals and stone patients.

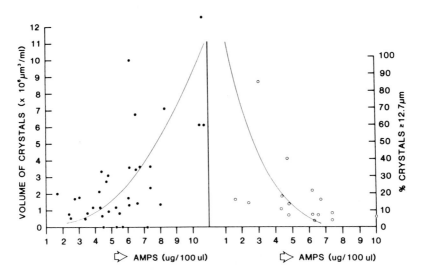

FIGURE 5. The relationship between the excretion of crystals and concentration of acid mucopolysacchride (AMPS) in urine.

increase in urine pyrophosphate.[62,63] In addition, orthophosphates bind calcium in the gut which results in a decrease in its intestinal absorption and urinary excretion, thus, resulting in an additional decrease in the urinary calcium oxalate concentration.

In a similar fashion, Thiazide diuretics decrease urine calcium[64-66] but also increase urinary pyrophosphate, magnesium,[67] and zinc,[46] all of which are strong inhibitors of crystallization. Magnesium can also be used by itself as a therapeutic agent either in the form of MgO or preferably Mg(OH)$_2$. During the administration of magnesium oxide orally,[68] urinary pH rises significantly and urinary calcium increases in some patients. Although a significant increase in the urinary magnesium concentration occurs, no change is found either in the formation products of calcium oxalate or brushite and no objective evidence for the beneficial effects of magnesium in calcium nephrolithiasis could be found.[24]

Oral administration of citrates or bicarbonate may lead to an increase in urine citrate and this treatment is indicated in those patients who have low urine citrates.

Administration of Methylene blue has been reported to benefit certain patients with kidney stones,[27] but no controlled trials have been done.

Finally, the presently available diphosphonates have not been recommended for use in patients with stones because of the risk of osteomalacia when used over long periods.

REFERENCES

1. **Finlayson, B.**, Physicochemical aspects of urolithiasis, *Kidney Int.*, 13(5), 344, 1978.
2. **Robertson, W. G.**, Measurement of ionized calcium in biological fluids, *Clin. Chem. Acta*, 24, 149, 1969.
3. **Crassweller, P. O., Oreopoulos, D. G., and Toguri, A.**, Studies of inhibitors of calcifications and levels of urine saturations of calcium salts in recurrent stone patients, *J. Urol.*, 120(6), 6, 1978.
4. **Oreopoulos, D. G., Wilson, D. R., Husdan, H., Polypchuk, G., and Rapoport, A.**, Comparison of two methods for measuring activity products of calcium salts in urine, *Urol Res,* Fleisch, H., Robertson, W. G., Smith, L. H., and Vahlensieck, W., Eds., Plenum Press, New York, 1976.
5. **Robertson, W. G., Peacock, M., and Nordin, B. E. C.**, Calcium oxalate crystalluria and urine saturation in recurrent renal stone formers, *Clin. Sci.*, 40, 365, 1971.
6. **Fleisch, H.**, Inhibitors and promoters of stone formation, *Kid. Int.*, 13(5), 361, 1978.
7. **Boyce, W. H. and Garvey, F. K.**, The amount and nature of the organic matrix in urinary calculi. A review., *J. Urol.*, 76, 231, 1956.
8. **Boyce, W. H., Garvey, F. K., and Norfleet, C. M.**, The metal chelate compounds of urine, *Am. J. Med.*, 19, 87, 1955.
9. **Boyce, W. H. and Swanson, M.**, Biocolloids of urine in health and in calculous disease. II. Electrophoretic and biochemical studies of a mucoprotein insoluble in molar sodium chloride, *J. Clin. Invest.*, 34, 1581, 1955.
10. **Boyce, W. H. and King, J. S.**, Crystal-matrix interrelation in calculi, *J. Urol.*, 81, 351, 1959.
11. **Boyce, W. H., Garvey, F. K., and Norfleet, C. M.**, Ion-binding properties of electrophoretically homogenous mucoproteins of urine in normal subjects and in patients with renal calculus disease, *J. Urol.*, 72, 1019, 1954.
12. **Boyce, W. H., Pool, C. S., Meschan, I., and King, J. S.**, Organic matrix of urinary calculi, *Acta Radiol.*, 50, 543, 1958.
13. **Lonsdale, K.**, Epitaxy as a growth factor in urinary calculi and gallstones, *Nature (London)*, 217, 56, 1968.
14. **Pak, C. Y. C., Hayashi, Y., and Arnold, L. H.**, Heterogenous nucleation with urate, calcium phosphate and calcium oxalate, *Proc. Soc. Exp. Biol. Med.*, 153, 83, 1976.
15. **Meyer, J. L., Bergert, J. H., and Smith, L. H.**, Epitaxial relationships in urolithiasis: the calcium oxalate monohydrate — hydroxyapatite system, *Clin. Sci. Mol. Med.*, 49, 369, 1975.
16. **Coe, F. L., Lawton, R. L., Goldstein, R. B., and Tembe, V.**, Sodium urate accelerates precipitation of calcium oxalate in vitro, *Proc. Soc. Exp. Biol. Med.*, 149, 962, 1975.
17. **Mandel, N. S. and Mandel G. S.**, Epitaxis between stone-forming crystals at the atomic level, presented at 4th Int. Symp. Urol. Res., Williamsburg, Virginia, June 22 to 26, 1980.

18. Coe, F. L., Hyperuricosuric calcium oxalate nephrolithiasis, *Kidney Int.*, 13(5), 418, 1978.
19. Thomas, W. C. and Howard, J. E., Studies on the mineralizing propensity of urine from patients with and without renal calculi, *Trans. Assoc. Am. Physicians*, 72, 181, 1959.
20. Pak, W. Y. and Buskin, B., Calcification of collagen by urine in vitro; dependence on the degree of saturation of urine with respect to brushite, *J. Clin. Invest.*, 49, 23, 1970.
21. Fleisch, H. and Bisaz, S., The inhibitory effect of pyrophosphate on calcium oxalate precipitation and its relation to urolithiasis, *Experientia*, 20, 276, 1964.
22. Gill, W. B. and Kavesh, J. W., Demonstration of protective (inhibitory) effects of urinary macromolecules on the crystallization of calcium oxalate, in *Urolithiasis Research*, Fleisch, H., Robertson, W. G., Smith, L. H., and Vahlensieck, W., Eds., Plenum Press, New York, 1976, 277.
23. Pak, C. Y. C., Ohata, M., and Holt, K., Effect of diphosphonate on crystallization of calcium oxalate in vitro, *Kidney Int.*, 7, 154, 1975.
24. Meyer, J. L. and Smith, H., Growth of calcium oxalate crystals. II. Inhibition by natural urinary crystal growth inhibitors, *Invest. Urol.*, 13, 36, 1975.
25. Ohata, M. and Pak, C. Y. C., The effect of diphosphonates on calcium phosphate crystallization in urine in vitro, *Kidney Int.*, 4, 401.
26. Meyer, J. L. and Smith, L. H., Growth of calcium oxalate crystals. I. A model for urinary stone growth, *Invest. Urol.*, 13, 31, 1975.
27. Meyer, J. L. and Smith, L. H., Growth of calcium oxalate crystals. II. Inhibition of by natural urinary crystal growth inhibitors, *Invest. Urol.*, 13, 36, 1975.
28. Robertson, W. G. and Peacock, M., Calcium oxalate crystalluria and inhibitors of crystallization in recurrent renal stone formers, *Clin. Sci.*, 43, 499, 1972.
29. Ryall, R. and Marshall, V. R., The effect of urine and other inhibitors on the growth and aggregation of calcium oxalate crystals in vitro, presented at 4th Int. Symp. Urol. Res., Williamsburg, Virginia, June 22 to 26, 1980.
30. Sheehan, H. E. and Nancolas, G. H., A constant composition method for modelling urinary stone formation, in press.
31. Finlayson, B., Where and how does urinary stone disease start? Int. Bladder Stone Conf. Proc., Bethesda, Maryland, 1976, 7.
32. Robertson, W. G., A method for measuring calcium crystalluria, *Clin. Chim. Acta*, 26, 105, 1969.
33. Crassweller, P. O., Brandes, L., Katirtzoglou, A., and Oreopoulos, D. G., Studies of crystalluria in recurrent calcium lithiasis, *Can. J. Surg.*, 22, 6, 1979.
34. Hansen, N. M., Felix, R., Bisaz, S., Fleisch M., Aggregation of hydroxyapatite crystals, *Biochim Biophys. Acta*, 451, 549, 1976.
35. Fleisch, H. and Monod, A., A new technique for measuring aggregation of calcium oxalate crystals in vitro: effect of urine, magnesium, pyrophosphate and diphosphonates, in *Urinary Calculi*, Cifuentes Delatte, L., Rapado, A., and Hodgkinson, A., Eds., S. Karger, Basel, 1973, 53.
36. Felix, R., Monod, A., Broge, L., Hansen, N. M., and Fleisch, H., Aggregation of calcium oxalate crystals: effect of urine and various inhibitors, *Urol. Res.*, 5, 21, 1977.
37. Fleisch, H. and Neuman, W. F., Mechanism of calcification: role of collagen, polyphosphate, and phosphatase, *Am. J. Physiol.*, 200, 1296, 1961.
38. Brandes, L., Oreopoulos, D. G., Husdan, H., and Crassweller, P., Study of calcium crystals in patients with kidney stones, presented at 4th Int. Symp. Urol. Res., Williamsburg, Virginia, June 22 to 26, 1980.
39. Fleisch, H. and Bisaz, S., Isolation from urine of pyrophosphate, a calcification inhibitor, *Am. J. Physiol.*, 203, 671, 1962.
40. Marshall, R. W. and Nancollas, G. M., The kinetics of crystal growth of dicalcium phosphate dihydrate, *J. Phys. Chem.*, 73, 3535, 1969.
41. Smith, L. H., Meyer, J. L., and McCall, J. T., Chemical nature of crystal inhibitors isolated from human urine, in *Urinary Calculi*, Cifuentes Delatte, L., Rapado A., and Hodgkinson, A., Eds., S. Karger, Basel, 1973, 318.
42. Oreopoulos, D. G., Walker, D., Akriotis, D. J., Roncari, D. A. K., Husdan, H., Symvoilidis, A, deVeber, G. A., and Rapoport, A., Excretion of inhibitors of calcification in urine. I. Findings in control subjects and patients with renal stones, *CMAJ*, 112, 827, 1975.
43. Fleisch, H., Russell, R. G. G., Bisaz, S., Termine, J. D., and Posner, A. S., Influence of pyrophosphate on the transformation of amorphus to crystalluria calcium phosphate, *Calcif. Tissue Res.*, 2, 49, 1968.
44. Francis, M. D., The inhibition of calcium hydroxyapatite crystal growth by polyphosphonates and polyphosphates, *Calcif. Tissue Res.*, 3, 151, 1969.
45. Bachra, B. N. and Fischer, H. R. A., The effect of some inhibitors on the nucleation and crystal growth of apatite, *Calcif. Tissue Res.*, 3, 348, 1969.
46. Bird, E. D. and Thomas, W. G., Effect of various metals on mineralization in vitro, *Soc. Exp. Biol. Mol.*, 112, 640, 1963.

47. Bisaz, S, Felix, R., Neuman, W. F., and Fleisch, H., Quantitative determination of inhibitors of calcium phosphate precipitation in whole urine, *Miner. Electroly. Metab.*, 1, 74, 1978.
48. Thomas, W. C. J., Use of phosphates in patients with calcereous renal calculi, *Kidney Int.*, 13, 390, 1978.
49. Williams, G. and Sallis, J. D., The sources of phosphocitrate and its influential role in inhibiting calcium phosphate and calcium oxalate crystallization, presented at 4th Int. Symp. Urol. Res., Williamsburg, Virginia, June 22 to 26, 1980.
50. Fleisch, H. and Neuman, W. F., Mechanisms of calcification: role of collagen, polyphosphates, and phosphatase, *Am. J. Physiol.*, 200, 1296, 1961.
51. Thomas, W. C. and Tilden, M. T., Inhibition of mineralization by hydrolysis of phytic acid, *Johns Hopkins Med. J.*, 131, 133, 1972.
52. Robertson, W. G. and Fleisch, H, The effect of imidodiphosphate (P-N-P) on precipitation and dissolution of calcium phosphate in vitro, *Biochim. Biophys. Acta.*, 222, 677, 1970.
53. Meyer, J. L. and Nancollas, G. H., The influence of multidentate organic phosphonates on the crystal growth of hydroxyapatite, *Calcif. Tissue Res.*, 13, 295, 1973.
54. Finlayson, B., Renal lithiasis in review, *Urol. Clin. North Am.*, 1, 2, 1974.
55. Van Triet, B., O'Lear, C. E., and Smith, M. J., Methylene blue and other agents as inhibitors of calcium oxalithiasis in vivo, *Invest. Urol.*, 16, 201, 1978.
56. Varadi, D. P., Ciffonelli, J. A., and Dorfam, A., The acid mucopolyssacharides in normal urine, *Biochim. Biophys. Acta*, 141, 103, 1967.
57. Leal, J. J. and Finlayson, B., Adsorption of naturally occuring polymers onto calcium oxalate crystal surfaces, *Invest., Urol.*, 14(4), 278, 1977.
58. Schrier, E. E., Lee, K. E., Rubin, J. L., Werness, P. G., and Smith, L. H., Characterization of the calcium oxalate crystal growth inhibitors in human urine, presented at 4th Int. Symp. Urol. Res., Williamsburg, Virginia, June 22 to 26, 1980.
59. Gill, W. B., Ruggiero, K. J., and Fromes, M. C., Calcium oxalate crystallization in urothelial-lined systems, presented at 4th Int. Symp. Urol. Res., Williamsburg, Virginia, June 22 to 26, 1980.
60. Pak, C. Y. C., DeLea, C. S., and Bartter, F. C., Successful treatment or recurrent nephrolithiasis (calcium stones) with cellulose phosphate, *N. Engl. J. Med.*, 290, 175, 1974.
61. Blaclock, M. J. and McLeod, M. A., The effect of cellulose phosphate on intestinal absorption and urinary excretion of calcium, *Br. J. Urol.*, 46, 385, 1974.
62. Pak, C. Y. C., Holt, K., Zerwelch, J. E., and Barilla, D. E., Effects of orthophosphate therapy on the crystallization of calcium salts in urine. *Miner. Electroly. Metabol.*, 1, 147, 1978.
63. Smith, L. H., Application of physical, chemical, and metabolic factors to the management of urolithiasis, in *Urolithiasis Research*, Fleisch, H., Robertson, W. G., Smith, L. H., and Vahlensieck, W., Eds. Plenum Press, New York, 1976.
64. Yendt, E. I. and Cohanim, M., Continuing experience with the use of thiazides in the prevention of kidney stones, *Trans. Am. Clin. Pharmacol. Assoc.*, 8, 1973.
65. Polypchuk, G., Ehig, V., and Wilson, D. R., Effect of hydrochlorothiazide on urine saturation with brushite — in vitro collagen calcification by urine and urinary inhibitors of collagen calcification, *JAMA*, 118, 792, 1978.
66. Yendt, E. R. and Cohanim, M., Prevention of calcium stones with thiazides, *Kidney Int.*, 13, 397, 1978.
67. Pak, C. Y. C., Hydrochlorothiazide therapy in nephrolithiasis, *Clin. Pharm. Ther.*, 14, 209, 1973.
68. Fetner, C. D., Barilla, D. E., Townsend, J., and Pak, C. Y. C., Effects of magnesium oxioce on the crystallization of calcium salts in urine in patients with recurrent nephrolithiasis, *J. Urol.*, 120, 399, 1978.
69. Howard, J. E., Studies of stone formation: a saga of clinical investigation, *Johns Hopkins Med. J.*, 139, 239, 1976.
70. Williams, G., and Sallis, J. D., The sources of phosphocitrate and its influencial role in inhibiting calcium phosphate and calcium oxalate crystallization, presented at 4th Int. Symp. Urol. Res., Williamsburg, Virginia, June 22 to 26, 1980.
71. Robertson, W. G., Physical chemical aspects of calcium stone formation in the urinary tract, in *Urolithiasis Research*, Fleisch, H., Robertson, W. G., Smith, L. H., and Vahlensieck, W., Eds., Plenum Press, New York, 1976, 25.
72. Pak, C. Y. C., and Chu, S., A simple technique for the determination of urinary state of supersaturation for brushite, *Invest. Urol.*, 11, 211, 1972.
73. Oreopoulos, D. G., Wilson, D. R., Husdan, H., Pylypchuk, G., and Rapoport, A., Comparison of two methods for measuring activity products of calcium salts in urine, in *Urolithiasis Research*, Fleisch, H., Robertson, W. G., Smith, L. H., and Vahlensieck, W., Eds., Plenum Press, New York, 1976, 325.
74. Robertson, W. G., Factors affecting the precipitation of calcium phosphate in vitro, *Calcif. Tissue Res.*, 11, 311, 1973.

75. **Nancollas, G. H. and Tomazic, B.**, Growth of calcium phosphate on hydroxyapatite crystals: effect of supersaturation and ionic medium, *J. Phys. Chem.*, 78, 2218, 1974.
76. **Oreopoulos, D. G., Soyannwo, M. A. O., and McGeown, M. G.**, Magnesium/calcium ratio in urine of patients with renal stone, *Lancet*, Augt. 24, 420, 1968.
77. **Boyce, W. H., McKinney, W. M., Long, T. T., and Drach, G. W.**, Oral administration of methylene blue to patients with renal calculi, *J. Urol.*, 97, 783, 1967
78. **Robertson, W. G., Peacock, D., Marshall, R. W., Marshall, D. H., and Nordin, C.**, Saturation — inhibition index as a measure of the risk of calcium oxalate stone formation in the urinary tract, *N. Engl. J. Med.*, 294(5), 249, 1976.
79. **Randolph, A. D., Kralijevich, Z. I., and Drach, G. W.**, Effect of urinary macromolecules on calcium oxalate dihydrate crystal growth and nucleation rates, presented at 4th Int. Symp. Urol. Res., Williamsburg, Virginia, June 22 to 26, 1980.
80. **Miller J. O., Randolph, A. D., and Drach, G. W.**, Observation upon calcium oxalate crystallization kinetics in simulated urine, *J. Urol.*, 117, 342, 1977.

Index

INDEX

A

A23187 ionophore, 6, 18, 218
Abacus-like bodies, 59
Abnormalities
  in bone metabolism, 144
  in chromosomes, 91
  in vitamin D metabolism, 127
Absorption
  cadmium, 224
  calcium, see Calcium absorption
Absorptive hypercalciuria, 113
AC, see Adenylate cyclase
Accumulation of calcium
  in β-cells, 18—19
  in mitochondria, 136
Acidic glycoproteins, 64
Acidic urine, 112
Acidosis, 92, 127—128
  renal tubular, 112
Acquired idiopathic hypoparathyroidism, 94
ACTH, see Corticotropin
Acute disuse hypercalcemia, 92
Acute disuse osteoporosis, 100
Acute hypercalcemia, 100
Adaptation, 48
Adenosine triphosphatase, 59
Adenylate cyclase (AC), 4, 7—9, 20
Adrenal cells, 4
Adrenal glucocorticoids, 90
Adrenal medulla, 6
  neomycin and, 211
Adrenergic transmission, 211
Adrenocortical tissue, 4
Adrenodoxin, 10
Adriamycin, 150
Adverse effects of calcitonin, 162
Affective illness, 158—163
Alcoholism, 165
Aldosterone, 5
Alkaline phosphatase
  cadmium and, 227
  calcium transport and, 59—60
Alkaline treatments, 34—36
Amino acids, 36
  hydrophobic, 37
Aminoglycosides, 150, 210—219
  ganglioplegic action of, 210
  medullar adrenal actions of, 211
  membrane structures and, 218—219
Amorphous calcium phosphate, 61
AMP, cyclic, see Cyclic AMP
Anilinonaphthalene-8-sulfonate (ANS), 34
Animals
  behavioral symptoms and CNS calcium in, 158
  virus effect on cells of, 200
Animal viruses, 200, 201
Anorexia, 162
ANS, see Anilinonaphthalene-8-sulfonate

Antagonists
  ACTH, 5
  calcium, 136, 210
  competitive, 209—222
Antibiotics, see also specific drugs, 150, 210
  aminoglycoside, see Aminoglycosides
  competitive antagonism between calcium and, 209—222
Anxiety, 163—164
Apples, 179
ATP, 59, 109, 240
Autoradiography, 58
Avian oviduct, 74
Avian shell glands, 73—83

B

Bacterial viruses, 200
Balance of calcium, 51
Behavior
  CNS calcium and, 158
  human calcitonin (HCT) and, 163
Benign hypercalcemia, 99
Bicarbonate, 242
  calcium relationship with, 80
  secretion of, 79—80
Binding, see also Calcium-binding protein, 219
  calcium-45, 39
  extra calcium, 39
  hormone, 7
  ion, 194—197
Biological roles of calcium ions, 210
Birds, 74—76
Bitter pit, 181
Blockers of calcium channels, 41
Blood coagulation, 88
Blossom-end rot, 180
Bone
  calcium content of, 46
  metabolism in, 125
  minerals in, 126
Bone disease
  classification of, 124—125
  high turnover, 124, 125
  low turnover, 125
  management of, 130
  pathogenesis of, 127—130
Bone metabolism abnormalities, 144
Bound cyclic AMP, 8
Breads, 98
Bridge, 234
Bursting of cells, 182—184
Bypass, 137

C

CaBP, see Calcium binding protein

Cadmium, 225, 227
  alkaline phosphatase and, 227
  changes induced by, 227—229
  dietary calcium and, 224
  intestinal absorption of, 224
  intestinal calcium transport and, 225—226
  poisoning from, 229, 230
  vitamin D-dependent calcium binding protein and, 227
  vitamin D in prevention of poisoning from, 230
Calcification, 56
  metastatic, 125
  soft tissue, 129—130
Calcitonin (CT), 88, 161—163
  anorexia and, 162
  CSF human, see CSF human calcitonin
  depression and, 162
  increase in levels of, 90
  malignant hyperpyrexia and, 162
  mechanism of action of, 163
  pain and, 165
  plasma, 128
  prolactin and, 163
  serum CPK and, 162
  synthetic salmon, 161—162
Calcium
  absorption of, see Calcium absorption
  balance of, 51
  binding of, see Binding
  biological roles of ions of, 210
  bone and, 46
  cells primed for, 6
  depletion of, 134
  dialysate, 128
  efflux of, 6
  exchange of, 6—7
  excretion of, see Excretion
  extra, 39
  extracellular, see Extracellular calcium
  extrusion of, 24—26
  flux of, 21—27, 201
  foods and, 46, 47, 95, 97
  glucose and, 21—27
  gradient of, 6
  influx of, see Influx
  intake of, see also Diet, 50, 89
  intracellular, see Intracellular calcium
  ionized, see Ionized calcium
  localization of during dentinogenesis, 58—59
  metabolism of, see Calcium metabolism
  removal of, see Removal of calcium
  repletion of, 135
  requirements for, 50—52
  responses to, see Responses
  restriction of in diet, 99, 159
  retention of, 48—50
  secretion of, see Calcium secretion
  serum, see Serum calcium
  skeleton and, 46
  taste receptor membranes and, 34—39
  translocation of, 6
  transport of, see Transport
  uptake of, 6—7
Calcium-45, 39
Calcium absorption, 46—48
  dietary, 89
  intestinal, 115—117
Calcium-antagonists, 5, 136, 210
Calcium-ATPase, 59
Calcium-bicarbonate relationship, 80
Calcium-binding protein (CaBP), 116
  soluble, 226
  vitamin D-dependent, 227
Calcium-calcium exchange, 22
Calcium channel blockers, 41
Calcium complex, 4
Calcium deficiency
  consequences of, 182—186
  in plants, 175—191
  symptoms of, 179—181
Calcium-free cardioplegic solutions, 137
Calcium homeostasis, 88—89, 144
  hormones associated with, 88
Calcium ionophores, see Ionophores
Calcium lactate, 160
Calcium lithiasis
  idiopathic, 108
  subgroups of, 113
Calcium metabolism, 55—71
  disorders of, see Disorders of calcium metabolism
  in kidney stone formation, 103—122
Calcium oxalate, 238
Calcium paradox, 134—138
Calcium phosphate, 56, 60, 63, 239
  amorphous, 61
Calcium-phosphate ratio in diet, 95
Calcium-phospholipid-sterol complex, 10
Calcium-phosphorus ratio, 49, 89
Calcium pump, 6, 60
Calcium secretion
  control of, 76—78
  diurnal timing of, 76
  egg distention and, 77
Calcium shifts, 158
Calcium urolithiasis, 113—115
Calcospherites, 60
Calculi, 234
Calmodulin, 8, 20
cAMP, see Cyclic AMP
Capillary, 234
Carbohydrates, 111
γ-Carboxyglutamate-containing proteins, 64—65
γ-Carboxyglutamic acid (Gla), 64, 105
  proteins containing, 58
Carcinogenesis, 144—145
Cardiac arrest, 137
Cardiac muscle, 215—216
Cardioplegic solutions, 137
Cardiopulmonary bypass, 137
Catalytic subunits, 4
Cavity spot, 180
Cells
  adrenal, 4

bursting of, 182—184
division of, 147
duodenal epithelial, 228
extension of wall of, 186
injury to, 145—150
isolated adrenal, 4
Leydig, 4
metabolism of, 149—150
priming of for calcium, secretory, 218
viruses and, 197—200
β-Cells
  accumulation of calcium in, 18—19
  pancreatic, 18
Cereals, 98
cGMP, see Cyclic GMP
Channels
  blockers of, 41
  slow, 136
Chemical transmitter, 43
Chemoreception, 34
Childhood disorders of calcium metabolism, 87—102
Chlortetracycline, 21
Cholecalciferol, 92
Cholera toxin, 4
Cholesterol, 4
Cholesterolester hydrolase, 9
Chondroitin sulfates, 66
Chromosomal abnormalities, 91
Chronic renal failure, 125—126
Circulating adrenal glucocorticoids, 90
Circumpulpal dentin, 57
Citrates, 239, 242
Citric acid, 239
Climate and stone formation, 107
Clinical relevance of calcium paradox, 137—138
CNS calcium in animals, 158
Coagulation of blood, 88
Coated vesicles, 59
Cobalt, 7
Colicins, 200
Collagen, 57, 63
Compartments, intracellular, 6
Competitive antagonism between calcium and antibiotics, 209—222
Concentrations
  dialysate, 128
  electrolyte, 79—80
  normal serum, 90
Contractibility of muscle, 88
Control of calcium secretion, 76—78
Corticotropin (ACTH), 4
  antagonist of, 5
Corticotropin (ACTH)-dependent phosphorylation, 9
Countertransport of sodium-calcium, 24
Coupling, excitation-contraction, 210, 213—218
Cows' milk, 46
CPK, 162
Cracking of fruit, 181
Crystalization, 57
  inhibitors of, 233—245

modifiers of, 235—237
CSF, 161, 163
CT, see Calcitonin
Cyclic AMP, 4, 19, 20, 42, 109
  bound, 8
  total intracellular, 8
Cyclic dependent protein kinase, 4
Cyclic GMP, 5, 42
Cytochrome P_{450}, 4
Cytoplasmic degeneration, 185
Cytostatics, 150

D

Daily intake of calcium, 50
Daily urinary excretion of calcium, 89
Dairy products, 98
Daunomycin, 150
Deficiency
  of calcium, 175—191
  of vitamin D, 92, 96
Degeneration of cytoplasma, 185
Delayed neonatal hypocalcemia, 90, 91
Denervation, 77
Dentin, 56
  circumpulpal, 57
  mantle, 57
Dentinogenesis, 55—71
  calcium localization during, 58—59
  defined, 56
  mineral formation in, 60
Dentin phosphoprotein, 65
Depletion of calcium, 134
Depolarization-induced release of calcium, 215
Depolymerization, 146
Depression and calcitonin, 162
Desserts, 98
1,25-DHCC, see 1,25-Dihydroxycholecalciferol
DHT, 161—162
Diabetes, 117
Diagnosis
  hypercalcemia, 96—99
  hypocalcemia, 93—95
  nutritional management and, 93—100
Dialysate calcium concentration, 128
Dialysis, 124—126, 127—130
  peritonal, 100
Diet, see also Intake; Nutrition, 89
  calcium in, see Dietary calcium
  calcium/phosphate ratio in, 95
  low calcium, 99, 100
  low vitamin, 100
Dietary calcium, 224
  absorption of, 89
  availability of, 46—48
  excretion of, 89
  intake of, 89
  intestinal absorption of cadmium and, 224
  requirements of, 50—52
  restriction of, 99, 159
Dietary fiber, 47

Dietary lead, 224
Dietary management of hypercalcemia, 100
Dihydrostreptomycin, 210
Dihydrotachysterol, 160—161
1,25-Dihydroxycholecalciferol (1,25-DHCC), 88, 127
1,25-Dihydroxy-vitamin D, 116, 117
Dimethylsulfoxide (DMSO), 137
Diphosphonates, 240, 242
Disease, see also specific diseases
   bone, see Bone disease
   high turnover bone, 124, 125
   Itai-Itai, 224
   liver, 96
   low turnover bone, 125
Disorders of calcium metabolism, 112
   in children, 87—102
   psychiatric symptoms of, 158
Dissociation, electromechanical, 135
Distention of egg, 77
Distribution of intracellular calcium, 26—27
Disuse hypercalcemia, 92
Disuse osteoporosis, 100
Diuretics, 99, 242
Diurnal timing of calcium secretion, 76
Division of cells, 147
DMSO, see Dimethylsulfoxide
Duodenal epithelial cells, 228

E

Early neonatal hypocalcemia, 90—91
ECC, see Excitation-contraction coupling
Ectopic PTH secretion from neoplasm, 99
Efflux of calcium, 6
Egg distention and calcium secretion, 77
Electrolyte concentration, 79—80
Electromechanical dissociation, 135
Electron microscopy
   scanning, 227
   transmission, 229
Electrostatic attraction, 234
Enhanced response suppression, 38—39
Environment
   ionic, 144—145
   stone formation and, 107—111
Enzymes, 9—10
   massive release of, 135
   regulation of, 7—10
   translocation of, 11—12
   viral, 198
Epitaxial growth, 235—237
Epitaxial induction, 236
Epithelial cells, 228
Ergosterol, 92
Error of metabolism, 92
Exchange
   calcium, 6—7
   calcium-calcium, 22
Excitation-contraction coupling (ECC), 210, 213—218

Excretion of calcium
   daily urinary, 89
   dietary, 89
   urinary, 49, 51, 89, 96
Extra calcium, 39
Extracellular calcium, 20, 21, 145—150, 202
   hormone binding and, 7
   insulin release and, 18
   role of, 4—6
Extrusion of calcium, 24—26

F

Facilitatory effect of streptomycin, 212
Familial benign hypercalcemia, 99
Fat malabsorption, 96
Fatty acids, 111
Fiber in diet, 47
Fish, 97
Flow of xylem, 177
Fluoride, 239
Fluxes
   calcium, 21—27, 201
   net electrolyte, 79
Foods
   calcium in, 46, 47, 95, 97, 99, 100
   components of, 116, 117
   minerals in, 108—109
   phosphorus in, 95, 97
Fruits, 97—98, 176, 178, 180—181
   cracking of, 181

G

GAGs, see Glycosaminoglycans
Ganglionic transmission, 210—211
Ganglioplegic action of aminoglycoside antibiotics, 210
Gastrin, 117
Genetic influences of kidney stone formation, 105—106
Gentamicin, 210, 215
Gla, see γ-Carboxyglutamic acid
Glucagon, 117
Glucocorticoids, 90
Glucose
   calcium extrusion and, 24—26
   calcium fluxes and, 21—27
   calcium inflow and, 21—23
   insulin induced by, 218
   insulinotropic effect of, 19
   intracellular calcium distribution and, 26—27
   metabolism of, 26
Glycoproteins, 11
   acidic, 64
Glycosaminoglycans (GAGs), 66, 239
GMP, 5, 42
Golgi complex, 58, 59
Gonadal tissue, 4

Gradients of calcium, 6
Growth, 235—237
GTP hydrolysis, 5
Guanylate cyclase, 8
Gustatory responses, 39

H

HCT, see Human calcitonin
Heavy metals, see also specific metals
  toxicity of, 223—231
Hemodialysis, 100, 124—126, 127—130
Heterogenous nucleation, 234
High-energy phosphates, 135
High potassium, 7
High turnover bone disease, 124, 125
Histamine, 218
Homeostasis of calcium, 88—89, 144
  hormones associated with, 88
Homogeneous nucleation, 234
Hormone receptors, 5
Hormones, see also specific hormones, 116
  calcium homeostasis and, 88
  extracellular calcium and, 7
  parathyroid, see Parathyroid hormone
  trophic, 4
Human calcitonin (HCT), 163
Human milk, 52
Hyaluroniadase polyacrylic acid ethylene, 240
Hydrolysis of GTP, 5
Hydrophobic substances, 37
Hydroxyapatite, 58, 60, 239
25-Hydroxy vitamin D (25-OHD), 92
Hygiene and stone formation, 107
Hypercalcemia, 92—93
  acute, 100
  acute disuse, 92
  benign, 99
  causes of, 93
  classes of, 93
  diagnosis of, 96—99
  dietary management of, 100
  disuse, 92
  familial benign, 99
  hypervitaminosis D, 93, 100
  idiopathic, see Idiopathic hypercalcemia
  immobilization, 99
  infantile, 96
  mild, 100
  predisposing factors of, 93
  sarcoid, 99
  treatment of, 99—100
  tumor-associated, 99
Hypercalciuria, 99, 108, 109, 112, 113
  absorptive, 113
  renal, 113
  resorptive, 113
Hypercalciuric rickets, 112
Hypernatremia, 92
Hyperostasis, 124
Hyperoxaluria, 112
Hyperparathyroidism, 99, 112, 117
Hyperphosphaturia, 112
Hyperplasia, 99
Hyperpyrexia, 162
Hypervitaminosis D, 92—93
Hypervitaminosis D hypercalcemia, 93, 100
Hypocalcemia, 90—92
  age of onset of, 91
  causes of, 91
  classes of, 91
  defined, 90
  delayed neonatal, 90, 91
  diagnosis of, 93—95
  early neonatal, 90—91
  infantile, 95
  in later life, 91—92
  neonatal, see Neonatal hypocalcemia
  physiological disturbance of, 91
  predisposing factors of, 91
  symptoms of, 90
  thyroidectomy and, 96
  treatment of, 95—96
  types of, 91
  vitamin D deficiency, 96
Hypocalcemic seizures, 95
Hypomagnesemia, 90, 92
Hypoparathyroidism, 91
  acquired idiopathic, 94
Hypophosphatemia, 116
Hypophosphatemic rickets, 100
Hypothermia, 136

I

Idiopathic calcium lithiasis, 108
Idiopathic hypercalcemia, 100
  of infancy, 93
Idiopathic hypoparathyroidism, 94
Idiopathic urolithiasis, 114—115
Imidodiphosphate, 240
Immobilization hypercalcemia, 99
Inborn error of metabolism, 92
Incisors of rat, 58, 59, 60
Infantile hypercalcemia, 93, 96
Infantile hypocalcemia, 95
Infection, see also specific infections
  lytic virus, 201
Influx of calcium, 145
  glucose and, 21—23
Inhibitors
  of calcium binding to synaptosomal
    membranes, 219
  of crystallization, 233—245
  stone formation and, 104
Injury to cells, 145—150
Inorganic pyrophosphate, 239
Inositols, 240
Inotropic response, 215
Insulin, 117
  basal release of, 21
  glucose-induced, 218
Insulinotropic effect of glucose, 19
Insulin release, 18—19

basal, 21
  calcium action and, 20—21
  extracellular calcium and, 18
Intake, see also Diet, 50
  calcium, 50, 89
  protein, 89
Interactions between viruses and calcium, 193—206
Intestinal absorption
  of cadmium, 224
  of calcium, 115—117
Intestinal microvilli cadmium-induced changes, 227—229
Intestinal transport of calcium, 225—226
Intestinal villi cadmium-induced changes, 227—229
Intracellular calcium, 145
  distribution of, 26—27
  glucose and, 26—27
Intracellular compartments, 6
Intracellular cyclic AMP, 8
Ion binding to viruses, 194—197
Ionic environment during carcinogenesis, 144—145
Ionized calcium, 88
  levels of, 92
  range of, 94
  serum, 90
Ionophores, 6, 150, 210
  A23187, 6, 18, 218
Ions of calcium, 210
Irradiated cholecalciferol, 92
Irradiated ergosterol, 92
Ischemia, 137
Ischemic cardiac arrest, 137
Isolated adrenal cells, 4
Itai-Itai disease, 224

K

Kanamycin, 210
Kidney
  failure, 125—126
  medullary sponge, 112
  transplants of, 99
Kidney hypercalciuria, 113
Kidney tubular acidosis, 112
Kidney stone formation, see also Stone formation, 103—122, 234

L

Labile protein, 4
Lactation, 52
Lactose, 47
Lanthium, 5
Later life hypocalcemia, 91—92
Laying birds, 74—76
Lead in diet, 224

Leaves, 180
Lesions, 179—181
Leydig cells, 4
LH, see Lutropin
Lipid composition change, 229
Liquid bridge, 234
Lithiasis, 112
  calcium, 108, 113
  idiopathic calcium, 108
  uric acid, 114
Lithium, 159
Liver disease, 96
Localization of calcium, 58—59
Low calcium diet, 99, 100
Low turnover bone disease, 125
Low vitamin diet, 100
Low vitamin D stores, 90
Luminal calcium, 78
Lutropin (LH), 4
Lysogeny, 202
Lytic virus infection, 201

M

Magnesium, 89, 109, 186, 212, 227, 239, 242
Malabsorption
  of fat, 96
  of vitamin D, 92
Maleic acid, 240
Malignant hyperpyrexia, 162
Manganese, 7
Mantle dentin, 57
Massive release of enzymes, 135
Matrix theory of stone formation, 105
Matrix vesicles, 57
Maximal response to calcium, 216
Measurement
  of inhibitors of crystallization, 237—239
  of intestinal calcium absorption, 115—116
Meats, 97
Mechanism of action of calcitonin, 163
Mechanism of action or inhibitors of crystallization, 239—240
Medullar adrenal actions of aminoglycoside antibiotics, 211
Medullary sponge kidney, 112
Membrane permeability, 184—185
  changes in, 200—201
Membrane receptors, 4
Membranes
  aminoglycosides and, 218—219
  mitochondrial, 4
  synaptosomal, 219
  taste receptor, 34—39
Mental retardation, 100
Metabolism
  abnormal vitamin D, 127
  bone, 144
  calcium, see Calcium metabolism
  cell, 149—150
  disorders of, see Disorders of metabolism

glucose, 26
inborn error of, 92
vitamin D, 90, 127
Metals, see also specific metals
heavy, 223—231
Metastable calcium complex, 4
Metastable saturation, 234
Metastatic calcification, 125
Methylene blue, 240, 242
Microfilaments, 4
Microincineration, 59
Microscopy
scanning electron, 227
transmission electron, 229
Microtubule depolymerization, 146
Microtubuli, 4
Microvilli
intestinal, 227—229
lipid composition of, 229
proteins associated with, 226
Mild hypercalcemia, 100
Milk
cow, 46
human, 52
Mineralization, 56—57
Mineralization front, 59
Mineralization vesicles, 57
Minerals, 49—50
bone and, 126
foodstuffs and, 108—109
formation of, 60
Mitochondria, 58, 59, 147—148
accumulation of calcium in, 136
calcium transport in, 201
membrane of, 4
Mixed function oxidase, 4
Modifiers of crystallization, 235—237
Molars of rat, 58
Monoamines, 163
Mood shifts, 158
Morbus Wilson, 112
Muscle
cardiac, 215—216
contractibility of, 88
skeletal, 213—215
smooth, 216—218
Myocardial function, 88
Myocardial high-energy phosphate decline, 135

N

Necrotic lesions, 179—181
Negative inotropic response, 215
Neomycin, 210, 212, 213, 218, 219
adrenal medulla and, 211
Neonatal hypocalcemia, 90—91, 93—95
delayed, 90, 91
early, 90—91
Neoplasm, 99
Net electrolyte fluxes, 79
Neuroactive peptide, 163

Neuroleptics, 164
Neuromuscular transmission, 212—213
Neurotransmitters, 210—213
Noncollagenous proteins, 63—64
Normal serum concentrations, 90
Normocalciuria, 114
Nucleation, 65
heterogenous, 234
homogenous, 234
sites of, 56
Nucleic acid, 198
Nucleotide, 7
Nutrition, see also Diet, 45—54
diagnosis and, 93—100
stone formation and, 108

O

Odontoblast-predentin region, 58
Odontoblasts, 56, 60
processes in, 58
$1,25-(OH)_2-D_3$
proteins induced by, 226
25-OHD, see 25-Hydroxy vitamin D
Olfactory reception, 43—44
Oncogenic viruses, 150, 202
Organic phosphate, 239
Orthophosphates, 240, 241
Osteofibrosis, 124
Osteogenesis, 56, 61
Osteogenesis imperfecta, 63
Osteomalacia, 124, 242
Osteoporosis, 46, 51, 52, 124
acute disuse, 100
Osteosclerosis, 124
Overdosage of vitamin D, 96
Oviduct, 74
Oxalate, 47, 109—111
Oxidase, 4

P

Pain, 164—165
Pancreatic β-cell, 18
Paradox, see Calcium paradox
Parathyroid glands 114, 117
suppressed function of, 90
Parathyroid hormone (PTH), 88, 117, 127, 160
ectopic secretion of, 99
increased amounts of, 92
levels of, 99
resistance to, 92
secretion of, 99
Parathyroid hyperplasia, 99
Pathogenesis of bone disease, 127—130
Pathophysiologic variables, 95
PDE, see Phosphodiesterase
Peppers, 180
Peptides, 163
Periodicity, 161—162

Peritoneal dialysis, 100
Permeability
    membrane, 184—185, 200—201
    sarcolemmal, 134
Phase transformation, 234
Phloem, 177
Phosphatases, 9
    alkaline, 59—60, 227
Phosphates, 109, 116, 117
    myocardial high-energy, 135
    organic, 239
    poisoning from, 92
Phosphocitrate, 239
Phosphodiesterase (PDE), 4, 7, 9
Phospholipase A, 10
Phosphoproteins, 63—66
    dentin, 65
Phosphorus, 89, 90, 125
    in foods, 95, 97
    poisoning from, 95
Phosphorylated inositols, 240
Phosphorylation, 9
Physiological stress, 90
Phytic acid, 47
Pits, 181
Plants
    calcium deficiency in, 175—191
    enhancing calcium levels in, 186
    transport of calcium within, 177—179
Plasma
    calcitonin in, 128
    membrane receptor in, 4
    proteins in, 64
Poisoning
    cadmium, 229, 230
    phosphate, 92
    phosphorus, 95
Polymyxines, 150, 210
Polyphosphates, 240
Potassium, 200
    high, 7
P-Proteins, 4
Precipitation-crystallization theory of stone
        formation, 104
Predentin, 58, 60, 66
Pregnancy, 52
Pregnenolone, 4
Primary mood shifts, 158
Priming of cells for calcium, 6
Product translocation, 11—12
Prolactin, 163
Propanolol, 136
Prostaglandins, 10
Protein, 4, 48—49, 111
    calcium binding, see Calcium binding protein
    γ-carboxyglutamate-containing, 64—65
    Gla-containing, 58
    intake of, 89
    labile, 4
    microvilli and, 226
    noncollagenous, 63—64
    1,25-$(OH)_2$-D_3-induced, 226
    plasma, 64
    synthesis of, 9
    vitamin D-dependent calcium binding, 227
Protein kinase, 4
Proteoglycans, 64, 66
Pseudohyperparathyroidism, 112
Pseudohypoparathyroidism, 91—92
Psychiatric illness, 157—173
PTH, see Parathyroid hormone
Purine, 111
Pyroantimonate method, 59
Pyrophosphate, 59, 241, 242
    inorganic, 239
Pyruvate dehydrogenase, 10

R

R_2, see Regulatory subunit
Rat incisors, 58—60
Rat molars, 58
Receptors
    hormone, 5
    plasma membrane, 4
    taste, 34—39
Recommended Dietary Allowances 89
Regulation of enzymes, 7—10
Regulatory subunit (R_2), 4
Release
    calcium-45, 39
    depolarization-induced, 215
    enzyme, 135
    insulin, 18—21
    steroid, 12
Removal of calcium, 34—36
    responses to salt and, 36
    responses to stimuli and, 36—38
Renal, see Kidney
Reperfusion damage, 137
Repletion of calcium, 135
Resorptive hypercalciuria, 113
Response to calcium, 216
    gustatory, 39
    suppression of, 38—39
Retardation, 100
Retention of calcium, 48—50
Rickets
    hypercalciuric, 112
    hypophosphatemic, 100
    vitamin D-dependent, 92, 96
Roots, 180
Rot, 180
Ruthenium red, 5

S

Salmon calcitonin, 161—162
Salts and calcium removal, 36
Sarcoid hypercalcemia, 99
Sarcoidosis, 93, 100
Sarcolemmal permeability to calcium, 134

Saturation, 234
Scanning electron microscopy, 227
Schizophrenia, 164
SCT, see Synthetic salmon calcitonin
Secondary calcium shifts, 158
Secretion
 bicarbonate, 79—80
 calcium, see Calcium secretion
 ectopic PTH, 99
Secretory cells, 218
Seizures, 95
Sendai virus, 200
Serum calcium, 90, 92, 96, 100, 128—129
 ionized, 90
 reduction in, 90
Serum CPK and calcitonin, 162
Sex and stone formation, 106
Sisomicin, 210, 218
Sites of nucleation, 56
Skeletal calcium content, 46
Skeletal muscle, 213—215
Slow channels, 136
Smooth muscle, 216—218
Sodium, 109
Sodium-calcium countertransport, 24
Soft tissue calcification, 129—130
Solid bridge, 234
Soluble calcium binding protein, 226
Solutions of calcium-free cardioplegic, 137
Sponge kidney, 112
Stems, 179—180
Steroidogenesis, 3—16
Steroid release, 12
Stimuli, 36—38
Stone formation, 112, 234, 240
 climate and, 107
 environmental factors in, 107—111
 hygiene and, 107
 nutrition and, 108
 sex and, 106
 theories of, 104—105
Stores of vitamin D, 90
Streptomycin, 210, 212, 218
 facilitatory effect of, 212
Streptozotocin 117
Stress, 90
Subunits, 4
Supersaturation theory of stone formation, 104
Suppression
 of enhanced responses, 38—39
 of parathyroid gland function, 90
Surface of duodenal epithelial cells, 228
Symptoms
 behavioral, 158
 of calcium deficiency, 179—181
 of hypocalcemia, 90
 psychiatric, 158
Synaptosomal membranes, 219
Synthesis of protein, 9
Synthetic salmon calcitonin (SCT), 161—162
Systolic arrest, 137

T

Taste receptor membranes, 34—39
Taste transduction, 40—43
TBNA, see Total body neutron activation
Tetany, 93
Tetracyclines, 150, 210
Theophylline, 19
Therapeutic use of inhibitors of crystallization, 240—242
Thiazide diuretics, 99, 242
Thioacetamide, 146
Thullium-171, 21
Thyroidectomy, 96
Tipburn, 180
Tissue
 adrenocortical, 4
 calcification of, 129—130
 gonadal, 4
 soft, 129—130
Tobramycin, 210, 214, 216
Tomatoes, 180
Topple, 179
Total body calcium, 125—126
Total body neutron activation (TBNA), 125
Total intracellular cyclic AMP, 8
Toxicity
 cholera, 4
 heavy metal, 223—231
 vitamin A, 100
 vitamin D, 100
Trace elements, see also specific elements, 49—50, 129
Transduction of taste, 40—43
Transformation
 phase, 234
 viral, 201—202
Translocation
 of calcium, 6
 of enzymes, 11—12
Transmission
 adrenergic, 211
 chemical, 43
 ganglionic, 210—211
 neuromuscular, 212—213
Transmission electron microscopy, 229
Transplants, 99
Transport of calcium, 59—60, 149, 150
 alkaline phosphatases and, 59—60
 intestinal, 225—226
 mitochondrial, 201
 plants and, 177—179
 vitamin D-stimulated, 227
Treatments, see also specific treatments
 alkaline, 34—36
 of hypercalcemia, 99—100
 of hypocalcemia, 95—96
Trimer of collagen Type I, 63
Trophic hormones, 4
Tubular acidosis, 112
Tumor-associated hypercalcemia, 99

Type I collagen, 63
Type III collagen, 63

U

Unstable saturation, 234
Uptake of calcium, 6—7
Uric acid lithiasis, 114
Urinary inhibitors of crystallization, 233—245
Urine
  acidic, 112
  calcium in, 100
  calcium excretion in, 49, 51, 89, 96
  calcium retention in, 48
Urolithiasis
  calcium, 113—115
  idiopathic, 114—115

V

Van der Waal forces, 234
Vasopressin, 218
Vegetables, 46, 97, 176
Verapamil, 5, 136
Vesicles
  coated, 59
  matrix, 57
  mineralization, 57
Villi, 227—229
Viruses, see also specific viruses
  animal, 200, 201
  bacterial, 200
  effects of on animal cells, 200—201
  entry of into cells, 197—199
  enzymes of, interactions between calcium and, 193—206
  ion binding to, 194—197
  lytic, 201
  multiplication of, 201
  nucleic acid of, 198
  oncogenic, 150, 202
  Sendai, 200
  takeover of, 201—202
  transformation of, 201—202
Viscous immersion, 234
Vitamin A toxicity, 100
Vitamin D, 88, 89
  abnormal metabolism of, 127
  calcium transport stimulated by, 227
  deficiency of, 92
  25-hydroxy, 92
  low stores of, 90
  malabsorption of, 92
  metabolism of, 90, 127
  overdosage of, 96
  prevention of cadmium poisoning and, 230
  toxicity due to, 100
Vitamin D_3, 47
Vitamin D deficiency hypocalcemia, 96
Vitamin D-dependent calcium binding protein, 227
Vitamin D-dependent rickets, 92, 96
Vitamins, see also specific vitamins
  diet low in, 100

W

Watercore, 181
Water-soaked condition, 182
Water-soaked lesions, 179—181

X

Xylem, 178
  flow of, 177

Z

Zinc, 239, 242
Zona fasciculata-reticularis, 4
Zona glomerulosa, 4

256381